3rd International Conference on Mechanical, Manufacturing, Modeling and Mechatronics (IC4M 2018) and 3rd International Conference on Design, Engineering and Science (ICDES 2018)

MATEC Web of Conferences Volume 167 (2018)

Barcelona, Spain
22 – 25 February 2018

Editors:

J. Detand
D. Ruxu

ISBN: 978-1-5108-6199-2

Printed from e-media with permission by:

Curran Associates, Inc.
57 Morehouse Lane
Red Hook, NY 12571

Some format issues inherent in the e-media version may also appear in this print version.

Printed by Curran Associates, Inc. (2018)

For additional information, please contact EDP Sciences – Web of Conferences at the address below.

EDP Sciences – Web of Conferences
17, Avenue du Hoggar
Parc d'Activité de Courtabœuf
BP 112
F-91944 Les Ulis Cedex A
France

Phone: +33 (0) 1 69 18 75 75
Fax: +33 (0) 1 69 28 84 91

contact-edps@webofconferences.org

Additional copies of this publication are available from:

Curran Associates, Inc.
57 Morehouse Lane
Red Hook, NY 12571 USA
Phone: 845-758-0400
Fax: 845-758-2633
Email: curran@proceedings.com
Web: www.proceedings.com

TABLE OF CONTENTS

PREFACE

IC4M 2018 & ICDES 2018 were held in Barcelona, Spain during Feb.22-25, 2018. IC4M 2018 & ICDES 2018 were organized by Hong Kong Society of Mechanical Engineers (HKSME) and IQS (Institut Químic de Sarrià), Barcelona, Spain. The conference provides a useful and wide platform both for display the latest research and for exchange of research results and thoughts in broad Mechanical, Manufacturing, Modeling, Mechatronics, Design, Engineering and Science. The participants of the conference were from almost every part of the world, with background of either academia or industry, even well-known enterprise. The success and prosperity of the conference is reflected high level of the papers received.

The proceedings are a compilation of the accepted papers and represent an interesting outcome of the conference. This book covers 3 chapters: Design of Engineering and Science; Mechanical and Manufacturing Engineering; Engineering Modeling and Mechatronics.

We would like to acknowledge all of those who supported IC4M 2018 & ICDES 2018. Each individual and institutional help were very important for the success of this conference. Especially we would like to thank the organizing committee for their valuable advices in the organization and helpful peer review of the papers.

We sincerely hope that IC4M 2018 & ICDES 2018 will be a forum for excellent discussions that will put forward new ideas and promote collaborative researches. We are sure that the proceedings will serve as an important research source of references and the knowledge, which will lead to not only scientific and engineering progress but also other new products and processes.

Prof. Jan Detand, Ghent University, Kortrijk, Belgium
Prof. Ruxu Du, The Chinese University of Hong Kong, Hong Kong

IC4M 2018&ICDES 2018 COMMITTEES

General Chair

Prof. Jan Detand, Ghent University, Belgium

Conference Committee

Prof. Mohd Hamdi Bin Abd Shukor, University of Malaya, Malaysia

Prof. Du Ruxu, The Chinese University of Hong Kong, Hong Kong

Prof. Cees de Bont , Hong Kong Polytechnic University, Hong Kong

Program Committee

Prof. Yong Zeng, Concordia University, Canada

Prof. Mohd Nasir Tamin, Universiti Teknologi Malaysia, Malaysia

Prof. Yan Wang, Georgia Institute of Technology, USA

Prof. James Andrew Self, Ulsan National Institute of Science and Technology (UNIST), South Korea

Local Chairs

Prof. Pascual Moya-Angeler Oriol, Director of IQS Tech Factory, Barcelona, Spain

Dr. Andrés-Amador García Granada, Industrial Engineering, IQS School of Engineering, Barcelona, Spain

Oriol Pascual, Director of IQS Tech Factory, Barcelona, Spain

Technical Committee

Prof. Yuliang Chen, National Defense University, Taiwan

Prof. Sim Kok Swee, Multimedia University, Malaysia

Prof. Amaresh Chakrabarti, Indian Institute of Science, India

edp sciences

MATEC Web of Conferences

Statement of Peer review

In submitting conference proceedings to MATEC Web of Conferences, I certify to the Publisher that I adhere to the **Policy on Publishing Integrity** of the journal in order to safeguard good scientific practice in publishing.

1. All articles have been subjected to peer review administered by the proceedings editors.
2. Reviews have been conducted by expert referees, who have been requested to provide unbiased and constructive comments aimed, whenever possible, at improving the work.
3. Proceedings editors have taken all reasonable steps to ensure the quality of the materials they publish and their decision to accept or reject a paper for publication has been based only on the merits of the work and the relevance to the journal.

Title, date and place of the conference

2018 3rd International Conference on Mechanical, Manufacturing, Modeling and Mechatronics (IC4M 2018)
2018 3rd International Conference on Design Engineering and Science (ICDES 2018)
February 22-25, 2018 Barcelona, Spain

Proceedings editor(s):

Prof. Jan Detand, Ghent University, Belgium
Prof. Du Ruxu,The Chinese University of Hong Kong, Hong Kong

Date and editor's signature

03-APR-2018

Ruxu Du
April 4, 2018

The design of a composite folding bike to improve the user experience of commuters

Thomas Vervisch[1], Yannick Christiaens[1] and Jan Detand[1, a]

[1]*Ghent University, Faculty of Engineering and Architecture, Department of Industrial System and Product Design, 8500 Kortrijk, Belgium*

Abstract. Over the last years, the popularity of folding bikes has been increasing as a result of the rise of multi-modal transport. They are used by commuters as a complement to public transport. Despite the increasing popularity, the current market offer of folding bikes still represents quite some restrictions and downsides which decrease their usability. This paper shows a user-centred process of designing and prototyping a composite folding bike with the aim of improving the user experience of folding bike using commuters. By improving the usability and ease of use of folding bikes, their full potential can be unlocked. The design process led to a disruptive folding bike design with front and rear single-sided offset wheel mounting. The concept excels in its intuitive and quick folding mechanism, superior riding performance and comfort, adjustability and overall ease of use. In addition to the design process and prototyping of the folding bike, this extended abstract elaborates on the performed user tests and its results. These tests range from the impact of offset wheels to the intuitiveness of the folding mechanism and were performed in order to prove different aspects of the design. This research shows how rethinking and redesigning a bike concept (product) from scratch, using a user-centred design process and taking into account the three aspects — business, technology and people — can lead to a disruptive design that improves usability and the overall user experience of the stakeholders.

1 Introduction

Despite the recent increasing research and innovation in the world of folding bikes [1-2], current market offer does not meet the desires of commuters using them for multimodal commute. The need for a folding bike that is designed to fulfil the demands and wishes of present-day commuters initiated this design process. The project started with initial user tests in which test subjects had to execute each aspect of a folding bike commuter's customer journey. These tests revealed several shortcomings of the current market offer of folding bikes. The first main deficiency is the complexity of the folding mechanisms with safety locks. This leads to a lack of intuitiveness of the folding mechanisms. In addition, the need for skilled and precise handling leads to a bike that does not fold fast and easy enough for the rushed daily use while commuting. A second important drawback is that most folding bikes are chunky and too heavy, making them uncomfortable and a big burden to carry. This is because they are made out of traditional bike-building materials, such as aluminium and steel alloys.

[a] Corresponding author: Jan.Detand@UGent.be

A third flaw is the abundance of exposed or poorly incorporated parts, such as exposed brake cables or externally mounted lights and reflectors. These elements obstruct the ease of use. A last main weakness of folding bikes is the small wheels. Folding bikes often have 20" or even 16" wheels. This results in a lack of comfort and even sketchy riding characteristics, especially at higher speeds.

This extended abstract explores solutions to these shortcomings and the user-centered design process of a concept folding bike, while taking into account the three design thinking aspects: business, technology and people. It furthermore elaborates on comparative user tests of the intuitiveness, the folding speed and the riding experience of the concept. These tests were performed in order to improve the usability of the folding bike and the user experience of the commuter, but also to prove various aspects and hypotheses of the concept [3-4].

2 State-of-the-Art

2.1 Multi-modal folding bike commuting

When it comes to multi-modal transport, cycling as a complement to public transport has various environmental, health, and traffic-reduction benefits [5-7], but it also has its challenges. For example, the relatively insecure parking at the station has its drawbacks [8]. It encourages the use of low-value, expendable bikes with often poor riding characteristics and a lack of safety and comfort [8]. In most cases, a commuter is required to have two bikes, one on both the departure location and one on the arrival location [6]. The perfect solution to these drawbacks is the use of folding bikes, enabling a commuter to retain the bicycle at all times as an item of personal luggage.

However, besides the clear benefits of folding bikes, the current market offer still has several downsides, such as the uncomfortable small wheel size [8], the oversized volume when folded and their excessive weight [9]. In literature, only very few studies were found concerning the user friendliness of folding bikes [9-10] or the user experience of folding bike using commuters. Nevertheless, the downsides mentioned above were demonstrated and reported by the multiple test subjects during the initial user tests.

2.2 Folding bike design

As regards the innovation and design of folding bikes, some literature studies can be found concerning folding bikes that have been commercialised, such as the design process of the Strida folding bike [11], concept studies that remained concepts, such as 'The Intermodal Bike' project [9] and more specific studies, such as the correct use of materials and production techniques for folding bikes [12].

An interesting study was conducted on the optimal direction of the pivot axis of the hinge of a folding bike [13]. An alternative was sought for the classic single lateral pivot type folding mechanism, which causes significant troubles in folding dimensions and the front and rear wheels end up arranged in a slanted way. The research shows for the need of a folding mechanism that aligns both the front and rear wheel in parallel, which would result in excellent usability and smaller folding width. A commonly occurring and remarkable phenomenon in patent research is the use of single-sided suspension, both for the front and rear wheel. In this way both wheels are positioned adjacently to each other with minimal width when the bike is folded [14-17].

3 Method

3.1 Methodology

Although inspiration was gained through several design methodologies, such as design thinking and user-centred design, an own method and structure was followed throughout the design process. First of all, the three aspects of design thinking (Figure 1), necessary for innovative design: business, technology and people were all constantly taken into account throughout the entire process [18].

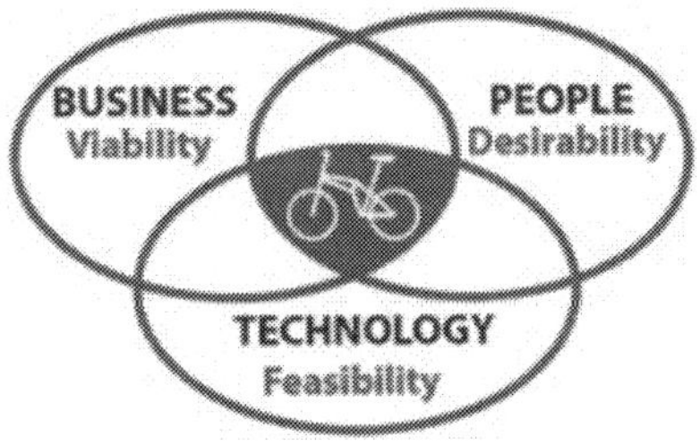

Figure 1. Three phases for innovative design.

The executed design process can be divided into three main phases. First there is the problem definition phase, in which market research and initial user tests are performed and the scope and research question of the thesis is defined. This is then followed by the developing phase that includes various iterative steps of ideation and prototyping. The final phase of the design process is the testing phase, in which several user tests were performed.

3.2 Problem definition phase

The problem definition was initiated with market research on the current offer of folding bikes, from which the market opportunity was derived (Figure 2).

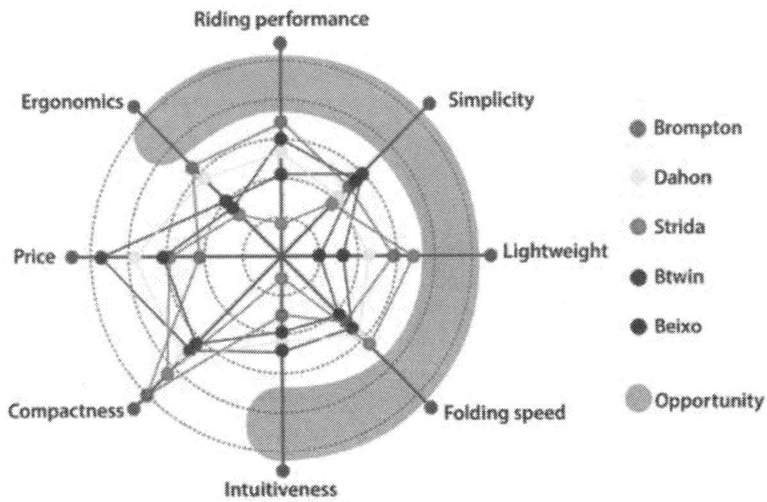

Figure 2. Three phases for innovative design.

Besides, a specific target group was chosen, namely young commuters, aged between 25 and 45, since this is the target group that matches best with the found market potential. In order to achieve a folding bike that works best for these users, User Experience Design techniques were used [19-20]. Personas were made to summarise the findings and research about the stakeholders [21]. A customer journey was drafted to get a better understanding of the specific context in which folding bikes are used by commuters and the actions that are performed in each part of this customer journey (Figure 3) [22-23].

Figure 3. The customer journey of a folding bike using commuter.

Both the persona and customer journey were used as guidelines throughout the whole design process and the user tests. This contextual design lead to a better implementation of the specific demands and wishes of the stakeholders during each aspect of the customer journey [24].

All findings of the market research and the initial user tests were then summarized and converted to the focus of the concept bike to be designed.

3.3 Developing phase

The first ideation phase started by brain-sketching ideas and solutions for sub-problems found in the problem definition phase. From there several concepts were derived, with the focus on the overall shapes of the bikes and a corresponding rough idea of compacting/folding the bike. Through further sketching and the use of quick and dirty scale models, the abundance of ideas was narrowed down to three main concepts, of which one main concept was chosen in consultation with the stakeholders. After further developing, prototyping and optimising the folding mechanism, a first quick and dirty full-size prototype was made as a proof of concept. The goal of this prototype was to get a better way of testing and communicating the complete folding mechanism, dimensions and the usability with the stakeholders (Figure 4). The development of the full-size prototypes always started in CAD, in which first the standard components and the overall geometry were drawn and from there the folding bike's frame and fork were designed.

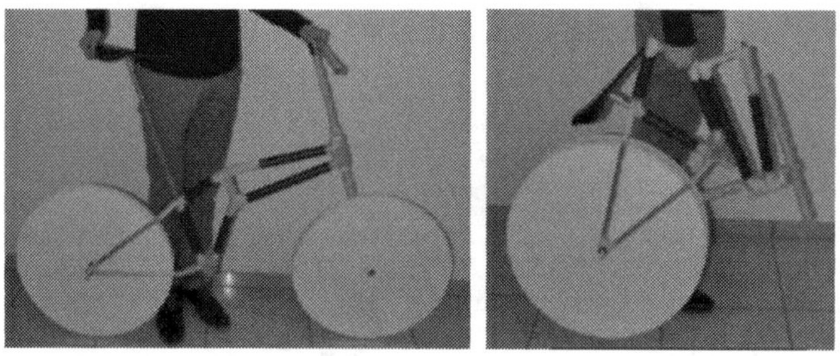

Figure 4. First quick and dirty full-size proof of concept.

Based on the learning outcomes of this first prototype, a second full-size prototype was made. The focus of this prototype was to optimise the dimensions and shape of the folding bike for structural rigidity, the incorporation of standard components and the overall ergonomics. Special attention was paid to make sure crucial bike geometry, such as the fork rake, the trail, the head tube angle and the bottom bracket height, are as optimal as possible in order to achieve the best bike handling characteristics. The goal of this prototype was to have a functioning bike, containing all the necessary standard components, enabling usability and user tests. Doing so, the strived improvement of the user experience of commuters and the usability can be researched. With the intent of fast, low budget easily adaptable prototyping, the frame and front fork were built out of carbon fibre tubes connected with 3D printed lugs.

3.4 Testing phase

<u>Test offset wheels:</u> One main aspect of the concept is that the front and rear wheels are both mounted with an opposing offset, so the two wheels are not perfectly aligned, but positioned in parallel to one another. This results in the two wheels coming right next to and against each other when folding the

bike perfectly vertically, keeping the overall width minimal and enabling the folded bike to be dragged like a trolley on its two wheels. The influence of the opposing offset of the two wheels on the riding experience was tested through several pilot tests and a user test.

Pilot test 1: In the first test the offset between the front and rear wheel was obtained by mounting the front wheel of a bike next to the front fork, resulting in an offset of 90mm (Figure 5).

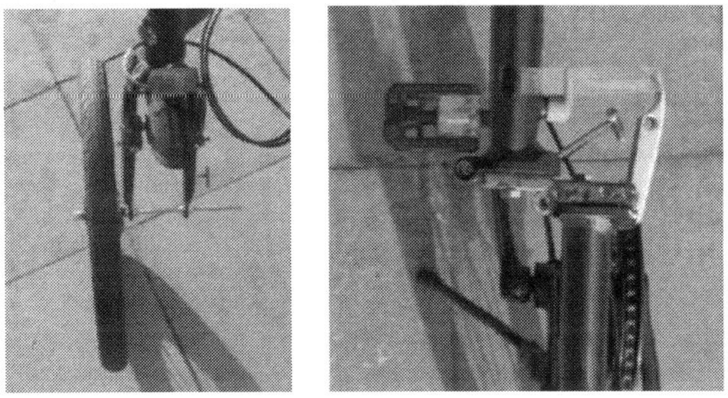

Figure 5. Pilot test 1 and 2 offset wheels

Pilot test 2: The goal of the second test was to create offset wheels, but with the head tube and thus the pivot point of the fork straight above the front wheel, as is normally the case in a bicycle. Now the parallel offset wheels are achieved by splitting a bike frame in half and mounting it with the desired 55mm offset (Figure 5).

User test: After the successful initial findings of the second pilot test, a blind user test was conducted on 25 random test subjects without any foreknowledge of the design project (Figure 6). The user test was a comparative test between two very similar bikes, of which one bike was not adapted and one was adapted with parallel offset wheels. The frames of both bikes were identically wrapped so the offset, located in the frame of one of the bikes was not visually noticeable. The test subjects were asked to ride a given circuit with both bikes. Which bike they started with was randomly chosen. Afterwards a short interview was conducted with each participant on their findings about the riding experience with both bikes.

Figure 6. User test offset wheels.

One of the main aspects of improving the user experience of the commuters is to enhance the usability and ease of use of folding and unfolding the bike. This was tested with two new user tests.

Intuitiveness test: The first way this was analysed was by testing the intuitiveness of folding and unfolding the bike. This test is a comparative user test in which 20 test subjects, who had never seen the folding concept bike and had no association with it, have to fold and unfold three folding bikes for the first time, without any instructions or outside help (Figure 7). The three folding bikes consist of the concept folding bike and two commercial bikes (Brompton and Dahon) that represent the market offer. To get a broader view on the intuitiveness, half of the test subjects started with folded bikes and the other half started with unfolded bikes. The time it took to fold and unfold each bike was timed to be able to compare and analyse the test results. Besides that, the folding and unfolding of the concept folding bike was analysed during the test to gain insights in how users interpret the concept folding bike and to be able to improve intuitiveness and usability of the folding mechanism afterwards. A time limit of three minutes was determined through pilot tests.

Figure 7. Intuitiveness and folding speed user tests.

Folding speed test: The second way to test the usability of folding and unfolding the bike was by performing a comparative test on the speed of folding and unfolding the bike. This test was executed by 20 test subjects after they had been demonstrated the correct way of folding each bike and once they got the hang of the folding motion (Figure 7). This test was also a comparative test between two commercially available folding bikes (Brompton and Dahon) and the concept folding bike. The folding and unfolding was timed to be able to compare and analyse the test.

Research was done to define which aspects determine the user experience of using a folding bike for commuting. The user experience and usability and all the different aspects combined, appeared to result in the overall complete user experience. This cannot be tested or judged through individual aspects alone; therefore, a fifth test was conducted.

Customer journey test: The fifth test is a user test focusing on the overall user experience of the stakeholder using the folding bike throughout the complete customer journey. In this test, three test subjects went through their complete commute, using the folding bike prototype (Figure 8). The full test was observed to analyse the test subjects' proceedings and the think-aloud protocol was applied to gain insights in the test subjects' thoughts [25]. Afterwards an interview was conducted to get a better understanding of participants' overall user experience of the folding bike during the commute.

Figure 8. Usability test during customer journey commuter.

4 Results

4.1 Developing results

The result of the design project is a functional prototype with which the usability, the handling and the folding of the concept bike can be tested and the user tests can be executed (Figure 9). Despite the fact that cycling with the bike is not possible due to the production method used for this prototype, this aspect of the concept has been thoroughly tested in Test offset wheels.

Figure 9. Final prototype of the concept folding bike.

A first main feature is the 24 inch ETRTO 507mm sized wheels, which are an intermediate size between the traditional 20-inch folding bike wheels and the classical 26-inch city bike wheels. This results in the ultimate balance between comfort, stable handling characteristics on rough roads, high speeds and compactness after folding.

A second main feature is the simplicity and speed of frame folding. This is achieved by lifting the bike vertically on the provided handle. Thanks to this lifting motion, the lever is being squeezed and the locking mechanism unlocks. Once the locking mechanism is unlocked, the same upwards lifting motion causes the wheels of the bike to ride towards each other and the bike frame to fold without having to lift the complete weight of the bike (Figure 10). Unfolding the bike frame is operated by taking the handlebar and the seat post and by moving them towards each other. This results in the wheels of the bike riding away from each other and the bike locking mechanism closing and automatically locking. An additional advantage of this folding mechanism is the resilience. The weight of the bike and the weight of the rider push the locking mechanism closed. This way there is no need for an additional safety lock on the locking mechanism, resulting in a faster, more user-friendly and intuitive locking mechanism.

Figure 10. Final prototype of the concept folding bike.

A third main feature is the trolley function. After folding the bike frame and if desired also sliding in the seat post, the bike can be pulled like a trolley on the wheels, using the handlebar as a handle. That means the bike does not have to be carried when transported by foot.

A fourth main feature is the adjustability of the handlebar height that is integrated in the compacting mechanism of the handlebar and stem. This is done by sliding the telescopic stem into the front fork. In the first part of the movement, the stem slides straight into the fork. This makes the handlebar adjustable in height. In the second part of sliding in the stem, the handlebar makes a 90° turn, causing the handlebar to end up against and parallel to the frame (Figure 10). This keeps the width of the folded bike as minimal as possible. A pin on the top of the front fork and the slot in the stem is used to guide the stem straight in the first section and then into a spiral curve.

4.2 Testing results

Pilot test 1 offset wheels: The first test shows that if the wheel is mounted on the right side of the fork, the bike has a tendency to turn left and vice versa. This can be explained by the bikes head tube, and thus the steering pivot point that is not straight above the front wheel. This results in the stable position not being straight forward, but a bike that constantly pulls to one side. Despite this, the test shows positive results regarding the riding experience, both riding straight and cornering.

Pilot test 2 offset wheels: This pilot test shows a perfectly stable bike, the offset of which does not impact the riding characteristics; it is even possible to ride no handed.

User test offset wheels: None of the 25 test subjects experienced a difference in riding experience or characteristics between the two test bikes caused by the 55mm offset located in the frame of one of the test bikes. Furthermore, this offset was only noticed by 1 out of the 25 test subjects because of the offset being located in the middle of the frame, resulting in a slightly crooked riding position.

Intuitiveness test: The intuitiveness test shows an average folding speed of 105s for the Brompton, 96s for the Dahon and 61s for the concept and an average unfolding speed of 116s for the Brompton, 72s for the Dahon and 33s for the concept bike. During this intuitiveness test 3 out of the 20 test subjects succeeded in folding the Brompton correctly, 6 out of 20 folded the Dahon correctly and 19 out of 20 folded the concept bike correctly (Figure 11).

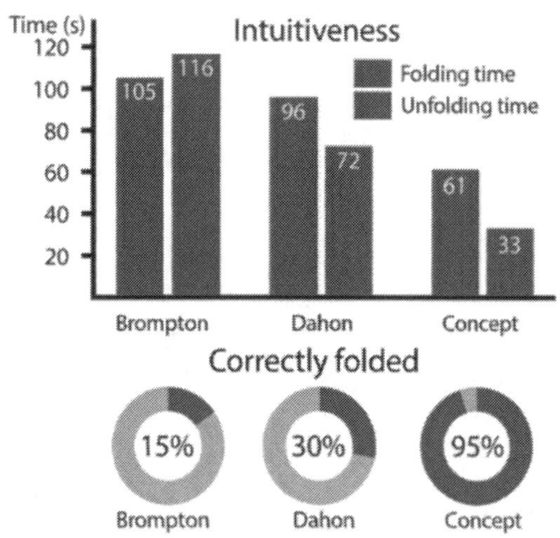

Figure 11. Test results user test intuitiveness.

Folding speed test: The folding speed test results in an average folding speed of 37s for the Brompton, 37s for the Dahon and 17s for the concept bike and an average unfolding speed of 38s, 33s and 12s respectively (Figure 12). All test subjects experienced the concept bike to fold and unfold the quickest. 11 out of the 20 test subjects experienced the Dahon to fold and unfold second quickest and the Brompton the slowest, while the other 9 test subjects experienced the opposite.

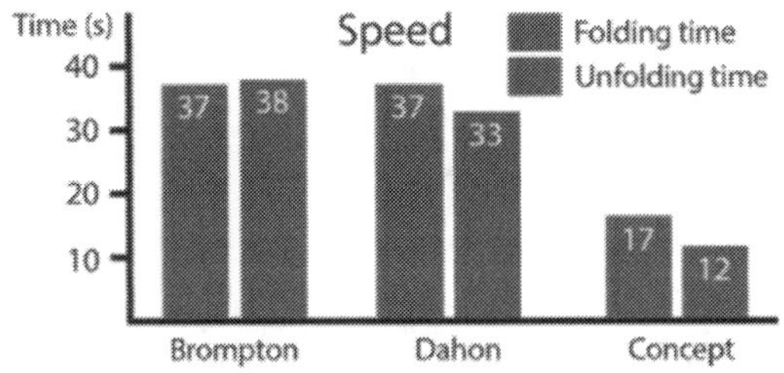

Figure 12. Test results user test folding speed.

Customer journey test: During the usability test in which the customer journey of a folding bike commuter was completed, some of the aspects allowing room for improvement were highlighted. For example, the magnet that keeps the wheels closed must be a lot stronger and custom wheels, of which the non-supported side is completely flat, are required for improved compactness and overall rigidity when folded. As expected, a telescopic seat post is required for smaller folding. Resting the bike on the seat instead of the extended handlebar when folded proved to be a more intuitive and better way to put down the bike when folded. Therefore, a bracket was designed and added to the seat. Moreover, the unique selling propositions of this bike, such as the fast and easy folding and the minimal weight, were valued by the test subjects as expected. Compared to the test results of the initial user test, an overall improvement of the user experience and the usability of using the folding bike was experienced by the stakeholders (Figure 13).

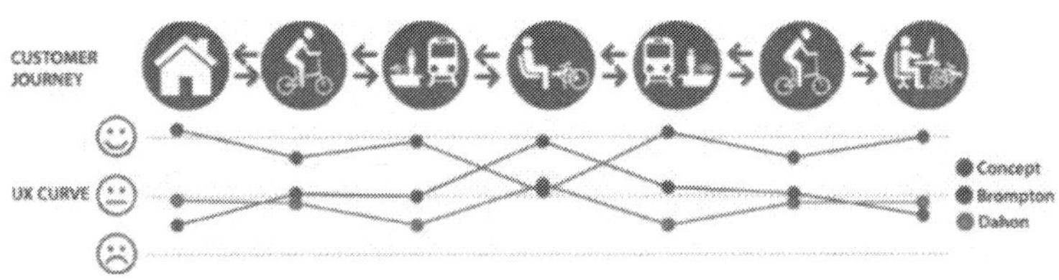

Figure 13. UX curve throughout the customer journey.

5 Discussion

User tests show that wheels aligned in parallel have no noticeable impact on the riding characteristics of a bike, provided the opposite offset mounted wheels are properly integrated in the bike's overall design. The offset was noticed by one test subject due to the crooked riding position caused by the offset being located in the frame. In the designed folding bike concept, the offset is located in the wheel mounting instead of in the bike frame and the riding position of the rider is thus straight and the offset unnoticeable.

The intuitiveness test shows that folding and especially unfolding the concept bike is a lot more intuitive than its two benchmarks. In addition, more participants succeeded in correctly folding the concept bike, which shows that it is more error-proof. The folding speed test shows that the concept bike is a lot quicker to fold and unfold than both benchmarks. All participants also experienced the concept bike to fold quicker. The participants however had limited practice time to master the folding movement. The folding speed of each bike can increase a little with more practice.

The results of both the intuitiveness test and the folding speed test indicate the folding and unfolding of the concept to be more user friendly and improve the user experience of folding the bike. The fragile prototype with 3D printed PLA lugs had a negative impact on the intuitiveness test, the folding speed test and the usability test since the participants were afraid of handling the prototype incorrectly and of damaging the frame. Apart from that, the prototype also presents room for improvement. Therefore, the executed user tests are no complete statistical tests, yet rather explorative tests performed in order to orient the design process. Further statistical tests require fully functional and properly constructed prototypes, of which the prototyping effort and costs should not be underestimated

6 Conclusion and future work

The design of an innovative folding bike concept can be seen as a big puzzle of standard components and self-designed components, in which certain traditional bicycle geometry aspects and rules have to be taken into account. For each standard component, an enormous amount of size and brand variations exist. A lot of foreknowledge is required to be able to select the right combination of standard components or universal dimensions. Not collaborating with a bicycle or folding bike company worked to the detriment of this aspect, but it enabled the design to be a lot more disruptive. During the design process, the consideration between custom wheels and standardised wheels was made. Due to time limitations of the thesis and the ease of use of standardised components, both for prototyping and in real life for bike repair, the preference was given to standardised wheels. The final prototype and design process however shows that a standard bicycle hub is still too wide for this application with offset wheels and single-sided suspension, resulting in the need for a very wide and thus custom bottom bracket and crank shaft. This, combined with the wide front fork and rear wheel mount, the overall folded prototype bike is a lot wider than initially intended. Therefore, custom bike rims with very narrow hubs and an inside mounted rear sprocket are a must for further development and improvement of this bike concept. Custom rims will also enable the inside of the bike wheel to be completely flat, so no hub protrudes. This will enable the two wheels to rest completely flat against each other without space for motion when the bike is folded.

A second suggestion for future work is researching and exploring the alternatives for production methods for the folding bike concept. Aspects such as the durability, weight and cost of production of different production techniques have to be compared. Within the field of composite manufacturing, different techniques such as SMC/BMC and traditional prepreg monocoque lay-ups may be considered and combined. Also, the weigh-up has to be done between composites or other bike building materials such as aluminium, titanium or steel. The material choice also influences the folding bike design, which will have to be fine-tuned according to the chosen material and production technique. There are two main suggestions for material choice and production technique to produce the concept bike's frame.

The first possibility is a traditional monocoque frame, such as present-day racing bikes. To apply this production technique to the bike concept, the monocoque has to be divided in a front part and a rear part, isolated by the central locking mechanism. This central hinge and locking mechanism could be an aluminium milled or forged insert of a composite BMC insert in the monocoque frame. Moulding this monocoque frame will be easier than traditional bike frames due to the single sided rear wheel mount. This means only two mould halves are required for the rear part of the bike frame.

A second and cheaper variant to produce the bike frame, involving less manual labour and cheaper material costs, is a welded aluminium bike frame. The individual parts could be a combination of hydroformed panels with CNC-milled or forged aluminium inserts for the central locking mechanism. The main selection criterion will be the production cost versus the weight of the frame.

There are also several potential materials to produce the custom bike rims, such as magnesium alloys, carbon fibre rims or other fibre reinforced plastics. In all variants, the bike rim would be bolted to a CNC-lathed wheel hub.

In this design process, the focus was on the user aspect of the folding bike design. When further developing this bike, attention will have to be paid to the safety standards and regulations. This will necessitate additional tests, such as the standardised fatigue tests, strength tests, corrosion tests etc. These mechanical tests require fully functional and correctly constructed prototypes. This entails high prototyping effort, costs and foreknowledge. To conclude, for these next steps in the development process, a collaboration with an established bicycle manufacturer would be highly recommended.

References

1. The Folding Cyclist, "Folding Bike News," (2017). Available: http://www.foldingcyclist.com/. [Accessed: 03-May-2017].
2. L. Cooke, J. Roderiguez, M. Chino, L. Wang, J. Fehrenbacher, C. Cameron, J. Dloffa, K. Lofgren, L. Cowan, P. Rebecca, L. Evelyn, A. Schwartz, M. Chino, and J. Chapa, "Folding Bike | Inhabitat - Green Design, Innovation, Architecture, Green Building." Available: http://inhabitat.com/tag/folding-bike/. [Accessed: 03-May-2017].
3. M. F. Story, "Maximizing Usability: The Principles of Universal Design," *Assist. Technology*, **10**, 1, pp. 4–12, (1998).
4. [4] D. J. Mayhew, "The Usability Engineering Lifecycle," in CHI '99 *Extended Abstracts on Human Factors in Computing Systems*, pp. 147–148. (1999)
5. K. J. Krizek and E. W. Stonebraker, "A Marriage Unrealized", pp. 161–167, (2010).
6. P. Rietveld, "Non-motorised modes in transport systems: a multimodal chain perspective for The Netherlands," *Transp. Res. Part D Transp. Environ.*, **vol. 5**, no. 1, pp. 31–36, (2000).
7. Y.-H. Cheng and K.-C. Liu, "Evaluating bicycle-transit users' perceptions of intermodal inconvenience," *Transp. Res. Part A Policy Pract.*, **vol. 46**, no. 10, pp. 1690–1706, (2012).
8. P. Cox, "The Role of Human Powered Vehicles in Sustainable Mobility," vol. **34**, no. 2, (2005).
9. F. Tosi, A. Belli, A. Rinaldi, and G. Tucci, "The Intermodal Bike : multi-modal integration of cycling mobility through product and process innovations in bicycle design," vol. **41**, pp. 1501–1506, (2012).
10. M. Arunachalam, "A Typical Approach in Conceptual and Embodiment Design of Foldable Bicycle," vol. **87**, no. 19, pp. 9–16, (2014).
11. R. Roy, "Case studies of creativity in innovative product development," *Des. Stud.*, vol. **14**, no. 4, pp. 423–443, (1993).
12. M. A. Maleque, M. S. Hossain, and S. Dyuti, "Material Properties and Design Aspects of Folding Bicycle Frame," vol. **265**, pp. 777–782, (2011).
13. [13]J. Hyeong, J. Roh, and S. Kim, "Optimization of pivot obliquity for a foldable bicycle," Int. J. Precis. Eng. Manuf., vol. 17, no. 7, pp. 931–936, 2016.
14. [14]M. C. Chen, "Bicycle foldable to align front and rear wheels along a transverse direction of the bicycle." *Google Patents*, (2005).
15. P. P. M. }ik, "Universal folding bicycle." Google Patents, (2004).
16. M. A. Jayasuriya, J. C. Cheng, and K. M. Schondorf, "Klapprad Folding bike." Google Patents, (2015).
17. D. K. De Jong, "Foldable bicycle." Google Patents, (2001).
18. T. W. Brown Jocelyn, "Design Thinking for Social Innovation," *Dev. Outreach*, vol. **12**, no. 1, pp. 29–43, (2010).
19. D. Norman, "Emotion & Design: Attractive Things Work Better," *interactions*, vol. **9**, no. 4, pp. 36–42, (2002).
20. M. Maguire, "Methods to support human-centred design," *Int. J. Hum. Comput. Stud.*, vol. **55**, no. 4, pp. 587–634, (2001).
21. J. E. Nieters, S. Ivaturi, and I. Ahmed, "Making Personas Memorable," in *CHI '07 Extended Abstracts on Human Factors in Computing Systems*, pp. 1817–1824, (2007).
22. S. Nenonen, H. Rasila, and J. Junnonen, "Customer Journey – a method to investigate user experience," *no. Schmitt* (1999).
23. "Experience-Maps-Using-Customer-Journey-Maps-to-Improve-Customer-Experience.pdf."
24. H. Beyer and K. Holtzblatt, "Contextual Design," *interactions*, vol. **6**, no. 1, pp. 32–42, (1999).
25. J. Nielsen, T. Clemmensen, and C. Yssing, "Getting Access to What Goes on in People's Heads?: Reflections on the Think-aloud Technique," in *Proceedings of the Second Nordic Conference on Human-computer Interaction*, pp. 101–110, (2002.

Mobile application tool for individual maintenance users on high-rise residential buildings in South Korea

Hee-Sung Cha, Ph.D.[1,a], Jun Kim[1], Do-Hee Kim[1], Jia Shin and Kun-Hee Lee[1]
[1]*Department of Architectural Engineering, Ajou University, Suwon, South Korea*

Abstract. Since 1980's, the rapid economic growth resulted in so many aged apartment buildings in South Korea. Nevertheless, there is insufficient maintenance practice of buildings. In this study, to facilitate the building maintenance the authors classified the building defects into three levels according to their level of performance and developed a mobile application tool based on each level's appropriate feedback. The feedback structure consisted of 'Maintenance manual phase', 'Online feedback phase', 'Repair work phase of the specialty contractors'. In order to implement each phase the authors devised the necessary database for each phase and created a prototype system that can develop on its own. The authors expect that the building users can easily maintain their buildings by using this application.

1 Introduction

The construction industry in Korea has grown rapidly for decades. At this time, countless new towns have been constructed enormously in order to solve the housing shortage problems. In this stage, construction companies have just focused on new construction orders, whilst maintenance of existing buildings has been relatively disregarded from the business area. As a result, many apartment houses requiring maintenance are not maintained properly. Also, it is difficult to conclude a contract with a maintenance company because most of the apartment complexes have different owners in each housing unit. As a result, the home owners themselves become responsible for maintenance work. However, there are too many paper-based building maintenance manuals for the house owners. Also, although the house owner would like to use the maintenance manuals, there is a need for some expert advice. There are also maintenance work that experts have to deal with directly. From the standpoint of the home owners, it is necessary to distinguish what can be directly repaired, what needs advice, and what the expert should do because of the cost issues. Therefore, in this study, after accumulating information about maintenance data, the authors develop a system which is selectively connected to maintenance manuals and maintenance experts according to information input by homeowners.

2 Literature Review

In the mid-2000s, a number of building maintenance system proposals for maintenance work using PDAs were developed. [1],[2],[3] As the smartphone became widespread to the general public since

[a] Corresponding author : hscha@ajou.ac.kr

2010, some studies have been conducted to use the smartphone to perform maintenance work on the buildings. However, this study focuses on the maintenance work of the building managers, not the owners of the building.[4],[5] Therefore, there is a need for a new system that allows the home owners to directly perform maintenance work throughout the advice of professionals, if necessary. Given a smartphone application, the ordinary phone users can easily be accessible to the proposed system.

3 Concept Design of Maintenance Application

In order to analyze the maintenance work, the authors classify the defects that can occur in the buildings. The defect type is classified as whether the residents themselves can be maintained, whether expert advice is needed, and whether they should be done by a professional. In addition, repair methods when defects occurred are classified in the same way as defect classification as seen in Table 1.

Table 1. Maintenance defect significance and solutions

Sig. Level	Type of Defect	Solution
1	Residents can repair	Action of Residents
2	Residents can repair, but need professional advice	Advice of Expert
3	Residents can't repair, need direct professional repair	Action of Expert

In order for residents to carry out "level 1 Residents Action", the system needs to provide maintenance manual with the residents. The maintenance manual must also be continuously updated. In order for residents to carry out "level 2 Expert advice ", The system should have the function for the residents to carry out the maintenance work and then provide the feedback by the expert. The system also needs a function that can compensate the experts who provide the feedback. In order for the experts to carry out "level 2 Expert advice ", the ordering, bidding, and awarding process for selecting experts in the system should be included. In this process, residents should be provided with information to assist them in selecting the experts. In Level 3, the selected contractors are evaluated, and the rating score of the construction result is assigned so that it could be confirmed by home owners. Based on this concept, the system algorithm is developed as depicted in Figure 1.

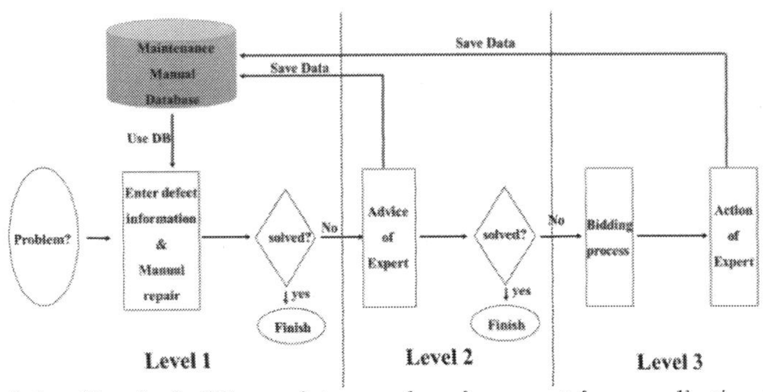

Figure 1. Conceptual algorithm for building maintenance based on smartphone application tool

The application uses three-step approach that progresses in sequence according to the defect classification. This concept consists of level 1 (Residents can repair), level 2 (Residents can repair, but need professional advice) and level 3 (Residents cannot repair, need direct professional repair). When a resident discovers a building defect, he/she carries out self-repair through the manual as the first step. If this method fails, the resident tries again with expert advice. If this method also fails, the resident selects the vendor and constructs the maintenance. In this process, the advice and maintenance contents are stored in the maintenance manual database. When this process is repeated, the maintenance database is updated.

4 Developing Application Tool for Building Maintenance

The proposed system begins with the user's running the application. The user enters the building information and creates a checklist based on whether or not the defect has been checked in advance. If it is proved that this is a problem according to the inspection, the first method, "repair according to the manual," is tried. If the 'repair by manual' fails, the user will enter the second level, 'level 2 expert advice'. At this stage, the user has to send the photographs and information about the defect to the experts. When the expert receives the information, the advice is sent back to the user in text format. If the user does maintenance work using the information received from the expert and fails to remove the defect, the next step will be taken. The third stage is the stage where the experts construct themselves. To do this, the user bids the construction work. After the experts are selected through the awarding, they will perform maintenance work. After the construction is completed, the user will rate an expert and end the entire procedure. The user follows the detailed procedures in each step to repair or exchange information.

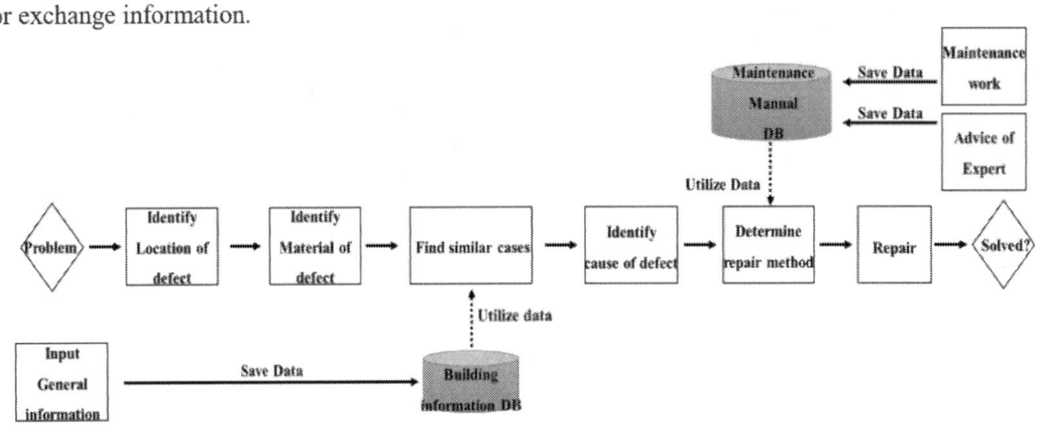

Figure 2. Detailed system design of level 1

Figure 2 shows a detail procedure for level 1. If a defect occurs, the system gets the information about the location and the building materials. The next step is to find a similar case to the defect and make a comparison. The user carries out repair work according to the maintenance manual. The general information entered in this process is stored in the building information DB, and the repair method is retrieved from the maintenance manual DB. Through this process, the user can perform the repair work in Level 1. The detailed procedure for Level 2&3 is shown in Figure 3.

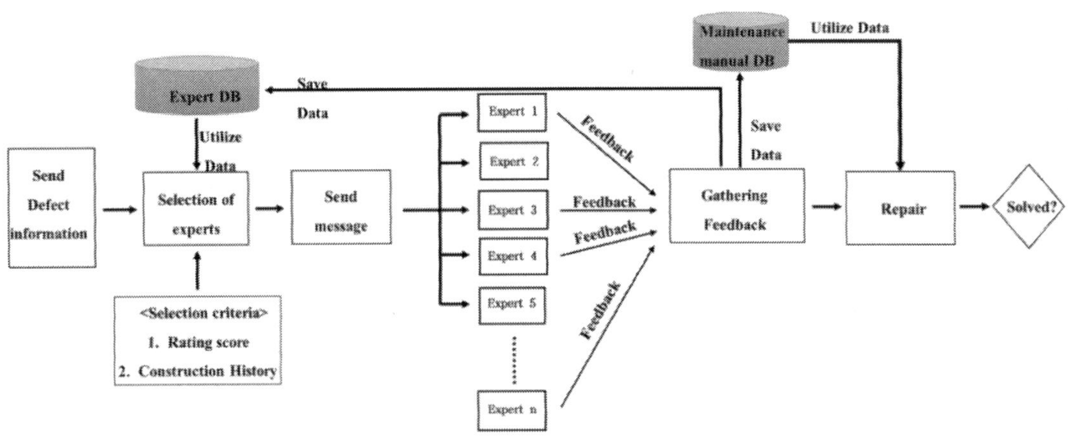

Figure 3. Detail design of level 2 &3

When the user sends the information from the level 1 to the specialist, the expert is selected according to the information sent. Expert DB is used in this process. Using the information provided by the specialist, the user will perform maintenance work. This information was also stored in the maintenance manual so that it could be used again in case of the next problem.

5 Detailed Design of Information Structure

Information generated in the whole process accumulates in four DBs (building information, defect cycle, maintenance manual, and expert information). The structure of each DB is organized as shown in Fig.4.

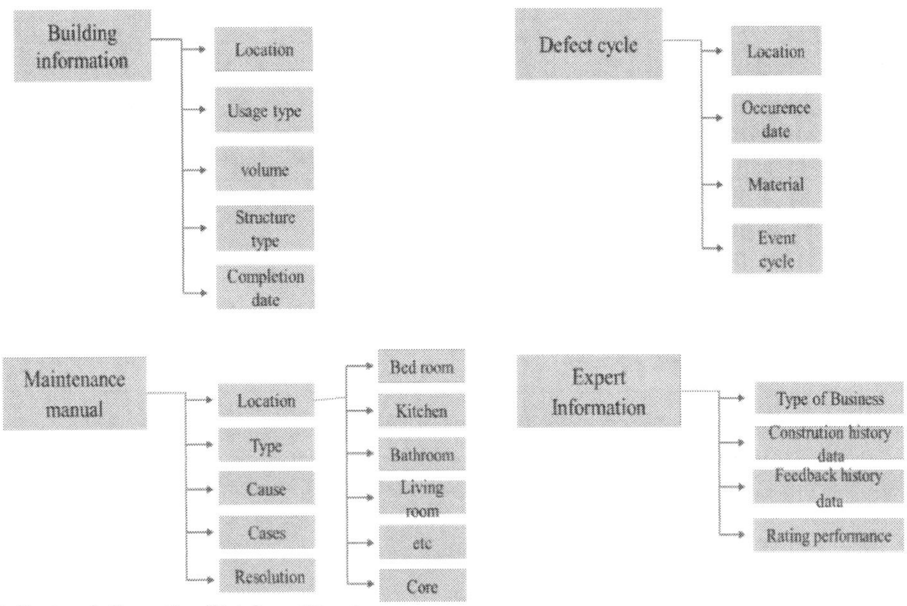

Figure 4. System Information Database Structure

Each DB includes detailed information as depicted in Figure 5. The examples of DBs developed in this study are provided in detail from Table 2 to Table 5.

Table 2. Building information DB

Location	Usage type	Volume (m^2)	Structure type	Completion date
316, Yeongtong-ro, Yeongtong-gu, Suwon	Apartment	3572.3	Wall structure	2010.11
96, Gyeongin-ro 3-gil, Guro-gu, Seoul	Villa	120.6	RC	2005.3
564, Nakdong-daero, Sasang-gu, Busan	House	82.64	RC	1999.01
102, Jungbu-daero, Paldal-gu, Suwon	Apartment	2500.23	Wall structure	2010.03
129, Jungbu-daero, Paldal-gu, Suwon	Office	12000	SRC	2012.4

Table 3. Defect cycle DB

Location	Event cycle(year)	Material
Kitchen-window	10	PVC window frame
Livingroom-finish material	5	Wall paper
Bathroom-Plumbing	3	Stainless pipe
Kitchen-HVAC	4	Exhaust hood
Livingroom-Electronic	1	Lamp

Table 4. Maintenance manual DB

Location	Type	Cause	Solution
Window frame	Can't open	PVC window frame	Need re-construction
Tile	Efflorescence	Poor construction	Grinding with sand paper
MEP	Water leaking at the toilet	Stainless pipe	Need Experts work
MEP	Puddle on the floor	Poor construction	Need re-construction
MEP	Don't work	End of life	Replacement

Table 5. Expert information DB

Expert	Type of business	Construction History(times)	Q & A History(times)	Rating
AAA	Tile	5	10	4.5
BBB	Window frame	10	12	4
CCC	Finish work	4	8	5
DDD	Masonry	5	15	3.5
EEE	MEP	8	10	5

6 Prototype smartphone application tool for building Maintenance

The smartphone application has been made by referring to the algorithms in Figures 2 to 4 by incorporating the information in Tables 2 to 5. The screenshots of the smartphone application are shown in Fig.5. There are four screens in the application. The initial screen shows a general information regarding building maintenance practice, such as ①Input general information, ② Use Checklist, ③ Q&A with Experts, ④Bidding, ⑤Counsel with specific Expert,⑥ Service center. The second screen shows an example of maintenance checklist (⑦ in Fig. 5) and in-situ solution items (⑧ in Fig. 5). The third one, maintenance manual screen, provides a cause-effect-solution relationship (⑨ in Fig. 5) in order to pinpoint the specified defect problems. Through menus ⑦, ⑧, and ⑨, the user is provided with a way to use the necessary checklists and to refer to the manuals. With functions Q&A screen (⑩ and ⑪ in Fig. 5), users can send photos and texts to experts and receive comments from them.

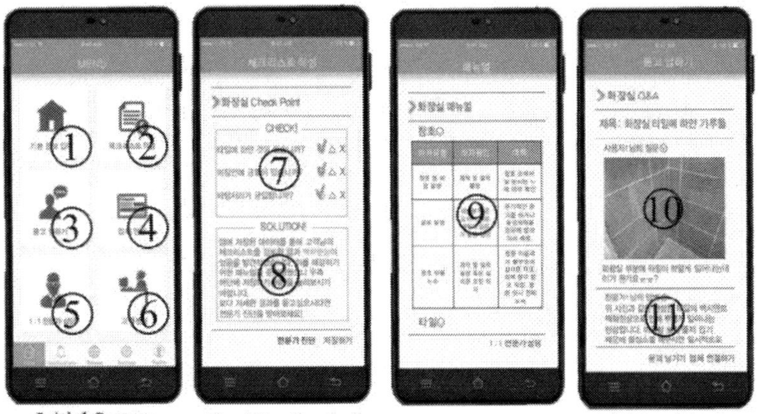

Initial Screen Checklist & solution Maintenance manual Advice of Expert

Figure 5. Exemplary Screenshots of Mobile Application Tool

7 Conclusion

Four positive effects are expected through this study. First, by using the easy-to-access application, it is expected that an active exchange of information among maintenance subjects can be achieved in a timely manner. Second, by providing users with a maintenance manual that is constantly updated, it can provide meaningful information for the general home owners who have little knowledge about the building maintenance. Thirdly, in case of defects, the repair work can be executed through acquaintance, personal contact, or using the Internet service. Using this application, the repair contractors can be selected based on the company's work performance and reliability. Fourth, since the information is structured in real time by using the application, the data about the defect cycle, time, and the location of the defect are continuously accumulated, so that preventive maintenance will be possible in the future. In this study, the authors classify the information into contents that the user can understand for the user's convenience. It is an advantage to be able to use easily by ordinary people who do not have enough knowledge of building maintenance through this method. However, it would be better to use a standard format, i.e, IFC (Industrial Foundation Class), which is used in BIM, to accumulate the relevant information. In this study, the authors have studied how to store and utilize information centered on the apartment, but future study need to be extended to the areas where information can be classified in various building types. Finally, in this study, the home owners have proposed a procedure to connect with experts individually, but a new method is needed for a new business format through collective bargaining agreement.

8 Acknowledgment

This research was supported by the Basic Science Research Program, through the National Research Foundation of Korea (NRF), funded by the Ministry of Education (No. 2015R1A2A2A01005242).

9 References

1. T. Kim, J. Joo,S. Kim, J. of Arch, *J. Architectural Institute of Korea*, **23**, (2003)
2. D. Kang, T. Kim, C. Lee,T. Park , *J. Architectural Institute of Korea*, **33**, (2013)
3. Y. C. Chen, HC. Chiu, M.D. Tsai, H. Chang, C.F. Chong, Compute Methods Programs Biomed, **85**, pp. 181-184, (2007)
4. S. Lee, J. Won, C. Kim, J. You, *J. Architrctural Institute of Korea*, **27**, (2007)
5. J. S. Kirwood, *Facilities*, **13**, pp. 8-12, (1995)

A Preliminary Study on the Priority of Business Evaluating Model for Design Office

Dae Young Kim(a)[1], Nak Ku Baek[2], Dong Heon Lee[3], Dae Young Kim(b)[4] and Hyoung Chul Lim[5, a].

[1]*Pusan National University, School of Architecture Engineering, Professor, Korea*
[2]*Changwon National University, School of Architecture Engineering, Master's Course, Korea*
[3]*Changwon National University, School of Architecture Engineering, PhD Candidate, Korea*
[4]*Kyungnam University, School of Architecture Engineering, Professor,Korea*
[5]*Changwon National University, School of Architecture Engineering, Professor, Korea*

Abstract. Considering that we are at the initial phase of application Business Process Management model, a sequential application plan must be taken for each unit task. Adapting Business Process Management into the enterprise recourse of the construction industry across all business process could cause problems due to rapid change in corporate operation systems. We finally suggested the definition of Business Process Management model by analyzing each business unit so that we could propose and generalize value chain and business process architecture suitable for design office. Furthermore, by evaluating each business unit (which is included in the business process architecture), based on quantitative standards, we presented a sequential application plan for business process model.

1 Introduction

Due to globalization of business environment and fierce competition, the construction industry requires management systems to increase management efficiency and corporate benefit[1]. At the same time, it is necessary to establish management strategy, increasing the value of intangible assets by collecting and analyzing real-time data so that we can take immediate and flexible reaction to cope with the changes.

Other industries noticed the changing environment in advance, actively engaging in research and discussions in order to apply BPM and to improve process-information technology separation[2]. While in the construction industry, however, only a few companies were reviewing the introduction of BPM and were limited to theoretical approach.

Considering that we are at the initial phase of application BPM model, a sequential application plan must be taken for each unit task. Adapting BPM into the enterprise recourse of the construction industry across all business process could cause problems due to rapid change in corporate operation systems. Further, it is unable to predict potential risks while BPM are being applied. Therefore, applying BPM model to the core strategy and task seems more effective than applying BPM model to the entire enterprise recourse planning. In that case, it is desirable to get

[a] Corresponding author : hclim@changwon.ac.kr
"This research was supported by Basic Science Research Program through the National Research Foundation of Korea(NRF) funded by the Ministry of Education(NRF-2016R1D1A1B01012129)."

feedbacks about the problem in the procedure and apply it to other operations accordingly.

This paper suggests a sequential application plan by quantitative criteria of each unit task which is included in business process architecture. The procedures are as follows:

- To consider and propose an evaluation criteria for BPM model application
- To research adoption priorities for BPM model application
- To analyze the importance of each evaluation criteria for BPM model application
- To propose adoption priorities for sequential BPM model

2 Definition BPM application

Us Smith (2003) defined BPM, the third wave, as a well-organized and designed activity, managing and improving business processes that raise productivity and its supporting methodology and system. Weskw (2004) agreed that BPM is a methodology used to design, execute, and control operating processes related to human, organization, application, document and other information sources, or it is something that supports business processes utilizing a system. As mentioned above, definitions of BPM in the existing research are as follows (Table 1.)[3] :

Table 1. Definitions of BPM

Author	Definitions of BPM
D. Jack (1995)	A structural method which analyzes, improves, controls and manages processes for enhancing quality of product or service
Zairi, M (1997)	A structural method which analyzes, improves essential elements such as production, and communication
M. Weskw (2004)	BPM is a methodology to design, execute, control operating process related man, organization, application, document and other information sources, or a thing that supports business processes applying a system
Smith (2003)	A managing methodology supporting business procedure by designing, managing and improving to raise business productivity, or its software system

The established definition of BPM allows the conception of enhanced effects such as a visibility of business processes, business monitoring.

However, difficulty in enterprise recourse planning and feedback followed by corporate core strategy, lack of sequential application plan and sustainable management plan through BPM were pointed as its problem.

Consequently, this paper defined BPM model application of the construction industry as:

A system to improve and visualize business process within and outside the company by applying the business process which is related to the strategy of the construction company; to execute and control human resources and systems (related to work performance) suitable to business process; to monitor and operate business processes continuously.

It is expected to realize business efficiency through business process improvement and maintaining the system by getting rid of inefficient factors.

Applying an BPM model to business strategy of construction industry requires a sequential adoption. The reason for this is that we have to get feedbacks from trial and error, and thus reduce potential risks while we apply it to other tasks.

In order to apply a sequential BPM model, analyzing the construction core strategy must be done ahead of all other procedures, and then analysis of core business tasks that could realize each company's core strategy should be followed. Then we visualize the process of each unit task to BPM's priority operations and get feedback by monitoring and measuring results so that we can get rid of the inessential factors.

3 How to evaluate business process

3.1 Analyzing business process criteria

Lee, Byoung-ok (2004) in his research 'the BPM building strategy of manufacturing industry' suggested long-term cycle operation and operations involved in many organizations[4]. Choi, Jin-ho (2006) suggested high-strategic priority work and those which need process improvement[5]. The BPM building report of 'L' company focused on the importance of business process, operations that involve number of organizations, and operations that need change and automation. The Korea Productivity Center presented work with typical disposal flow, familiar work done by large number of people, and work that is repeated across departments. Finally, the BPR report of 'B' company focused on dissatisfaction factors and operations that involve more than 30 people in one department. The business criterion for the application of BPM to the existing research is as below (Table 2.). After considering each criterion, there are two significant findings. One is efficiency when business process model adoptions occurred, and the other is something that suggests evaluation standard in easiness factor.

Table 2. Evaluation Criteria for the Application on BPM

Reference	Evaluation Criteria	Efficiency	Easiness
A study on the Case of Implementing BPM for a Manufacturing Company Byung-Ok Lee(2004)	An operation involved by many department	○	
	A long-term cycle operation to complete	○	
	An operation apt to frequent changes on guidelines and/or disciplines		○
	A core business of corporate	○	
	An operation which can be easily automated		○
	An operation apt to frequent changes in organization and management	○	
	An operation to be understood of its progress status	○	
	An operation with high-frequency of occurrence	○	
A Study on Process Selection and Implementation for Business Process Management Jin-ho, Choi(2006)	An operation of higher strategic importance (relation with a strategic purpose)	○	
	An operation requiring urgent improving of process (Business productivity, Accuracy, Adequacy of cost)		○
	An operation which needs to be managed and monitored	○	
	An operation which needs rapidity	○	
	An operation to be processed with multiplication: (Organization, complication of processing level)	○	
	Easiness of building BPM(clearness of organization, possibility of supporting legacy system)		○
	An operation of having less risk in application (relation with other process, cases)		○
Report of Implementing BPM of Company 'L'(2004)	Process importance, an operation involved by many departments	○	
	An operation which needs to be standardized		○
	Work needed to manage change and automation		○

	An operation which requires plenty of communication	○	
	An operation which needs to be enhanced for speedier processing	○	
	An operation of strict time limit and often forced control required	○	
	An operation requested frequently to be computerized and to be controlled in loose forms		○

3.2 How to evaluate sequential BPM application

Table 3. Conceptual methods of criteria for applying BPM

Criteria	Evaluation Point	Contents
Efficiency	Corporate value enhancement	The degree of effect on corporate value and importance of the process mean work of importance which enhance corporate value..
	Process complexity	It indicates visualization effect on structural and non-structural works. Level of involvement in the process by related departments affects complication of the Process Under the circumstances of a complicated process and having difficulty to grasp and monitor progressing status of work, we could manage operational process by adopting BPM. On the other hand, more operational process (which is more immanent than simple-repetitive work) visualizes BPM effects more significantly.
	Unit task teamwork frequency	A basic majority-led process performed by variant subjects \ When lots of cooperation is needed, the efficiency of the BPM application increases.
Easiness	IT support	Easiness of BPM application according to the level of information system built in the company
	Frequency of unit task change	Difficulty and irregularity of managing process due to frequent change of organization and management.
	Sensibility of external change	The amount of environmental effects on business process resulting in many changes in law or regulation.

Criteria for BPM application consists of Efficiency (EF) and Ease (EA). Efficiency is classified as improvement of corporate value (V), diversity of process (D), and teamwork frequency (T). Ease (EA) consists of IT support (I), frequency of unit task change (C), and sensibility of external change (S).

Conceptual methods of criteria for applying BPM are as in Table 3.

A system of Process criteria is as in Fig 1.

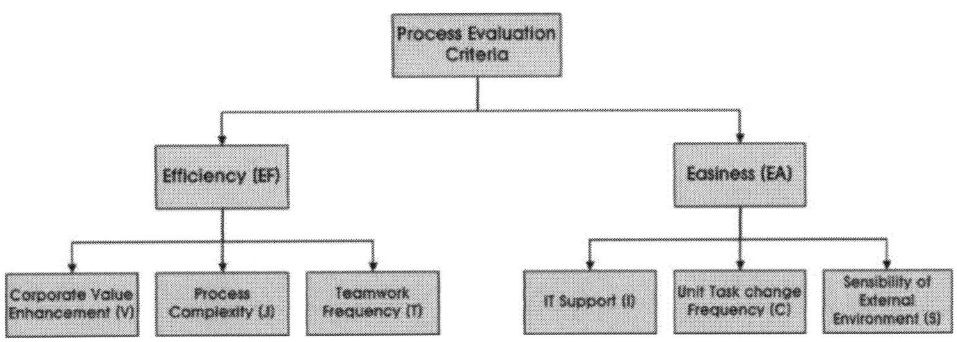

Figure 1. System of Process criteria

4 Analysis Method for Evaluating Business Process

4.1 Questionnaire outline

Relative comparison among each criteria and evaluation of the priorities of BPM model application for each unit task were performed during the week of Feb 28, 2007 through AHP (Analytic Hierarchy Process). These surveys were asked to staff from 11 companies who were in charge of Management, HR, Processing, Finances & economy, IT, or construction performance for at least 10 years. The return rate of the questionnaires was 75% out of 110 copies distributed and is as in Table 4.

Table 4. Questionnaires outline

Criteria	Contents
Purpose	(1) Analyzing the importance among each criteria choosing core area for BPM model application (2) Evaluation the priorities of BPM model application on each unit task
Period	2007~2008
Target	Staff who were in charge of Management, HR, Processing, Finances & economy, IT, and construction performance (11 domestic general contractors)
Research Methods	Step 1: To introduce the survey and to confirm the participant's consent by telephone. Step 2: To introduce the concept of the survey. To explain how to fill out the questionnaire and to distribute personal visit. Step 3: to collect it by mail or visiting.
Return rate	- The number of total copies : 110 - Contractors : 11 companies, each 10 copies - Replied copies : 83 - The return rate : 75%

4.2 Organizing questionnaire to analyze the importance of evaluation criteria

The questionnaire for analyzing the importance of evaluation criteria for applying an BPM model is organized as follows: In terms of Efficiency and Easiness, the criteria were written on the left and right end of the sheet in order to test relative comparison and to allow respondents to mark the importance.

Meanwhile, the form for analyzing Efficiency and Easiness of an BPM model application on each unit task is as in Fig.2. The targets of each unit are the hundred tasks in the Process chain level in the business process architecture. Quantitative evaluation was carried out under an eleven-point measurement.

5 Priority analysis result by applying sequential BPM model.

5.1 Analysis result of the importance of each criteria

Through the relative comparison of each evaluation criteria, we analyzed the importance of evaluation criteria and performed CR (Consistency Reliability) under the result suggested by Satty (1980).

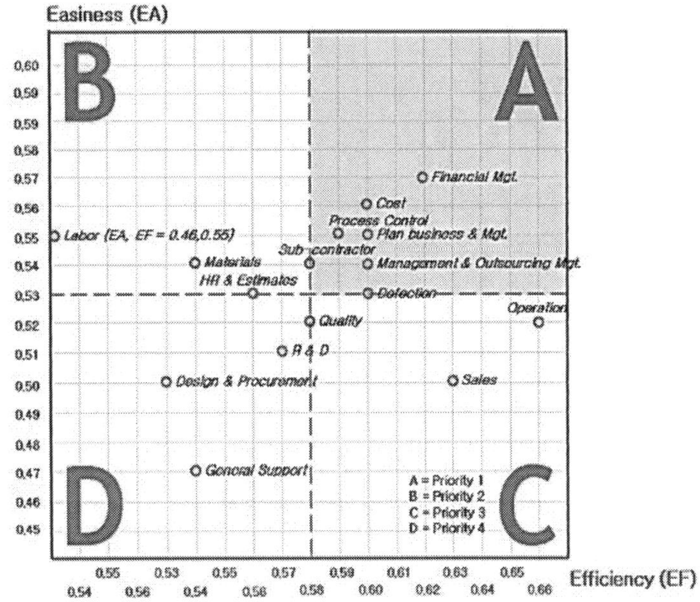

Figure 2. Application Priority for Sequential BPM Model.

6 Conclusion

We finally suggested the definition of BPM model and a sequential application plan for business process model. The details are as below:

In order to analyze the peculiarity of construction firms' business process and adopt business process management related to construction firm's strategy, we enhanced and visualized business process of internal and external environment. Then we suggested the definition of BPM model which executes and controls people and system, consistently monitors business process and continuously improves. It is expected to realize business efficiency through process improvement and also get rid of inefficient factors since the adoption of business process model.

Finally, we have suggested business priorities for a sequential BPM application by quantitative standards. Efficiency (EF) and Easiness (EA) of applying an BPM model.

References

1. Howard Smith and Peter Fingar. The Third Wave analysis. (2002)
2. Hammer. The Reengineering Revolution analysis. (1995)
3. Weskw. Design, management and improvement plan for productivity improvement. (2004)
4. B.O. Lee, A study on the Case of Implementing BPM of a Manufacturing Company analysis. (2004)
5. J. H. Choi, A Study of Process Selection and Implementation for Business Process Management analysis.(2006)

Throughput analysis in mobile adhoc network: a literature survey

Madhurya Kodialbail Divakara[1] and Dr. A. R. Abdul Rajak[2]

[1]*Department of Electronics and Communications Engineering, Bits-Pilani Dubai Campus, UAE*
[2]*HOD of Electronics and Communications Engineering, Bits-Pilani Dubai Campus, UAE*

Abstract. A rapid advancement in the area of mobile networks has been observed over the recent years due to the spread of reasonable, existing wireless devices. Hence it has given way to numerous opportunities to many researchers to work on Mobile Adhoc Networks. There are several key challenges in implementing the MANETS. It does not use any existing infrastructure or centralized administration. Each mobile node acts as a router. Due to its dynamic features, they are endangered. This paper provides a detailed study and throws light on the MANETS Topology in Wireless Adhoc Networks and focusses on performance parameters throughput, end to end delay and network load, its effect on the performance of the MANET network depending upon the number of users.

1 Introduction

Wireless Adhoc Networks are peer to peer wireless networks where wireless devices are able to communicate directly with each other. They have a Base Service Set (BSS) without an Access Point (AP) in order for communication to take place. They neither rely on any pre-existing infrastructure nor use any set of central administrations to follow.

There are 2 main types of Wireless Adhoc Networks i.e. Infrastructured and Infrastructureless. In Infrastructured Networks, a node can move while communicating and the base stations are fixed. As the node elopes out of range from the base station, it merges into the range of another station, whereas in Infrastructureless Networks, the base stations are not fixed. From the above mentioned they can be classified as MANETS (Mobile Adhoc Networks), WMNS (Wireless Mesh Networks), WSNS (Wireless Sensor Networks) and HWNS (Hybrid Wireless Networks).

MANET is a self-configuring infrastructure with fewer networks of mobile devices connected by wireless connections. Such networks may also be connected to the larger Internet. They are dynamically formed by an absolute system of mobile nodes and are made to move freely, randomly hence leading to the change of the network topology occasionally. Besides being in a routable networking setting, MANETS are on the top of a Link Layer model.

In MANETS, nodes within a wireless transmission range are able to communicate precisely ; however, nodes out of its range have to rely on alternative nodes to relay messages. Thus, a multi-hop scenario occurs, where several transitional hosts relay the packets sent by the source host to take it to its destination node. MANETS are capable of multi-hop routing. It is also able to allocate and maneuver security, routing and host configuration. A malicious behavior highly resembles another benign behavior.

Mobile nodes are evolved with lesser memory, power and light weight characteristics. Due to their impulsive behavior it burdens human involvement to configure the network. The nodes have exact features with related responsibilities and competences and thereby enable an entirely symmetric environment. They have immense user density and user mobility.

Types of MANETS includes : VANETS (Vehicular Adhoc Networks), In-VANETS (Intelligent Vehicular Adhoc Networks) and i-MANETS (Internet Based Mobile Adhoc Networks).

VANETS (Vehicular Adhoc Networks) facilitates an effective communication with another vehicle and is free to communicate with roadside equipment.

InVANETS (Intelligent Vehicular Adhoc Networks) uses artificial intelligence to challenge unexpected situations like vehicle collisions and accidents.

i-MANETS (Internet Based Mobile Adhoc Networks) provides a link for both fixed and mobile nodes.

Some challenges in MANETS encountered are that these wireless link features are time-fluctuating in nature; there are transmission impairments like fading, path loss, blockage and noise that increases the vulnerability of wireless channels, limited range of wireless transmissions causes the narrow radio band to reduce the data rates. Hence optimized usage of bandwidth is mandatory by keeping low overhead. Other challenges also include packet losses due to errors in transmission, route changes due to mobility and frequent network partitions.

There are 2 types of routing protocols in MANET. Proactive routing protocol is where nodes endlessly calculate and update routes. Periodic & triggered updates occur when a link modifies. This method is productive if the routes are often used, in order to keep up-to-date information about the network routes. It is quite similar to conventional routing protocols. Each node sustains steady and up-to-date routing information in the form of a table with the next-hop to reach all nodes in the network. These include DSDV (Destination Sequenced Distance Vector), CGSR (Cluster Head Gateway Switched Routing). Reactive routing protocol is where nodes evaluate and alter routes whenever needed. When a node has a packet to send, it checks to see if it has a valid route. If no valid route is known, then node must send out a route-request message to obtain a valid route. This method is efficient if routes are not used often. These include AODV (Adhoc On Demand Distance Vector), DSR (Dynamic Source Routing).

QOS is a depiction and measurement of the overall performance of a service, such as computer network, a cloud computing service and the performance seen by the network users. QOS also provides support to Mobile Adhoc Networks, by supporting internet directly. There is a difficulty in getting and managing the link states directly because they change with surrounding circumstances and also have certain bandwidth constraints.

The main QOS parameters in the network are :

Throughput- The rate of successful message delivery over a transmission.

Latency- The amount of time taken for the message to traverse a system.

Packet Loss- The number of packets lost during the transmission process.

Jitter- The variation in the delay of received packets.

Reliability - Percentage of packets discarded by the user.

Bit Rate- The number of bits per second that can be transmitted along a digital network.

End-to-end delay- The time taken for a packet to be transmitted across a network from source to destination.

2 Related Works

This section presents some previous works related to throughput and end to end delay performance analysis on MANETS. Adhoc Networking has become one of the most interesting research fields inside the networking.

H. Hallani, et al. [1] in 2005 proposed in their paper about creating a small physical network, gathered experimental results and compared those with simulations through OPNET. The simulation results showed good agreements with those obtained through the experiments conducted physically. It also showed that a linear drop in the throughput, with reference to the number of nodes simultaneously connected to the same destination.

There are also some interesting references based on real applications, so Monika Rajput [2], et al. have highlighted in their research paper about the various reactive routing protocols such as Ad hoc On-demand Distance Vector (AODV), Dynamic Source Routing (DSR) and Temporally Ordered Routing Algorithm (TORA) and compared them on the basis of throughput by increasing number of nodes in the network.

There is an interesting approach that researchers are currently investigating in the application of reinforced learning in routing protocols [3]. For e.g. Mostafa Fazeli et al. presented in their research paper about throughput analysis using 10, 15 and 20 users without using any routing protocols through OPNET software. These researchers extensively used graphs to highlight their point of view. From the simulations it was observed that as the number of users keeps increasing then the throughput of the network increases.

Ikramullah Khosa et al. [4] evaluated in their paper by providing a detailed analysis of IEEE 802.11 medium access control (MAC) protocol. The performance of CSMA/CA protocol is studied in depth by using optimized network engineering Tool (OPNET) with different sets of WLAN parameters. Simulation results have been obtained in different network scenarios by changing the number of stations in the network. This was also analyzed for adhoc and infrastructure networks.

Fan Ya-qin et al. [5] suggested a better solution to overcome problems of inefficiency in large expansion of adhoc routing protocols through DSR routing protocol and concluded that DSR protocols are suitable for only small scale MANET networks. Simulation results thus have obtained references for further research & study.

Thodeti Srikanth, et al. [6] besides doing a bit of introduction in their research paper mainly considered different methods like routing load, end to end packet delivery and performance of protocols and used http and low load traffic. These researchers concluded that the efficiency of a network can be achieved by selecting efficient protocols based on the network requirement. This research article mainly explained about the simulator OPNET Modeler, its architecture and performance parameters to a certain extent.

In their paper N.S. Samshette, et al. [7] calculated about throughput, end to end delay, packet delivery ratio in order to evaluate the performance of different mobility models; Random Walk, Random Waypoint, Random direction, Gauss Markov, Manhattan Grid through routing protocols. By using Random Walk AODV model high throughput and packet delivery ratio were produced leading to low end to end delay as compared to Random Waypoint and Reference Point Group Mobility model. Hence this also led to the explanation of relevant points on hybrid on demand protocols by the researchers.

Beata Krupanek et al. [8] presented a detailed study on the quality of service for modern wireless networks and had also explained relevant details about multimode network simulations in the OPNET and discussed about the prime ways in ensuring proper transmission of quality measurement in wireless sensor networks.

P.Venkata Maheswara et al. [9] in their paper compared about the various routing protocols for MANETS i.e. AODV, DSR, TORA and OLSR. Its performance was evaluated by considering the parameters such as routing overhead, delay, throughput and network load by varying the mobile nodes.

Parulpreeth Singh et al. [10] highlighted the performance of the DSR reactive routing protocol under traffic http, video conferencing, ftp and email. On the basis of evaluation the performance parameters such as throughput and end to end delay were taken into consideration and they concluded which traffic delay would be best for those parameters.

Sukhwinder Kaur et al. [11] in their paper had shown a comparison regarding performance parameters in terms of delay, throughput and media access delay by using the routing protocols AODV, OLSR and DSR also studied the network performance. The application traffic taken into consideration were http and data base. They wanted to check which protocol gave the best result for their study.

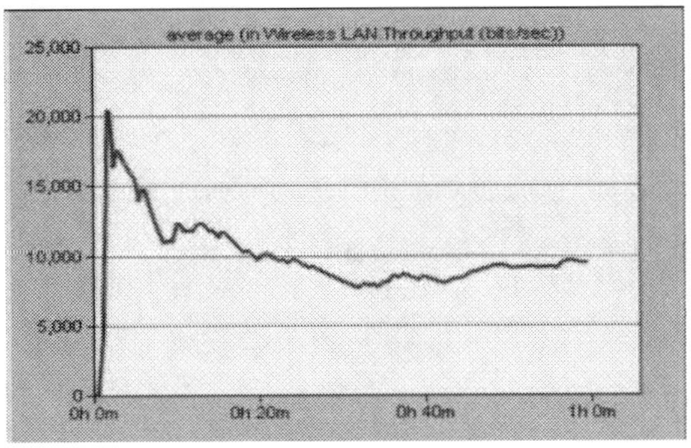

Figure. 1 Throughput Graph for 10 users. (Fig 1, [3])

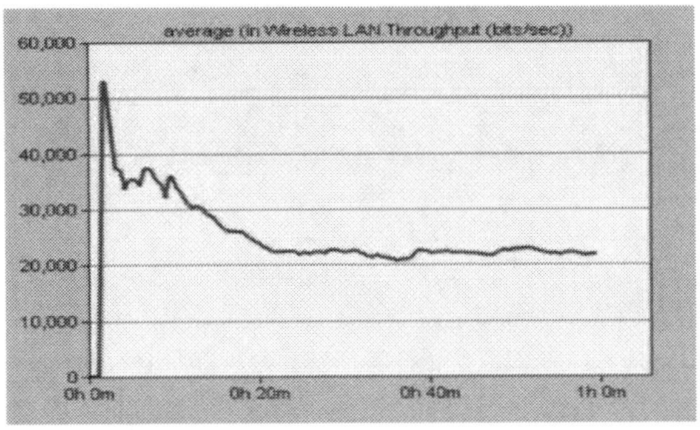

Figure. 2 Throughput Graph for 15 users. (Fig 2, [3])

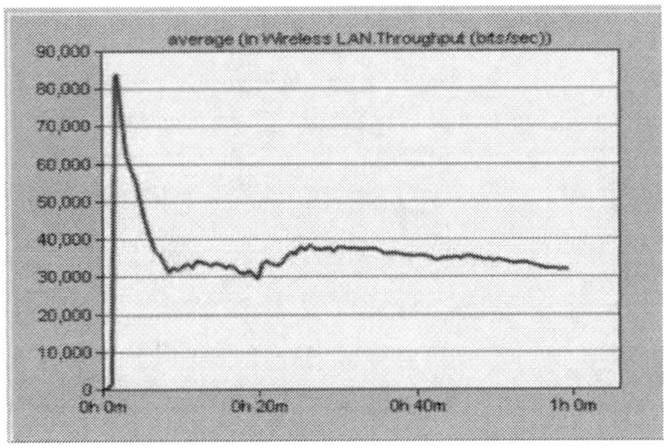

Figure. 3 Throughput Graph for 20 users. (Fig 3, [3])

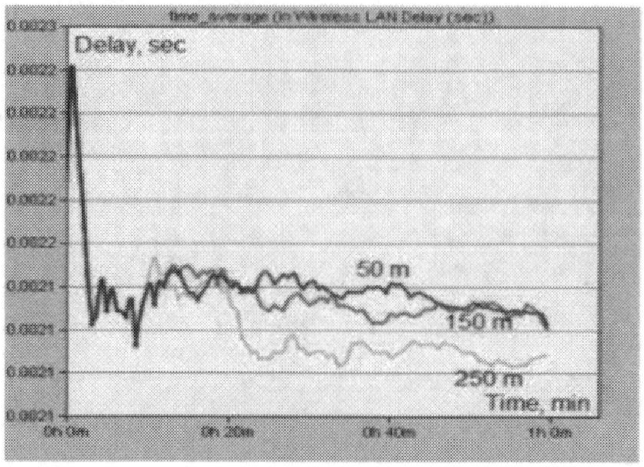

Figure. 4 Delay vs. node distance for DCF (Distributed Coordinated Function) algorithm. (Fig 4, [8])

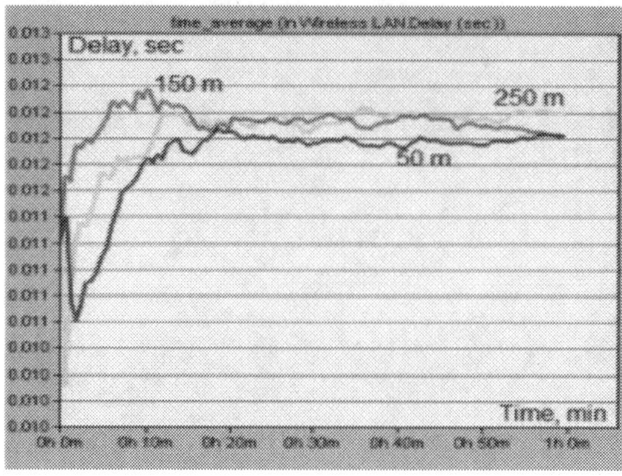

Figure. 5 Delay vs. node distance for PCF (Point Coordinated function) algorithm. (Fig 5, [8])

From the above figures it can be observed that the throughput keeps increasing as the number of users keep increasing as illustrated in Figure. 1, Figure. 2 and Figure. 3 but in Figure. 4 and Figure. 5 it can be seen that the node distance and delay are indirectly proportional in DCF algorithm, whereas in PCF algorithm distance and delay are proportional to each other. Hence we say that PCF mechanism affects delay much more than DCF.

3 Results

H. Hallani, et al [1] concluded by evaluating the performance of adhoc networks through simulation and analysis. Simulation results confirmed the results that matched with the ones obtained through physical experiments. It has also showed that when the signal strength between source and destination is weak, routing the traffic will take place through an interconnecting node and leads to greater values of throughput.

Monika Rajput [2], et al concluded by comparing all the protocols and through the results commented that TORA performs better than AODV and DSR as the number of nodes keeps increasing. The performance of the network was evaluated for throughput explicitly.

Mostafa Fazeli et al. [3] compared the throughput parameters for different scenarios when the number of users were 5, 10 and 15. Through the simulation graphs they observed and analyzed that as the number of users increase then throughput also increases.

Ikramullah Khosa et al. [4] concluded by analyzing the performance of channel access protocol under varied conditions of the network. By simulating through OPNET they provided good views that performance was better if packet size was small and it was better performing under Adhoc mode. The performance of the protocol can be upgraded by using adaptive algorithm.

Fan Ya-qin et al. [5] concluded by conducting simulations of DSR protocol through a broad range of distinctive networks. It was better to be used when the topology of the network was small and showed complexities when the topology of the network changed and when the network size increases continuously. DSR will show a number of networks performance problems. Some kinds of improvements are needed to be made to upgrade the DSR protocol to easily adapt to the changes in network topology.

Thodeti Srikanth, et al [6] completed their research by giving and ensuring an insight about the simulation based approach on MANET and Adhoc. They also explained about the software in great depth.

N.S. Samshette, et al [7] concluded by using different mobility models for analyzing different parameters such as throughput, end to end delay, packet ratio with different routing protocols as AODV, TORA, DSR and AOMDV. Through these results it was finally proved that Gauss Markov Model has better results as compared to Manhattan Grid Model.

Beata Krupanek et al. [8] summarized and concluded that if DCF (Distributed Coordination Function) was used then it would cause high latency depending on the type of traffic. If PCF (Point Coordinated Function) was used it would work for all nodes and provides a good overall performance. P.Venkata Maheswara et al. [9] evaluated and concluded by explaining the protocols AODV, DSR, TORA and OLSR ad-hoc routing protocols under different load and different number of users. The software used is OPNET Modeler 14.0 and simulations with varying traffic were run for 3600 sec.

Parulpreeth Singh et al. [10] concluded their research by considering various load conditions and checked which one performed the best for throughput and end to end delay. So it was found that http showed best performance in terms of both throughput and delay. In video conferencing throughput was the highest and was lowest in http.

Sukhwinder Kaur et al. [11] evaluated and concluded that in data base traffic, OLSR achieved highest throughput, least delay and least media access delay. In http traffic, OLSR performed well. So, it was finally observed from their research that OLSR performed the best compared to other two protocols.

4 Conclusions

From the above 11 surveys and researches carried out by various researchers it summarizes, an overview of Mobile Adhoc Networks (MANETS), its applications and characteristics that distinguishes it from other wireless networks. Due to its characteristics, there is a need of separate routing protocols for MANET. Classification of routing protocols for MANET has been done on the basis topology of the network i.e. proactive or table-driven and reactive or demand-driven. A summarized overview of routing protocols belonging to each type of classification has also been presented above hoping that it will be useful and helpful to students and researchers in this field. From this, it is concluded that MANET routing protocols are designed based on the application area and environment and it is not possible to design a single protocol, which is suitable for all MANETS in all situations.

References

1. Hallani, S.A. Shahrestani and F.H. Stootman, "Wireless Adhoc Networking: Analysis and Simulation of Validation Results," TENCON 2005 IEEE Region 10.
2. M. Rajput, P. Khatri, A. Shastri and K. Solanki, "Comparison of Ad-hoc Reactive Routing Protocols using OPNET Modeler", International Conference on Computer Information Systems and Industrial Management Applications (CISIM), 2010 IEEE, 530-534.
3. M.F.H. Vaziri, "Assessment of Throughput Performance under OPNET Modeler Simulation Tools In Mobile Adhoc Networks (MANETS)", 2011 Third International Conference on Computational Intelligence, Communication Systems and Networks, 328-331.

4. I. Khosa, U. Haider and H. Mosood, "Evaluating the Performance of IEEE 802.11 MAC Protocol using OpNET Modeler", 2010 International Conference on Electronics and Information Engineering, **2**, 91-95.
5. F. Ya-qin, F. Wen-yong and W. Lin-Zhu, "OPNET- Based Network Of MANET Routing Protocols DSR Computer Simulation", 2010 WASE International Conference and Information Engineering,46-49.
6. T. Srikanth and Dr.V.B. Narsimha, "Simulation- Based Approach to Performance Study of Routing Protocols in MANET and Adhoc Networks", IJCSNS, **11,**111, (September 2011).
7. N. S. Samshette, A.D.Bhoi and R.D.Kharadkar, "Analysis & Comparison of Mobility Models for Ad-hoc Network", IJRITCC, **2**, 1357 – 1362, (May 2014).
8. B. K. R. Bogacz, " OPNET Modeler Simulations of Performance for Multinodes Wireless Systems",IJMQE, 2-5.
9. P.V. Maheswara and K. B. Naik, "Routing Protocol Performance Issues and Evaluation Considerations in MANET", IJERA, **3**, 1-5, (Jul-Aug 2013).
10. P. Singh, E. Barkhodia, G. K. Walia, "Evaluation Of Various Traffic Loads in MANET with DSR Routing Protocols Through OPNET", IJDPS, **3**,75-83, (2012)
11. S. Kaur, R.K. Bansal and S. Bansal, "Performance Analysis Of AODV, OLSR and DSR Routing Techniques For Adhoc Mobile Networks", IJCSEIRT, **3**, 195-200, (Dec 2013).

A Study on the Design of Rural Material Flow System Based on the Traditional Chinese Twenty-four Solar Terms - A Case Study of Lifengjie in Liyang City, Jiangsu Province, China

Xiao-Yu Xie [1,a] , Sheng-Xi Fan[2]

[1]*Behavior Cognition Design Research Lab, Tongji University, No 1239 Si Ping Road, Shanghai, China*
[2]*Behavior Cognition Design Research Lab, Tongji University, No 1239 Si Ping Road, Shanghai, China*

Abstract. In the view of Chinese traditional twenty-four solar terms, the circulation of various substances in the material flow system of base production and life is systematically analyzed, and on the basis of the existing system problem point, as well as by combining with the concept of the twenty-four solar terms and the system design method, the author reconstructs each subsystem of the base material flow system, moreover, the author eventually connects the subsystems and reconstructs a material flow system conforming to the traditional Chinese solar terms and conforming to the requirements of ecological sustainability.

1 Introduction

Limit to Growth describes such a scene: the huge explosion in population, environmental degradation, industrial growth without restraint, and the loss of urban morality, causing the loss of earth's resources as well as environmental burdens in human economic activities, eventually leading to the environmental disruption [1]. Agricultural production is governed by the laws of industrial production, whose subversion of the laws of nature can affect the whole food production system, and such a single development model will have more serious environmental and social problems for developing countries, or less developed areas, thus causing immeasurable damage to the traditional customs and cultures that have been brought together in the most harmonious way with the ecosystem for hundreds of years [2]. The return of the twenty-four solar terms and the contemporary people's concern about the safety of diet are the background of this study, while because of its better preservation of the characteristics of Chinese traditional farming society, Lifengjie has become the experimental field for systematic design and research based on the traditional twenty-four solar terms culture in China. The understanding and the evaluation of the environmental, social and economic performances of complex agricultural food systems is probably the real challenge, and the design of more sustainable alternatives has been recognized as necessary for a correct territorial management [3]. However, the system design started late in China and lacked practice cases, so if we can use systematic design method to carry on systematic research to "twenty-four solar terms", and carry on the practice of system design, on the one hand, it provides a new idea for the study of the twenty-four solar terms,

[a] Corresponding author: xiexiaoyu_tongji@163.com

and on the other hand, it fills the gaps of the study of the traditional agricultural circulation model of the Orient in the design of the system, especially the ecological design.

2 Tools and Methods

System design is an applied design method combining system thinking and industrial design. In this study, the author uses the system design method to analyze the elements of the base material flow system [4].

2.1 Application range of system design

As a subject, design has both sensibility and rationality, and is widely used in art, anthropology, economics and management, technological progress and other aspects. And the system design is concerned with human development and economic benefits, and reflects the significance and value of design, moreover, it differs from artistic design from the inner emotional level, and requires a certain amount of scientific measurement and analysis to solve the problem rationally. Their relationship is shown in the following figure. 1:

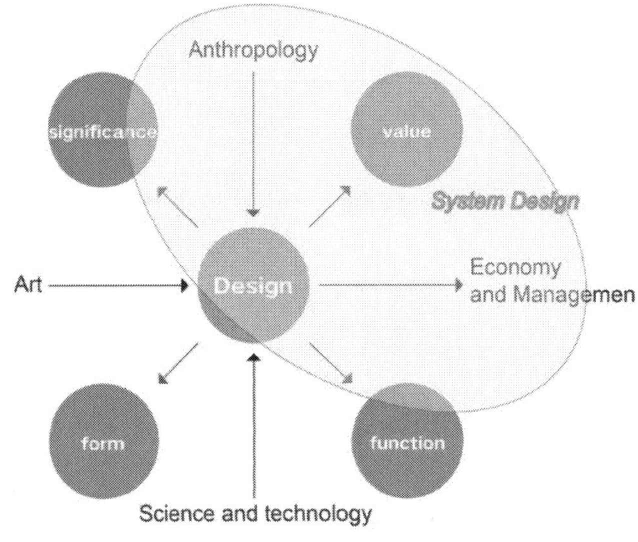

Figure 1. Range of application system design

2.2 System design method

Systematic boundaries should be delineated at the beginning of systematic studies, the qualitative and quantitative methods can be used to study the nature, definition and quantity of the input and output in the system according to the resources in the system boundary, moreover, the relationship between the activities in the system should also be paid attention to. At the same time, we should pay attention to the factors such as the ability of the participants and the action ability of the system, which will affect the efficiency of the system [4]. The relationship is Showed in figure. 2.

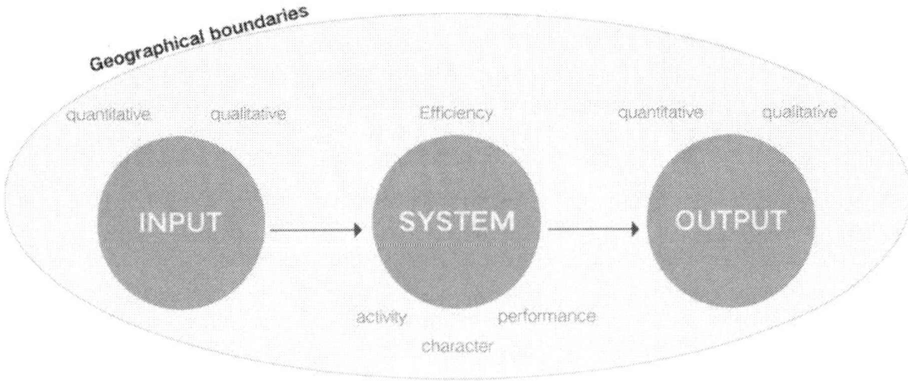

Figure 2. System design method

2.3 Guidelines for system design

2.3.1 Waste is resources

Mechanical linear thinking often results in excessive exploitation of resources and the production of large amounts of waste, while the rational planning system can improve the efficiency of the system, thereby enabling the output of a system to be transformed into another system's resources, then there will be no waste.

2.3.2 Open system

Various systems in nature, including objects as small a enzyme catalysis in cells, as well as the energy exchange between the earth and the universe, are all open systems [4]. The system design is an open system that allows material flow and energy flow to flow together, which is different from closed loop systems in industrial ecological design.

2.3.3 The original system is related to other systems

The open system repeats the functions of the system in dynamic equilibrium, while maintaining itself and evolving with other systems associated with it, and when an element of a system changes, the system adjusts itself, thereby affecting the output of the system, and the change of the output of this system will inevitably affect the input of another system [5]. Finally, each system is connected with each other and adjusted dynamically.

2.3.4 Geographical importance

The system has boundaries which can be physical boundaries, such as cell walls; and they can be cultural boundaries as well, such as Islamic cultural circle, etc., and the system is meaningful only if it is discussed within a certain boundary. Different geographical environment contains different resources, and breeds unique resources, culture and residents. Therefore, the emphasis on regional differences is the basis of system design, so there are no two identical system designs.

2.3.5 The importance of man in the system

The system has boundaries which can be physical boundaries, such as cell walls; and they can be cultural boundaries as well, such as Islamic cultural circle, etc., and the system is meaningful only if it is discussed within a certain boundary. Different geographical environment contains different resources, and breeds unique resources, culture and residents. Therefore, the emphasis on regional differences is the basis of system design, so there are no two identical system designs.

The idea of suiting measures to local conditions and waste utilization advocated by the system design coincides with the concept of fine agricultural production in the traditional twenty-four solar terms of China, in addition, the ecological farming experience, vegetable rotation for sericulture reeling in Chinese traditional agriculture is also praised by the scholars who use the system design method to carry on the circulation agriculture pattern research, and is often listed as a positive case in various books. Because of this, the system thinking and system design approach can work when we try to optimize the life system of the project base according to the twenty-four solar terms. The idea of suiting measures to local conditions and waste utilization advocated by the system design coincides with the concept of fine agricultural production in the traditional twenty-four solar terms of China, in addition, the ecological farming experience, vegetable rotation for sericulture reeling in Chinese traditional agriculture is also praised by the scholars who use the system design method to carry on the circulation agriculture pattern research, and is often listed as a positive case in various books[2]. Because of this, the system thinking and system design approach can work when we try to optimize the life system of the project base according to the twenty-four solar terms.

3 Achievements

3.1 Linear

In administrative relations, Lifengjie belongs to Pingqiao village, Tianmu Lake Town, Liyang City, which has 103 households, more than 300 resident population, and whose north-south length is about 1000m, east-west span is about 500 meters, the lowest point is 100 meters, and the hills are around 230 meters above sea level. The total area of cultivated land is 279 mu, of which 143 are paddy fields and 145 are dry land. The bamboo forests around the mountains are about 900 mu, and the area of the ponds is 23 mu. Moreover, agriculture and forestry are the main economic fields, paddy fields mainly produce two season rice, dry land grows various kinds of vegetables and chestnuts and other economic crops; and the forest provides bamboo shoots, bamboo shoots, bamboo and other raw materials, therefore, it maintains the production and life style of the family units in the traditional farming society of China.

Lifengjie is surrounded by the Lifeng Mountain to form natural geographical boundaries, therefore, the author's research will also take this as the research boundary. The villagers there open up the flat areas in the middle of the depression into their fields, and their residential areas are located on the foothills of the depressions, thus their main sources of water resources are natural rainfall and mountain spring water, and villagers set up several ponds to function as water storage. In addition, according to the different types of land, the main output areas of various ingredients are also reflected in the Figure. 3.

Based on the analysis of the material flow direction, rhythm and economic benefits of the existing life production system, the author sums up the systematic characteristics and problems of the current material flow in Lifengjie:

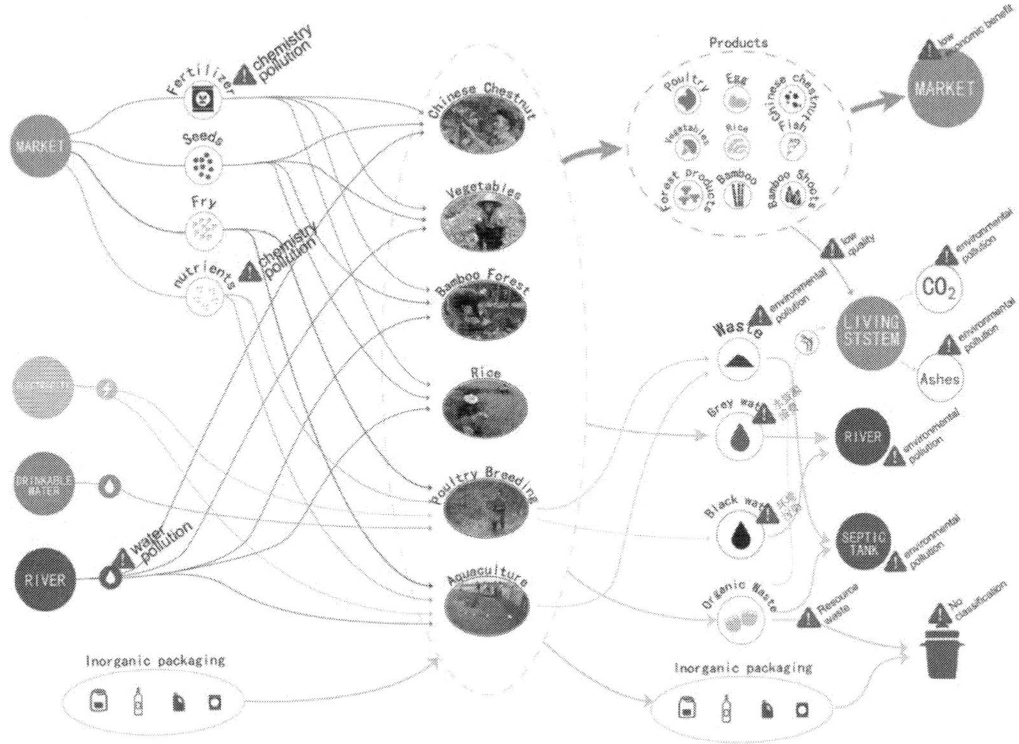

Figure 3. Linear agriculture system and problem points

3.1.1 The whole is in line with the twenty-four solar terms

Lifengjie keeps many production experiences of Chinese traditional agriculture in agricultural production, such as plant crops in rotation, and raise chickens in bamboo forests, etc.. In selecting crop types, in addition to the economic benefits, seasonal crop suitability is also considered, therefore, it has done better in accordance with local conditions, moreover, it makes full use of local weather, light and heat resources, and because bamboo is suitable for production in this area, it energetically develops the related industries of bamboo forest industry.

3.1.2 The diet of the residents does not meet the requirements of the solar terms

After systematic analysis, there are two main sources of food ingredients for residents in Lifengjie: First of all, the ingredients come from stores in Pingqiao village, which include all kinds of anti season foods, vacuum canned food and so on. These foods share a common feature: after being preserved in the processing area, they are transported to Pingqiao village, and then reached the villagers' hands through the stores. And the ingredients are the products of over industrialized production, which are co opposed by two authors in Slow Food and Blue Economy. Whether GM crops affect the body is not yet conclusive, however, the damage done to humans by a large amount of chemical agents have been paid attention to by an increasing number of people, what's more, long-distance transportation often destroys the nutrition of the ingredients, therefore, the author believes that the quality of such ingredients is not high. The other source of food ingredients for residents in Lifengjie is local product, and this approach seems to be much better in quality than the above approach, however, when the author systematically analyze, it is found that a large number of local crops are produced by using chemical fertilizers and chemical herbicides, insecticides and so on, and these chemicals remain on the surface of the crop and enter the soil and water and enter the crops, so the ingredients are not of high quality.

3.1.3 The waste of part resources

When we systematically analyze the production process of its ingredients, we can find another point of pain in the system: a large amount of "resources" are wasted. The "resources" mentioned here are actually treated as waste by villagers, which include not only organic substances, including abandoned limbs and chestnut shells in agricultural production, and food scraps and poultry feathers in poultry, etc., as well as inorganic substances, which include food packing, raw material packing of crops, etc.. When dealing with waste, the villagers cannot achieve the sorting of organic and inorganic substances, so that waste cannot be used efficiently, and a large number of "resources" are simply buried, even burned outdoors, resulting in air and dust pollution. Another kind of "waste of resources" is the waste of energy, first of all, the use of abandoned limbs for incineration is a very low rate of conversion. The waste of water resources is also a thorny problem. In the theory of system design, water resources are divided into several levels: drinking water, grey water (available water) and black water (the water that causes pollution), and during the daily life of the villagers there, a large number of water with the drinking water level is directly used for activities that do not require high water level water, which in itself is a waste of water resources.

3.1.4 Environmental pollution factors

A great deal of data tells us that there is a limit to the carrying capacity of the environment which can purify itself within a certain range. Through the analysis of the dietetic cultural system in Lifengjie, the author finds that a great deal of chemical fertilizer and chemical herbicides are the source of environmental pollution in its agricultural production, while chemical elements cannot be quickly degraded naturally, which move into the soil and water sources and enter the human body with the plant chain, and after accumulating to a certain amount, they can affect the health of the human body. Another source of pollution comes from the burning of plant stems, which produce large amounts of carbon dioxide gas and harmful gases that do not burn enough, in addition, when dust enters the respiratory tract of the human body, it can also cause damage to the human body. Therefore, in the subsequent systematic transformation, it is necessary to reduce the access of chemical fertilizers, and avoid the method of burning plants is used again in life and fertilization.

3.2 Systemic

In the recycled material recycling system, clean energy is used more often in residents' lives, for example, they mainly use electricity to control the temperature of the environment and to clean their daily lives, and compared with the burning of dried bamboo and dead branches, it can reduce a large number of harmful greenhouse gas emissions, and reduce dust pollution, thereby improving the living environment of residents and environmental benefits; in the use of detergents, the author advocates to reduce the use of chemical detergent and washing by physical methods or using biodegradable detergent, and residents can even make lemon and sodium bicarbonate detergent at home, thus reducing the chemical pollution to the base soil and water; in the water cycle, when the drinking water enters the diet, the residue of the solid residue after washing can be removed from the grey water, and the grey water can be used for the cleaning of the individual or the living environment to realize the hierarchical utilization of the water quality, finally, the black water which cannot be used directly can be introduced into the septic tank for purification, in addition, the air conditioning water when using air conditioning can also be used for environmental cleaning. Through the reconstruction of the water cycle of the residents' living system in Lifengjie, the author realizes the multi-level utilization of water resources, thus reducing the waste of water resources; and the residual organic matter in the residents' lives can be input into poultry breeding and renewable energy systems to achieve multi-level trophic stratification.

3.2.1 Agricultural production system

In the "food production" system in Lifengjie, rice planting and bamboo forest industry is the largest, and thus the use of water resources is also the most, therefore, rice planting and bamboo forest industry can be taken as the starting point, and water stratification can be used in the two quarter of spring and autumn, water can be used as a medium, and the nutrients in water can be layered and utilized at many levels, thus saving water resources and saving nutrients. The system flow to implement this idea is: Use the river water to irrigate paddy fields, remove evaporation water, and use the rest of the water for vegetable and chestnut farming, and grey water with nutrients is used in poultry farming and aquaculture, until the water contains harmful substances and turns into black water and concentrates on its water purification system, while the bamboo forest industry uses natural rainfall to irrigate the surface water and consciously guide the pond for aquaculture. Farmers can arrange crops to select crops according to the information in their solar terms, so as to ensure every solar term can provide the appropriate solar term food for the dietary system in the Lifengjie scenic spots. In its agricultural production system, the surplus of rice, straw and other organic materials are no longer converted to fertilizer by burning, but are concentrated in its renewable energy system, and then compost through biogas digesters, while producing biological fertilizer, the biogas provided can be used for the living system of the residents and the solar diet and commercial system. In addition, the remaining organic matter such as rice can be used as a substitute for the chemical nutrients for poultry farming and aquaculture. At the same time, some of the raw materials in the agricultural production system can be kept for breeding and returned to the system.

3.2.2 Commercial system

In the past, because it was not developed for a large scale, it did not have commercial diet. With the completion of the twenty-four solar tourism development, the solar commercial system will become a major source of income for residents in Lifengjie and an important component of its system transformation. Sources of ingredients for the solar business system is partly from its own solar term ingredients, including rice, seasonal vegetables, spring bamboo shoots, wild mushroom etc., the other source is from the featured food around Liyang, such as Tianmu Lake fish head, Liyang white tea and all kinds of fruits and so on. The black-water produced in the throttle restaurant is introduced into the integrated water purification system of Lifengjie. The organic residue in the solar energy restaurant can be introduced into the poultry breeding program to achieve multi-level trophic stratification.

3.2.3 Water treatment system

In the past, it did not have a comprehensive water purification system. In the future, the use of water resources, which had not been sufficient, would have been more intense as its population and visitor population increased, therefore, under the premise of ensuring the rational use of water resources, which is of great significance to carry out the complete planning and rational use of water resources. The future water treatment system will transform the grey water from the "food production" system, the living system of residents and the commercial diet of the solar system into recycled water and return it to the system through foam separation, septic tank solid-liquid separation, water purification, plant filtration and other stages, and the solid waste separated from the septic tank can enter the renewable energy system and can be converted into bio fertilizer and clean energy.

3.2.4 Renewable energy system

Now, some families have methane pools, in the new system, the large-scale renewable energy system will be built with Lifengjie as a whole, collect solid waste that can be converted into biogas and organic fertilizer, and thus the organic fertilizer produced by the renewable energy system flows to the agricultural production system, and biogas can be used for the food and cooking of the residents' living system and the solar business system. The designed material flow system is showed in Figure. 4.

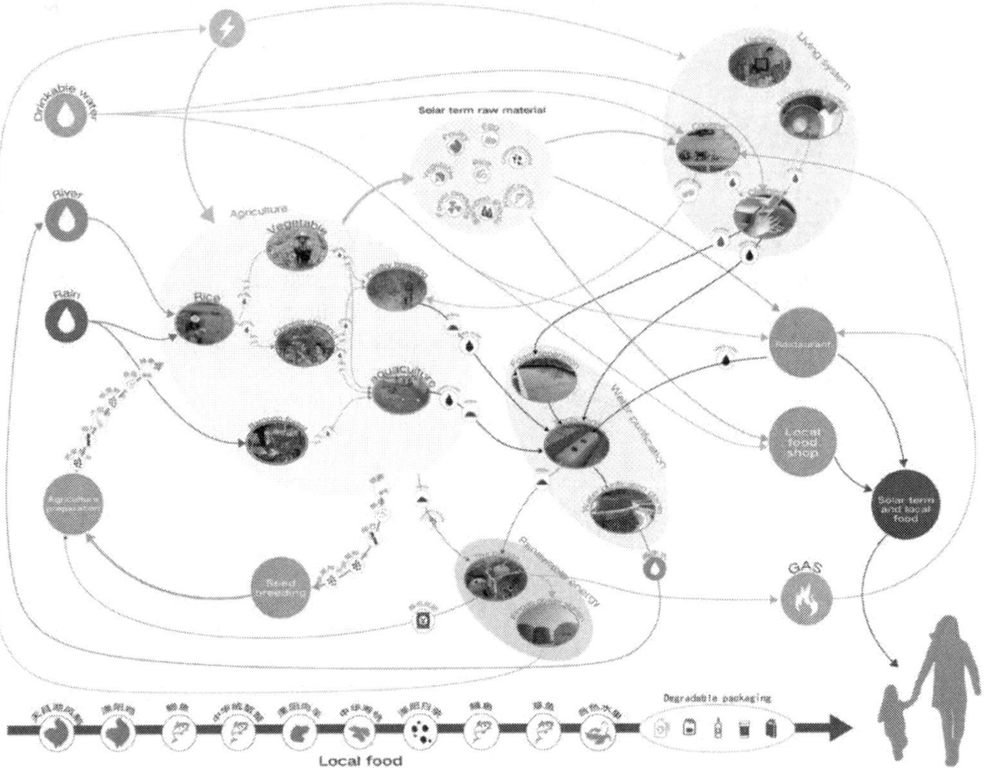

Figure 4. Solar terms ecosystem

On the basis of subsystem reconfiguration, the author combs the relations among subsystems, in general, when reconstructing the circulation of agricultural production related substances from the perspective of the dietary culture of the solar terms, the concept of solar energy ingredients is emphasized, and a new material cycle is built around the solar energy ingredients, and in the new system, the solar feed is provided by the agricultural production system and flows into the solar energy commercial system and the living system of the residents. As the material basis of constructing the system of solar term food culture, other material circulation should follow the concept of twenty-four solar terms, which adopts measures suiting local conditions, and does farm work in the right season, and study modern ecological agriculture and recycle agriculture to plan, so as to reduce the material needs of the system, and to produce less waste of system materials, and thus improve the overall efficiency of the use of resources.

Then we start each subsystem in the system, and finally get its material flow of dietetic culture of the twenty-four solar terms, and system material cycle diagram. From the perspective of tourists, they buy food ingredients from its own seasonal ingredients, which mainly include rice, vegetables, chestnut, spring bamboo shoots, wild mushroom products etc., on the other hand, it is provided by the nearby areas, such as Tianmu Lake fish head, seasonal fruits, meat, Liyang white tea and other products. The support system for self-produced seasonal ingredients can not only ensure the output of seasonal ingredients, but also provide seasonal agricultural tourism experience, and because the system meets the requirements of circular agriculture, it is also an important part of its natural environment.

4 Conclusion

4.1 Economic performance

On the whole, we re-plan Lifengjie based on the diet of the twenty-four solar terms, and add a new commercial diet to it, changes in the traditional population can only rely on the wholesale of crops to the status of crop wholesalers, thus raising the possibility of increasing income for residents. And in accordance with the twenty-four solar terms, their production, festivals, food habits can be regarded as a tourist feature to attract more tourists, and more tourists will therefore bring more economic benefits.

4.2 Environmental benefit

The concept of adopting measures suiting local conditions and twenty-four solar term in local conditions, and doing farm work in the right season can serve as a guiding principle in the reconstruction of rural material flow system, and based on the idealized transformation of the twenty-four solar terms in Lifengjie, in the new system, nutrients, energy and water resources can be classified and utilized to improve the utilization of resources, establish a water purification system and a renewable energy system, improve the recycling efficiency of waste and black water and improve the overall efficiency of the system; on the other hand, the new system opposes the use of chemical cleaning agents, fertilizers and so on, and advocate the use of biological detergents, biological fertilizer, physical, insect and other means to reduce environmental pollution sources into the system. Ideally, Lifengjie can benefit from the environment.

4.3 Cultural benefit

In the survey, the author can appreciate the sense of pride in the local culinary culture, in this topic, the author has long established the design principle of enriching the connotation of solar terms on the basis of protecting local culture, and in the design, based on the twenty-four solar terms, the author combs the connotation of the base diet culture, and links each element with the solar terms, change its eating habits from the concept, and thus strengthen the soft power of Liyang food culture.

References

1. P. Cristiana, P. Migliorini, F. Sottile, Eur. *Ecology and Society*. J. E **19,** 4 (2014)
2. G. Pauli, The Blue Economy (Academic Foundation, 2016)
3. C. Peano, N. Tecco, E. dasero, V. girgenti, F. Sottile, Eur. *Sustainability*. J. E **7**, 6741 (2015)
4. L. Bistagnino, *System Design: designing the productive and environmental sustainability*(Slow food, 2011)
5. F. Capra, The Web of Life (Anchor, 1997)

The Study on Heuristic and Algorithmic Processing Creative Training on Product Design Education: Based on the Bags & Luggage design Course

Ziran Zhang[1], Ying Yu[2] , Shengxi Fan[3] and Yujing Tian[4]

[1]*Shanghai University of Engineering Science, School of Art and Design, 201620 Shanghai, P.R. China*
[1]*Tongji University Shanghai Institute of Design and Innovation, School of Design, 200092 Shanghai, P.R. China*
[1]*Tongji University, College of Design and Innovation, 200092 Shanghai, P.R. China*
[2]*Donghua University, Fashion and Design College, 200051 Shanghai, P.R. China*
[3]*Tongji University, College of Design and Innovation, 200092 Shanghai, P.R. China*
[4]*Donghua University, Fashion and Design College, 200051 Shanghai, P.R. China*

Abstract. In the study, context-creativity of Teresa M. Amabile was used as the foundation to apply it in the bags & luggage design course. Moreover, the sectional creative training education mode of prior heuristic task and postpositional algorithmic task was proposed. 26 junior students in product design were used as the trial objects. The Consensual Technique for Creativity (CAT) was considered as the scoring standard of creative performance. In the end, the sectional theoretical framework of effective creative training in product design was finally proposed.

1 The relationship between context-creativity theory and product design

1.1 Based on context-creativity theory

Differing from relevant psychological studies that creativeness or creativity is equal to *divergent thinking* ability test[1], respondents participate in actual activities in the experiment conducted by Teresa M. Amabile, such as barbola, prose creation and storytelling. This is consistent with the final purpose-ending of practice problems in design subject. The achievement is *really visible and evaluable*[1]. In early studies, Amabile created a test technology to evaluate whether works could own some creativity. And this is called as Consensual Technique for Creativity (CAT)[2](It is a relative science and repeatedly verified subjective evaluation test). This actually has a great similarity with subjective judgment for works. Even if most of design teachers don't verify scoring and evaluation form for works creativity, CAT has more practical value for the *quantized creative* test in psychological study conducted in other labs(such as TTCT)[3]. Moreover, it is closer to subjective evaluation method for single works in artistic design subject. The research conclusions are more suitable for theoretical foundation in artistic design subject education methods.

1.2 Algorithmic Processing and Heuristic Processing in Product Design Process

[1] Corresponding author: zhangziran_080@126.com

T.M. Amabile repeatedly proposes an idea that creative performance of external limitations in the algorithmic task and heuristic task is entirely different in Context Creativity. *Artistic creation lays particular stress on algorithm in technologies, while creativity focuses on heuristic*[2] (The algorithmic and heuristic concepts will be stated in details as follows). Similar to creative process of artistic works, designers control technical knowledge and creative design of products in product design process, so as to create a new product. As a result, training on "technologies" and "creativity" is the teaching target in product design major. The commonly used basic design method and process of professional studies in integrated product design course is shown in the right side of Figure 1[4]. It can be clearly observed from the figure that as training students in product major, creative inspiration(conceptual extraction and creativity in the figure) is included, and practical operation of relevant technologies(product formation and achievement publication) is also contained. After studying every part, it can be clearly shown that creative inspiration also contains knowledge in technical levels, such as market investigation, sketch and analysis of functional structure. Based on the context-creativity theory, these contents should belong to the algorithmic processing mode. The influence factor is totally different from the heuristic processing mode, thus the training method is also diverse.

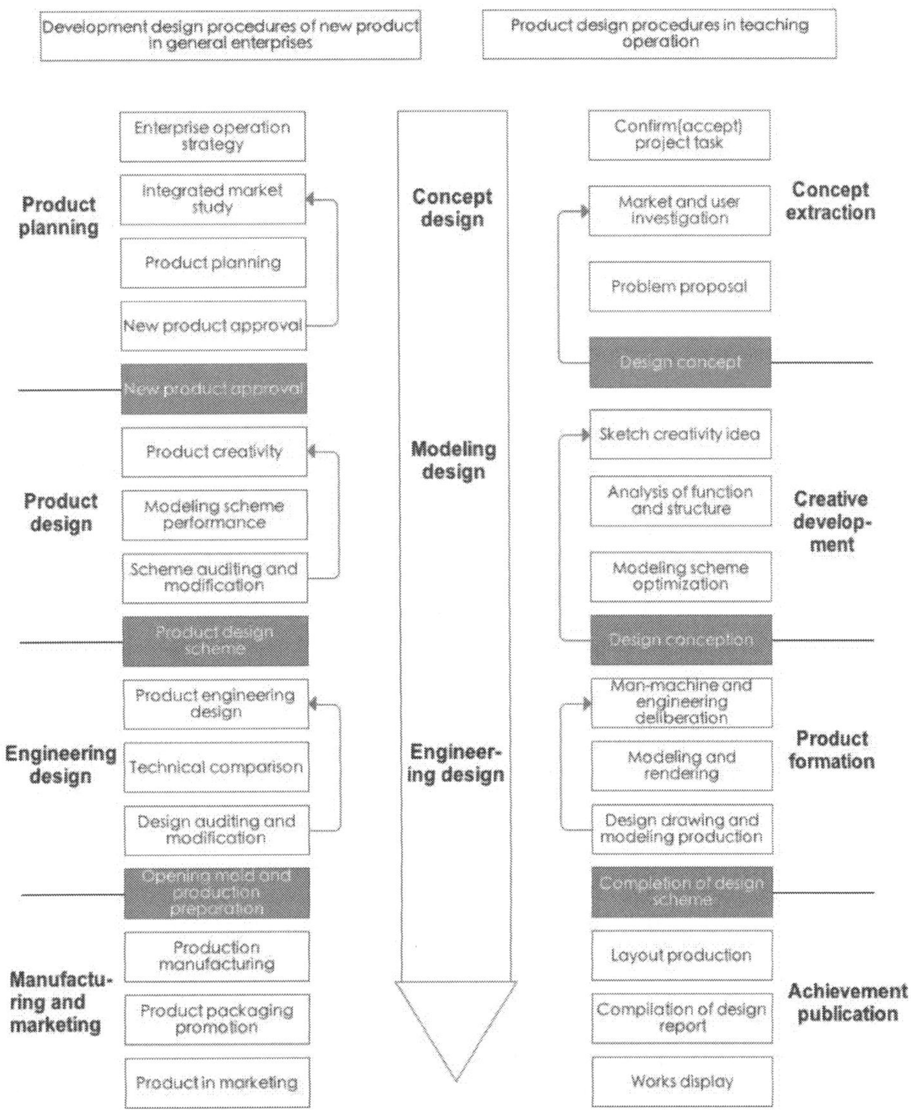

Figure 1. Comparison between Development Design Process of New Products in General Enterprises and Product Design Procedures in Teaching Operation [4]

1.3 The Research Object of the Bags & Luggage Design Course

There is a large dispute in professional attribution of the bags & luggage design course. Some colleges in Chinese mainland classify it in clothing major. Some of them are divided into the product major, because this is the impression on general classification of products(such as supermarket, market and online shopping platform). Generally speaking, these products are displayed to consumers as *garment accessories*. From the perspective of processing technology, it is exactly the same with relevant technical knowledge of products in ordinary meanings for luggage structure parts and production technologies, especially for hard cover luggage. As a result, it has no need to doubt whether bags &luggage course has representativeness in the product design course system. The truth that we design in the study means that heuristic processing and algorithmic processing in product design task implementation are relative key points on how to cultivate creativity. Therefore, the differences of diverse courses in technical knowledge, creative training contents, degree and depth won't affect preliminary conclusion of the empirical study. Relative to furniture, automobiles and household appliances, technical knowledge contents of luggage design are more intuitive and easier to be mastered, so that students are able to reach the required baseline level in a short time(32 hours in the study)[2]. According to three basic requirements in CAT[5] proposed by Ambile, the third one means that *research objects shouldn't have the large individual difference in baseline performances*. As a result, the trail should improve research objects to the same baseline level in the shortest time of the course teaching. Thus, it can be observed that relative to other products that have the relatively complicated technologies and materials, the bags & luggage design course conforms to the required conditions of the trial.

2 Necessity of Sectional Training

2.1 Differences between "Algorithmic" Processing and "Heuristic" Processing

Amabile found in the pasteup task in 1979 that when the "creative" solution in the task can't figure out steps in task development, the creative performance of respondents will be easier to be affected by external limitation and form the negative correlation. In other words, the more obvious limitation is, the lower creative scores of respondents will be. On the contrary, for respondents of technical scoring, when they are provided with definite information of technical concern and generate evaluation expectations, the technical level scoring will be higher than relevant control group[2], showing that technical level of artistic works will be slightly improved in definite scoring concern and evaluation expectation on respondents. Such a process is called as the "algorithmic" processing. In the "heuristic" processing, task type must be an open method without standard answers or obvious solutions. Secondly, external limitation in task implementation should be prevented or weakened. For example, respondents shouldn't generate evaluate expectation and avoid from passive effects of awards on creative performance.

2.2 Sectional Training and Definite Teaching Methods

Educators often insert thinking training and relevant technical knowledge in previous design education according to their experience(shown in Figure 1), but it is not systemic and scientific. Moreover, sectional teaching method points are not definite. Homework setting has no strong purposefulness to how to enhance creative performance. Reward mechanism is also applied at random. Evaluation proposal is also mastered by virtue of personal experience. There is no scientificity and demonstration support. As teachers, we often show the clear marking criterion and rewarding mode to students at the beginning of courses and then start the specific step implementation of courses. Specific contents will be based on market investigation and divergent thinking training shown in Figure 1(especially for brainstorming [6]). Then, comments and technical knowledge are inserted to carry out the specific design procedures. The research contents of Amabile show that rewards and external limitations will be helpful to enhancement of technical level. As the creative conception dominated by the heuristic processing, too early or overemphasis reward mechanism will reduce the creativity. By aiming at totally different conclusions, the sectional teaching method was put forward in the study to divide product design teaching process into three steps: (1) prior heuristic task; (2) postpositional algorithmic task; (3) concrete implementation of heuristic task achievements. At the same time, by aiming at two types of task process, people formulate the relevant teaching mode in accordance with the different creative reaction mode performed by external limitation.

2.3 Universality of the Sectional Teaching Method in the Product Design Course

The teaching method proposed in the paper was proposed in line with the context-creativity theory by Amabile. In the trial, 26 undergraduates in product design major were selected as observational and interview objects of the empirical study. Due to limitation of sample quantity and implementation in the same school, the preliminary exploration was only formed in the theoretical framework. Later, the more extensive empirical study on different courses of product design in different colleges should be conducted, so as to prove universality of the method.

3 Trail

3.1 Trail subjects and contents

This trial is a preliminary attempt of the distributed teaching method. By selecting 26 undergraduate students of Grade Three majoring in product design from a famous university of Shanghai, China as the research subjects, and dividing the content of the bag product design course into three stages: the heuristic task stage, the algorithmic task stage and the implementation stage of heuristic task outcomes, it aims to improve the creative performance of students when designing works through the application of Amabile's theory of Creativity in Context.

26 students in the trial were divided into 2 groups and they were sampled according to their student number. Students with odd student number were the control group, and students with even student number were the trial group. Two groups of students participated in the first two stages. Only in the final concrete implementation stage of task outcomes, the content of the trial was different, so two groups of subjects were tested separately. In the final stage, the trial group was asked to further design for the creativity in the heuristic stage and put it into a specific design plan. The subjects in the control group were asked to design an arbitrary concept suitcase or bag and show the design plan. Two different instructions were used to test whether the segmented education was effective in enhancing the creativity of the subjects. The former was the effective segmented education method that we assumed, and the latter was a proposition designed to weaken the influence of the heuristic process in the final product design. Finally, works of the two groups were merged and randomly sorted and handed over to the judges for the creativity of the works. (The judges didn't know the difference between the groups).

3.2 The heuristic task first

3.2.1 Heuristic Task

The heuristic task is a type of task that is quite different from the algorithmic task. The results of such a task should be sufficiently open to *ensure that the response has considerable flexibility and innovation.* [2] T*he task must have certain openness, and it needs some exploration to get the solution*. Moreover, th*e idea of getting the right solution is less clear.* At the same time, the task should be designed to prevent the task executor from being transformed into an algorithmic task. *When a heuristic task is converted to an algorithmic task, subjects perform better under external constraints.* [2] Since heuristic tasks are sometimes transformed into algorithmic tasks, how to design propositions for heuristic tasks and how to reduce the generation of such transformations has become the most important focus.

3.2.2 Heuristic task first

Baseline level of subjects. Since the selected subjects are third-grade students majoring in product design, they have not studied any design method courses related to the design of bag products before taking this course. Therefore, they can be regarded as subjects at the same baseline level, with a certain way of thinking theory, computer production skills and hand-painted techniques that are not aimed at bag products.

The necessity of the heuristic task first. As mentioned above, we have made a clear explanation of the "heuristic" task, so in order to ensure the proposition that the idea of getting the right solution is less clear, we should weaken students on some issues, such as "how to make out a bag"; "what are the structures of a usual bag?"; "what are the features of the bag?"; "What functions does a bag need?", so as to prevent the subjects from getting a set of solutions for themselves by combining their own experience with the focus in the instructions, thereby leading to low levels of creativity in the final results.

Task contents. In a heuristic task, the task is executed as follows:

(1) The teacher prompts the task, and the concrete instructions are as follows:

I am the keynote teacher of this course. Before the beginning of the course, I would like to ask you to participate in an interactive game in order to experience the pleasure of being a designer. Please design a bag according to your preferences, and the form is unlimited. Please note that this task will not record your results. My focus is not on the outcome of the design, but for the freedom to experience this casual thinking. Please fill out a questionnaire about the design process at the end of the task.

(2) When the students make the design plan, the teachers also make their own plans together

(3) Teachers and students participate in an informal program discussion: All people (including teachers) are divided into two groups: one group is the suggestion group, and the other is the design presentation group. Each subject in the presentation group can communicate with any of the members of the suggestion group, while members of the suggestion group can only give positive comments or affirmation on the strengths of the works of the presentation group members. All exchanges are free to choose locations and exchange roles at the end. The recommended subjects need to be evaluated and documented.

(4) According to the suggestion of others and their new ideas, the subjects need to revise the plan again to form the new final plan, or they can choose not to modify the plan.

Matters needing attention and theoretical basis in the task.

(1) When the task is put forward, the teacher must make sure that the answer to the task is open and that attention should be given to avoid restrictive conditions, such as the steps and design procedure, etc.

(2) In the course of the task, it is necessary to ensure that there are no evaluation expectations for participants as described in the instructions above: "I'm not focusing on the final design, and I'm not going to score it."

(3) In the course of the task, teachers should also participate in the discussion as a design participant. There are two purposes for doing so: 1. Set an example. According to the theory of Creativity in Context, this will help drive the intrinsic motivation of the subjects. 2. Create a more "open" classroom atmosphere [7-8], encourage curiosity, exploration and self-directed learning, but not advocate scoring and authoritative instruction [2].

(4) This course is a specialized elective course, so students have the right to choose for the participation of the course. According to the conclusion of the experimental study of Amabile in 1982, the creativity score under this type of task would be higher if "no reward" was applied. As a result, teachers cannot refer to anything such as "reward" and "scoring criteria" when conducting a heuristic task [2]. If the course is a compulsory course, and then the situation is opposite (Figure 2).

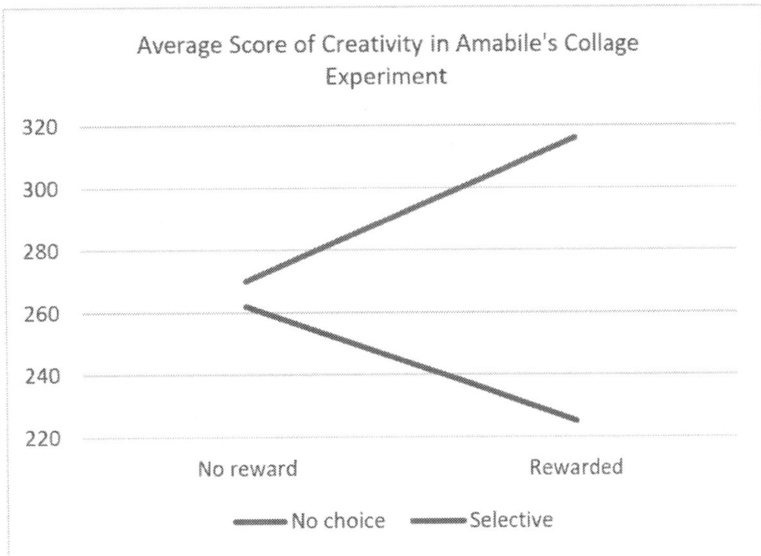

Figure 2. Average Score of Creativity in Collage Experiment[2]

(5) Students must be free to choose their creative expressions. *The individual's intrinsic motivation will be elevated if he/she perceives that he/she has the right to choose some aspect of the task.* [9] This conclusion is well proved in an experiment by Amabile and Judy Gitomer. [2]

(6) At the end of the task, when giving specific evaluation of each creative plan, the evaluation must be targeted and affirmative or positive comments on the individual's ability to do the task, and some unexpected rewards may also be presented. For example, the reward used in this study is that students with good performance can get internship opportunities in related enterprises. *Unexpected rewards will allow participants to enhance their internal drive to the design of bags. Research shows that the individual's interest in an activity is positively related to the creativity shown.* [2]

3.3 Followed by the algorithmic task

3.3.1 Algorithmic task

Amabile has repeatedly stressed in her book that theory of Creativity in Context applies only to heuristic tasks [2]. Algorithmic task processes are more appropriate to describe techniques in art creation. Although there is no explicit definition of algorithmic tasks in the text, some of the specific descriptions can also help us understand the characteristics of this task type. *It is a "script" (Schank&Abelson, 1977) or an algorithm that performs a task, that is, a fixed step that performs a task or solves a problem in a certain field* [2]. *If you can clearly and directly get the solution, and then the issue is an "algorithmic task".* [2] It is certain that the "arithmetic" task means that when the task is executed, it can draw clear and direct steps to implement the task or solve the problem according to the previous knowledge and experience.

3.3.2 Followed by the algorithmic task and its main contents

The reasons for the baseline level and followed by the algorithmic task have been discussed in the above, so only the main contents of the algorithmic task will be explained here.

As shown in Figure 1, the product design process used in general teaching, namely, a series of teaching contents used to strengthen *domain-relevant skills*,[2] such as the sketch, functional structure analysis, man-machine engineering knowledge, computer modeling, drawing design and model production and graphic display production, are the algorithmic tasks described in this article, because the methods you need to learn these skills are step-by-step, and you have to connect them gradually with specific steps to achieve the desired learning requirements. "Market research" in the initial stage, the design, the implementation and subsequent statistical work of a questionnaire are tasks with a fixed format. Only some qualitative analysis (User Studies) processes are likely to be closer to what we call heuristic task processes.

3.3.3 Algorithmic tasks in the bag design course

In algorithmic tasks, we mainly adopt the task-oriented curriculum teaching model, which is commonly used in the traditional design education field, to bring the task of solving specific problems into the whole design process, and we can achieve the special training purpose of the students' performance techniques of bags, structure and manufacturing process of bags and computer modeling through the step-by-step implementation of the task.

This approach is designed to enable students to have certain domain-relevant skills in real time in the third stage of concrete implementation of outcomes, and allow students not reduce the design creativity performance because of the lack of technology in terms of the expression of the design plan.

Specific implementation steps are divided into: (1) Select an existing product on the market and present its problems (2) Put forward a variety of solutions according to the problem (3) Report the program and get the teacher's evaluation (4) Research on materials, man-machine engineering and other knowledge (5) Choose one of the options and optimize the final solution (6) The teacher evaluates the work. Different from heuristic tasks, at the beginning of the algorithmic task, the instructor will make a specific explanation and emphasis on the evaluation criteria, make a targeted positive evaluation in the middle and late phases of the task execution, or give some suggestions for improvement on the part of the work which is insufficient. According to Amabile's conclusion in the experiment of assessment-creativity in 1979, such evaluation expectations and technical focus will enable the subjects to perform well in technical scoring [2].

4 Results

4.1 Results of heuristic task stage

The results of the heuristic task stage can be summarized as follows:

(1) The subjects showed strong initiative and enthusiasm in the process of interactive review. According to the playback after class, the students generally reflected that it was interesting to work with teachers and comment on each other.

(2) Open propositional way enabled the students to put forward various ideas

(3) At the end of the heuristic task, many students showed a strong interest in becoming a bag designer because of the unexpected rewards, and several students expressed the idea of getting into my bag design studio as a volunteer.

4.2 Results of algorithm task stage

The results of the algorithm task stage can be summarized as follows:

(1) The subjects showed stable learning state when they have evaluation expectations, and they could learn and master the skills in related fields in accordance with the guidance procedure.

(2) The mid-term evaluation recommendations were helpful to improve the skills of the subjects.

4.3 Results of heuristic task outcomes in concrete implementation phase

The results of the heuristic task outcomes in concrete implementation can be summarized as follows:

(1) After the final creative evaluation, the average score of the trial group was 83 points (the 560 point system of the CAT scale was rounded and translated into 100 points), which was slightly higher than the 76 points of the control group.

(2) One of the students in the trial group was below average, with a score of 64, and the reason for this was still to be further studied.

5 Conclusions

Based on Amabile's theory of Creativity in Context, we made a tentative application in the bag product design course and conduct a trial with samples. Moreover, we proposed a segmented education model of "the heuristic task" first and followed by "the algorithmic task" to enhance product design and students' creativity level. A Consensual Technique for Creativity was applied to measure students' creativity. Results effectively proved that the segmented teaching method could promote the creativity of students in the bag design course. Thus, we summarized this method into a preliminary theoretical framework, as showed in figure 3.

The art design education can be divided into three main stages: (1) Heuristic task first (2) Followed by the algorithmic task (3) Implementation of heuristic task outcomes. The specific steps at each stage are shown in the green squares in the table, while the red word shows the corresponding theoretical basis at this stage.

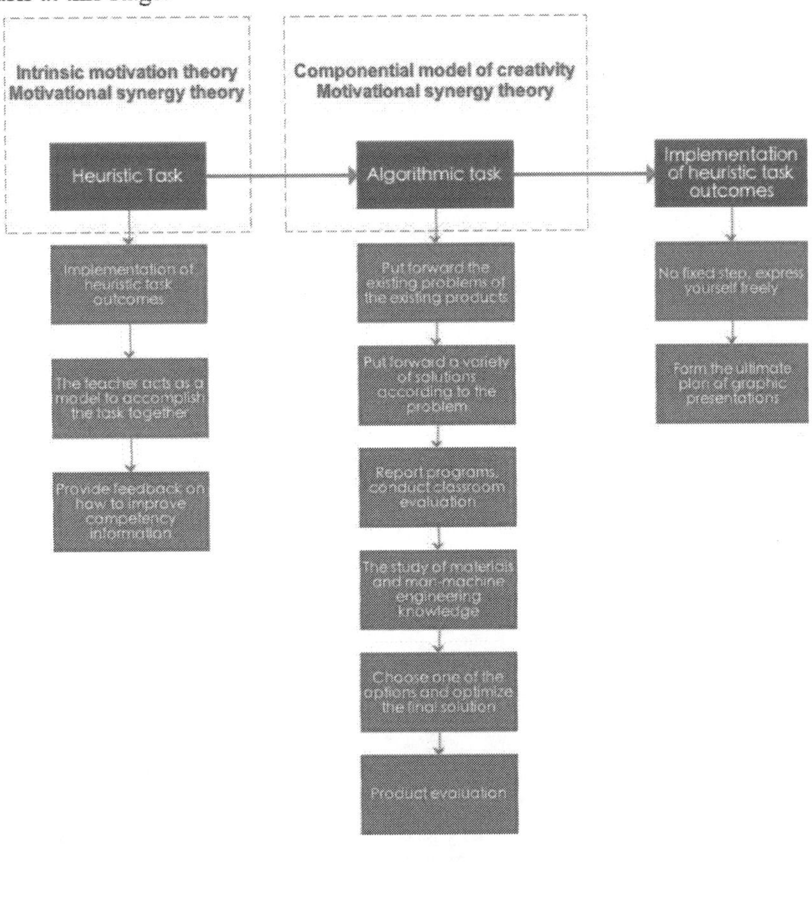

Figure 3. A segmented educational theory framework for effective creativity training in product design courses

Acknowledgement

Project Research Team of Intelligence Sustainable Package Design Support by Shanghai Summit Discipline in Design; Foundation item: Chenguang Plan (13CG74) sponsor

References

1. T. E. Chang, *Psychology of Creativity: Theories and Methods of Creation*, (Hebei People's Publishing House, Hebei, 2000)
2. T. M. Amabile, Creativity in context: *Update to the Social Psychology of Creativity* (Westview, Boulder, CO, 1996)
3. H. Y. Sun, *Creativity Psychology* (Beijing Normal University Publishing Group, Beijing, 2016)
4. Y. L. Gui, C. Yang, *Product Design* (China Light Industry Press, Beijing, 2016)
5. T. M. Amabile, Creativity in context: *Update to the Social Psychology of Creativity* (Sichuan People's Publishing House, Sichuan, 2017)
6. A. Osborn, *Applied imagination: Principles and Procedures of creative thinking* (Scribner's, New York,1963)
7. C. T. Ramey, V. Piper, *Child Development*, **45**, 557-560 (1974)
8. M. Hickey, *JRME*, **49**, 234-244 (2001)
9. M. Zuckerman, J. Porac, D. Lathin, R. Smith, E. L. Deci, *Personality and Social Psychology Bulletin*, **4**, 443-446 (1978)

Survey, Research And Prospect Of Signage Systems In National Parks In Yunnan Province

XU wenjuan[1,a] and FAN shengxi[2]

[1]*Department of Architecture and Urban Planning, Tongji University, 200092 No. 1239, Siping Road, Shanghai, College of Art and Design, sanda University, Shanghai*
[2]*College of Design and Innovation, Tongji University, 200092 No. 1239, Siping Road, Shanghai*

Abstract: The Signage System is essential to establishing a national park, The authors conducted surveys, analysis and research of identification signs, informational sign, directional signs and functional signs from the signage systems adopted by five national parks in Yunnan Province. Relying on the results, with reference to industry experience, years of research related to national park, successful cases of overseas national parks and the current signage systems across China's national parks, the paper aims to explore future development strategies of national park signage systems that are suitable for China.

1. Introduction

National parks are established to conserve the most characteristic, valuable, representative natural and cultural landscapes and ecological systems in a country [1]. In a way, they are regarded as national symbols. The signage system is the direct manifestation of how well a national park is established and managed and an important part of its brand identification and promotion.

Started late in building national parks, China is still exploring and developing national park management systems, policies, and laws and regulations [2]. The authors carried out surveys and analysis of five pilot national parks in Yunnan Province and developed the design principles, guidelines and evaluation models for the signage system used in China's national parks, thereby making contribution to creating a distinctive national park image and facilitating development of national park systems and policies.

2. Surveys on Signage Systems in Yunnan's National Parks

2.1 Overview

2.1.1 Purpose

From the perspective of the national parks' brand-building efforts, the survey, analysis and research focused on various categories of signs, such as identification signs, directional signs and functional

[a] Corresponding author: 34425965@qq.com

signs, from the signage systems adopted by the national parks in Yunnan Province. The ultimate goal is to explore strategies and roadmaps to systematically develop the national park signage systems that suite China's situation.

2.1.2 Survey Methods

(1) Observation: This is a non-participating survey approach. The observation activities for this survey were divided into two categories. First, all signs in the parks were catalogued, photographed and sorted. Second, by following the tourists visiting the parks, the researchers observed and objectively recorded all their activities and conversations from the point when they arrived at the parks to the point when they left, down to details such as when and where they stopped, made turns, looked for directions, tried to find ways and asked for information [3].

(2) Interview: Executives of the management companies in charge of the five national parks were interviewed to gather information about the pilot signage systems in the parks.

(3) Experiment: One respondent was selected at every point of observation to carry out the specified tour tasks as planned. The respondents were asked to find ways using the signs currently in use in the parks. All their activities and experience were objectively recorded.

2.2 Implementation

2.2.1 Highway Traffic Signs

The first two stops were Podatso National Park and Meili Snow Mountain National Park, The researchers drove from Lijiang City to the parks and observed the signage systems along the way. Both the national highways and the provincial highways have complete and well-rounded signage systems, including traffic signs, scenic site signs, warning signs, prohibitive signs, instructional signs, directional signs, road construction and safety signs, and other signs, with comprehensive coverage and explicit directions. Particularly, the signs for Meili Snow Mountain, Podatso, White Horse Snow Mountain, and other scenic sites were placed at intervals along the national highway and clearly indicate the direction and distance to the destinations. Besides, noticeable billboards were usually erected 5 kilometers away from the scenic spots for the publicity purpose.

The third stop was Laojun Mountain National Park in Lijiang. From the interviews with related park management executives, it developed and built a signage system by itself, rather than using the standard signage system from the Tourism Administration Bureau of Yunnan Province. Due to limited funding, the signage designs fell short in many aspects, from styles, sizes, materials and visual effects, to reasonable distribution, quantity and design expertise. In addition, the scenic site signs along the way to the national park, used the name "Lijiang Laojun Mountain Liming Scenic Zone", without mentioning its national park status. Evidently, after being titled as a national park, Laojun Mountain had not re-planned and re-designed its signs in the scenic spot.

Xishuangbanna Tropical Rain Forest National Park's Parashorea Scenic Zone was the last stop. The scenic site signs in remote regions of the prefecture and different districts of the tropical rain forest national park were not comprehensive and widely distributed. The travel almost entirely relied on the navigation system. Although there were scenic site signs along the way indicating there was a tropical rain forest national park, they did not indicate which districts were nearby. Tourists had to ask the national park staff members for more information. Parashorea Scenic Zone is situated in a remote county, Mengla, with harsh road conditions and underdeveloped traffic. Without sufficient guidance of scenic site signs, the researchers took wrong ways from time to time, even assisted by the navigation system.

2.2.2 In-park Signage Systems

(1) Podatso National Park

Podatso is the first pilot national park established in Mainland China upon approval by China's State Forestry Administration [4]. At the time of the survey, the Tourist Service Center was under reconstruction. However, tourists visiting the park still could use the comprehensive signage system. The signs for different facilities, such as parking lots, the Tourist Service Center and different businesses in the park, were well distributed and structured. Tourists can choose the single-direction route or the destination-specific routes. There were a great variety of the signs, such as the identification signs at the entrances, informational signs, directional signs, and functional signs. Meanwhile, the signs were widely and sufficiently distributed. To facilitate on-going maintenance and easy locating, every informational sign was numbered, helping tourists easily locate themselves and find ways in the park and staff members maintain the signs. Podatso National Park has a reasonable overall planning and design of the in-park signage system. But still, across the park there was no sign set up for fire-fighting safety and related facilities.

(2) Meili Snow Mountain National Park

Each zone in Meili Snow Mountain National Park covers a large area and has a well-planned and reasonably-organized signage system. as the examples in Figure 1, did an outstanding work in narrating the culture of Meili Snow Mountain. Along the hiking trail in Meili Snow Mountain, there were rest facilities and directional signs at one-kilometer interval, and serial numbers and altitudes were marked on the electric poles.

Figure 1. The signage system in Meili Snow Mountain National Park

(3) Laojun Mountain National Park and Xishuangbanna Tropical Rain Forest National Park

In Laojun Mountain National Park, the cultural promotion and park management were quite obsolete. The identification sign at the park's entrance still showed the former name, "Liming Scenic Site, a Key National Scenery Area for Three Parallel Rivers". Inside the park, the identification signs indicated that the park was "Laojun Mountain Geological Park", without any text or logo about its national park status. The in-park signage system used two distinct logos for the park. Identification signs were not consistent and introduced the park as "geological park", "famous scenery area", "World Heritage", or "national park". The park's logo was not consistent in design: some were using three symbols, and some were four patterns with a national park mark.

Signs of the tourist service center at Parashorea Scenic Zone, Xishuangbanna Tropical Rain Forest National Park only showed its name "Parashorea Scenic Zone" and indicated that the park was a "famous state scenery area", without any mention of its national park status. Riding a boat into the scenic zone, the researchers saw bulletin boards that showed the text "tropical rain forest national park" and identification signs in the zone. However, such signs were not placed at the entrance. Generally speaking, the zone mostly used its original out-of-date signage system.

(4) Pu'er National Park

Reconstruction of Pu'er National Park was approved in 2011 and it reopened to the public in 2013. However, the park lacks consistency in naming, as it now uses two names, Pu'er National Park and Pu'er Sun River National Park. Its Xiniuping Scenic Site did not have any sign to indicate its national park status at the time of the survey, its signage system underwent re-planning and re-construction as part of its overall planning efforts after gained the national park title. The signs and facilities were considerate, reasonably planned, well designed, and fully functional, perfectly fitted into the park environment.

3 Survey Results and Analysis

On the basis of the survey and analysis of the five pilot national parks in Yunnan Province, it can be concluded that the signage systems in China's national parks vary in maturity: some are well planned, while other remain underdeveloped. In this case, planning and design of the overall signage systems in Podatso National Park and Meili Snow Mountain National Park are relatively complete and comprehensive. In comparison, Laojun Mountain National Park, Xishuangbanna Tropical Rain Forest National Park and Pu'er National Park mostly made some adjustments to their existing scenic spot signage systems and come short of consistent management, guidelines and standards.

3.1 Naming of National Park

Every one of the five national parks in this survey was named differently, even the same national park uses distinct names in different places. There is no standard naming approach.

3.2 Out-of-park Signs and Support Information

The out-of-park signs and support information mainly refer to the traffic signs, scenic site signs, warning signs, prohibitive signs, instructional signs, directional signs, road construction safety signs, secondary signs and billboards. Planning of the signs and support information reflects primarily the overall management of a national park, and can directly impact tourists' visual perception and psychological feeling [5].

3.3 In-park Signage Systems

The survey results show that Podatso National Park and Meili Snow Mountain National Park were pilot parks established earlier than others and the planning and design of their signage systems were reasonable and widely distributed. Laojun Mountain National Park and Xishuangbanna Tropical Rain Forest National Park were using their existing signage systems with some adjustments and additions, and there was information inconsistency in some aspects. Pu'er National Park's signage system, after it gained its national park status, underwent reasonable planning and design as a whole. Table 1 analyzes the in-park signs of the five national parks.

Table 1. In-park Signage Systems of the five National Parks

National park names and logos					

Identification signs		No identification signs	No identification signs		
Informational signs					
	No general index, but the regional index is numbered	General index Regional index	Both information signs and informational signs	General index	General index Regional index
Directional signs					One-direction tour route, no directional signs
Introduction signs	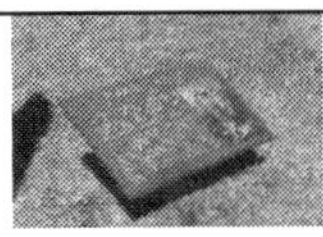	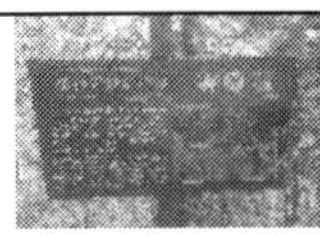			
Fire fighting signs	No such signs	No such signs;	No such signs;	No such signs;	No such signs; only fire
Other signs					

4 Long Way ahead in Developing China's National Park Signage System

New hope is born out of pains, and transition comes with confusion. The first national parks in China are challenged in every way. There is still a long way ahead to develop and mature China's signage systems. On the basis of industry experience in signage design, years of research related to national parks, successful cases of signage systems in overseas national parks, and the current situation of signage systems in China's national parks, the authors put forward a few ideas about future development of the systems in China.

4.1 National Park Naming

When it comes to brand building, national parks first need to develop and follow standard naming conventions or fixed name structures, such as "Provincial name + City name + Park name + National

Park", or "Park name + National Park", or other structures. A national park must use one name and keep it consistent for various purposes, such as signs, advertisements, tickets and websites.

4.2 Develop a Visual Identification System for China's National Park Brands

China has to establish standards and regulations on planning and design of a VI (visual identity) system for national parks, including logos, standard fonts, standard colors, supporting graphics, mascots, and so on. Various VI systems, particularly signage system, must remain consistent when being used for different purposes.

4.3 Principles and Rules of Signage Design

The signage system is a complete and independent system. In the same way, China's national parks also need separate principles and standards to guide planning and design of the signage system.

4.3.1 Graphical design standards

Develop and implement the Standards of Public Information Graphics and Symbols for National Parks in China, to meet the specific needs of national parks.

4.3.2 Out-of-part road signs

Standardize the out-of-park scenic site signs along the highways and road that lead to the national parks. The parks can choose to continue using their existing road sign standards, or further develop and adjust the old standards, which should cover interval distance, dimensions, materials, colors, and so on. The design and location of billboards also should follow related standards, such as sizes, advertising messages, and locations.

4.3.3 Design standards on sign categories (specific design standards for every category of signs)

The signs should be primarily categorized by functions, such as identification signs, informational signs, directional signs and functional signs.

4.3.4 Other signage systems

When adopting new categories of signs, like touch-sensitive signs, auditory signs, and induction signs, national parks should follow the specific design guidelines and design standards.

4.3.5 Sign making and installation standards

Sign making and installation should follow related national and industry standards.

References

1. Forest Park Administration Office of State Forestry Administration, Central South University of Forestry and Technology, Comparative Study of National Park Systems [M], Beijing: China Forestry Press, 2015:8-9.
2. LI Rusheng, Administration Systems of American National Parks [M], Beijing: China Architecture and Building Press, 2005.
3. National Park Service. (2008). The National Park Service Organic Act.

4. Tourism research and planning center of Peking University , Tourism Planning & Design [J], Beijing: China Architecture and Building Press, 2005.
5. WANG Mo, Study of Establishing Scenic Site Signage Systems [J], Chinese Landscape Architecture, 2013, 12 (2):38-42.

The Influence of Technoethics on Industrial Design

Zhong Fan[1,a], Yanming Ge[2]

[1]*Lecturer, College of Design and Innovation, Tongji university, China*
[2]*Lecturer, College of Design and Innovation, Tongji university, China*

Abstract. Technoethics is an interdisciplinary research area that means ethics in technology. Technology is transformed into the products in our daily life by industrial design and the negative effects of technology abuse makes ethics issues cannot be ignored. But design ethics research rarely from the perspective of technoethics, which is worth reviewing and summarizing. This review focused on the influence of technoethics on industrial design in the context of technological development. Through studying technoethics and industrial design from the 19th century, we find they have similar developing processes and the early 20th century and 1970s are two key point in time. This article aims to present the development of technoethics and industrial design intuitively on a timeline to discuss the impact on industrial design from the perspective of technoethics.

1 Introduction

1.1 Definition

Technoethics means ethics in technology and it could be found in Dewey and Peirce's pragmatism firstly. The term "technoethics" was coined in 1977 by the philosopher Mario Bunge in order to describe the responsibilities of technologists and scientists to develop ethics as a branch of technology.

1.2 Background

The development of technology has revolutionized our way of life, but it also brings serious problems of population, social problems and environmental problems. In the face of increasingly severe demographic, social and environmental issues, ethical dimensions are becoming very important. According to the results of the sixth China census, November 1, 2010 China's population aged 60 years and over reached 178 million, accounting for 13.6% of the total population. The growth of modern industry since the late 18th century produced massive urbanization and the rise of new great cities in other regions as well as Europe, as unprecedented opportunities brought enormous migrants from rural communities into urban areas.

[a] Zhong Fan: fanzhong@tongji.edu.cn

In general, environmental impact comes from (excessive) consumption of natural resources and emissions of pollutants to air, water, and land [1]. The emergence of great factories and the concomitant immense growth in coal consumption gave rise to an unprecedented level of air pollution in industrial centres; after 1900 the large volume of industrial chemical discharges added to the growing load of untreated human waste [2].

In the context of technical intelligence, technoethics is becoming more and more important, especially the artificial intelligence and robotics communities facing an important ethical decision [3]. Repetitious work is being replaced by algorithms that control databases and robots. The scope of these algorithms continues to expand. There are arguments about the limits of machines, but it is fair to say that machines cannot currently perform "creative" tasks [4]. However, robots may always have limitations on perception, cognition and actuation. There are still hurdles to overcome to enable robots and humans to co-exist safely and productively [5].

Technoethics enriched the meaning of technology. Recently, many instances in the world culture are pushing towards the rediscovery of the ethical dimension of technology [6]. As the inherent dimension of technology, ethics has evolved with the development of technology. Technology demanding for ethics is the inherent requirements for itself, and it mainly manifested in three aspects: First, the quality of technology is ultimately tested by nature and technology should be harmony with nature. Second, the freedom of technology determines the technology must be self-discipline. Third, ethics intentionality of technology is pre-existence comparing with technology practice. Therefore, the ethical dimension is the inherent dimension of technology, and the demand for ethics is the inevitable requirement of technology developing [7].

Technology is the critical factor in the industrial design. Industrial design originated in the era of industrial civilization and the root cause is the mechanized mass production and the division of labor by the industrial revolution in the West [8]. From steam rea to digital era, technology is the main driving force for social development and is the main service role for design. In the field of design itself, technology also has made great revolution in design process, like Computer Aided Design. Digital technologies are influencing the way that design project work and practical studio work are taught [9]. Neil Gershenfeld argues that digital fabrication is being regarded purely in terms of conventional thinking as an additional production process by many engineers and designers and one of the intellectual frontiers associated with the development of digital fabrication is the development of an associated design theory [10].

However, at present, most of research about design ethics does not involve technoethics issues. In the context of technology-based service development, design encompasses those elements of the development process which enhance and communicate the value of services [11]. Design is the process of trying to make the world friendlier to us clumsy humans; it is the effort to make the world more caring toward us, more accepting of us and so more morally acceptable to us [12]. The questions raised for ethics by the objects of technology are defined by the major areas of their impact and thus fall into such fields of knowledge as ecology (with all its biospheric subdivisions of land, sea, and air), demography economics, biomedical and behavioral sciences (even the psychology of mind pollution by television), and so forth [13].

1.3 Motivation

Technology is the key factor in industrial design, while the development of technology is also critical factor in our social progress. But the energy crisis of the 1970s allowed us to realize the negative impact of blind pursuit of technology and some philosophers began to discuss the ethics issues of technology. Of course, this crisis has spread to the field of industrial design and designers began to think about the value of design. In particular, Victor Papanek referred to the design ethics for the first time in 'design for the real world'. However, there is little literatures has discussed the relationship between technoethics and industrial design. In our review study, we focus on the influence of technoethics on the industrial design to do this literature review.

In this section, we discuss the definition of technoethics and why we try to do a literature review exploring the influence of technoethics on the industrial design. In the coming section, we explain the methods of filtrating the literatures and analyzing them. Subsequently, we will sort out the development of technoethics and the impact of industrial design in accordance with the timeline. Finally, we will discuss and summarize the information obtained from these literatures.

2 Methods

In this section, we try to present the method we have followed in research to conduct the literature review. We begin with the illustration about how we selected the articles on which we based the literature review. Then, we describe how we analyzed the articles by drawing on the historical timeline of the information.

For our review, we have selected thirty journals from three important databases, which were Elsevier, JSTOR, IEEE using keywords "technoethics", "philosophy of technology", "history of industrial design" and "design ethics". There are three main aspects of the literature: 'technoethics', 'the history of industrial design' and 'design ethics'. At the same time, in order to have a comprehensive systemic understanding of technoethics, we have read the important chapters of six books briefly.

After review many technoethics and industrial design articles, we find technoethics and industrial design have similar processes and there are two critical points in time that they both had great changes. Finally, based on the outcome of above, the detailed analysis and findings are as follow.

3 Analysis and Findings

The history of industrial design is not long, it formed after the human into the industrial civilization gradually, and the mechanized large-scale production and division of labor from the modern Western industrial revolution is the root cause of its appearance [8].

So in this section, we divide out three times, with two key points, and each part has been described based on the technical background on the technoethics and its impact on industrial design. After that we briefly outlined the situation today with the same structural approach.

3.1 In the 19th century

Europe's industrial revolution is an important turning point in human civilization in the 18th century. It is a sign that the Western world truly enters the capitalist society from feudal society and the beginning of human civilization from the era of industrial civilization. This revolution is the most direct cause of industrial design [8].

At the end of 18th century, the invention of the steam engine marks the era of human civilization into 'the steam power era', and then appeared in the factory, steam locomotive and the British became the first industrial country in the world. In the 19th century, with the discovery of electricity, manufacturing and processing and people's lifestyle have undergone a radical change. With inventing telegraph, cable, telephone, electrified railway and then to the production of electricity, electricity began into people's daily life from the factory. Electricity era is a more efficient and more convenient than the era of steam.

The great changes brought about by the industrial revolution have had an unprecedented impact on the design field. At the same time, the traditional artisans have no way to keep up with the increasingly complex product technology issues. However, the serious disengagement of new technology and art has become the most important and most acute contradiction in the field after the outbreak of the industrial revolution. In this context, the early industrial design was born.

3.2 At the end of the 19th century and early 20th century: The first key point in time.

The outbreak of the First World War in 1914 lasted for four years and 1939 to 1945 the Second World War broke out. These two World War have brought destructive blows to mankind, and social development has been stagnant. But their strong demand for new technologies and materials has allowed technology and industrial design to develop unparalleled.

At the beginning of the 20th century, the standardization consciousness was based on the national scope, and some industrial powers established standard institutions to publish production standards. In 1902, the British established the 'British Engineering Standards Association, and later developed into 'British Standards Institution'. In 1916, the German Standards Commission was established. In 1918, the American Standards Association was officially established with the support of Hoover.

In the aspect of materials, steel die technology was used to manufacture the shell of household appliances. Followed by refining pure aluminium technology and electricity become cheap, allowed aluminium products began mass production. In addition, the more important material is plastic discovered in 1869 in American, and from the 1930s it was widely used in product design and the technical civilization went into the plastic era.

In this period, technology theory has begun to develop, and the most important theory is the technology determinism by Thorstein Veblen. 'Technology marches in seven-league boots from one ruthless, revolutionary conquest to another, tearing down old factories and industries, flinging up new processes with terrifying rapidity.' [14]. In the middle of the 19th century, the development of new technologies was growing rapidly, but the problems of mechanization and the marked decline in product quality were more and more prominent.

From 1810 to 1897, the usage of the word "design" declined. After 1897, it increased—and notably so after 1936 and again after 1967 [4]. Just this period happens to be the beginning of the modern movement in the history of design, after the industrial design into the unprecedented prosperity. Almost all of the basic ideas of industrial design, methods and its role in the socio-economic development are formed in the twenty years after the First World War.

In this period, the impact of technical and philosophy of technology on the industrial is to make rude new technology elegant and to find the right way to visualize the products. And the second aspect is the change in the way of manufacturing has had a profound impact on industrial design, which is the core of the manufacturing standardization. Ford T-car is the success of automobile production standardization model in the beginning of the 20th century. The establishment of standardization is the inevitable result of the change in the way of production, but for the designers it also means that the new product aesthetics and the corresponding design language changes.

Der Deutsche Werkbund was established in 1907 and its core was 'quality'. They were actively concerned about the progress of new technologies and product design related to the invention and maintain an unprecedented industrial spirit. Peter Behrens designed Turbine Workshop was the closest to modern architecture and he designed a fan for general companies as a model of functional and technical unity. In 1914, the result of Cologne debate was that the historical development is not to avoid or reject the inevitability of technology, but how to grasp and transcend. In 1919, Weimar Art Institute and Weimar Arts and Crafts School merged into Der Staatliches Bauhaus, and Walter Groupius as its principal. Bauhaus laid the foundation of modern design education and was also the main force of modern movement. They argued that technical and artistic barriers should be eliminated and design aesthetics designed to adapt to new production methods and technical levels. The products designed should reflect the technical characteristics of the production mode.

A typical example is the famous S-shaped steel pipe cantilever chair designed by Brewer in 1928. The structural relationship between the steel pipe material is developed in the form of largest simplified. In addition, there are Teague designed Bantam Special camera for Kodak and Alto designed bending wooden chair by steam bending forming technology. They all tried to unify aesthetics and technology to express the characteristics of new technology. They thought handicrafts were only the means to reach the target, and the role of industry was the basis for future production [15].

In this period, technology was developing rapidly and designers helped create a lot of new products without hesitation. We invented and designed a large number of products and all kinds of new materials and new products appeared every day. Unprecedented prosperity in the areas of industrial and industrial design covered up the possible negative crisis. May be due to mechanized large-scale production has not yet shown a negative impact in the short time, the mainstream of technical philosophy theories still focused on the technical determinism theory.

3.3 In 1970s: The second key point in time.

After the Second World War, the economies of Europe began to recover, especially in the 1960s, when the economic development of the capitalist world entered a period of victory. On the on hand the benefits of industrial production methods have been the greatest degree of expression, on the other hand, all the crises of industrial civilization in this golden age began to appear. In 1950s, Planned Obsolescence in the rise of Detroit Motor City supported that the life of the product should be limited, or by reducing the durability of the use, or making it quickly out of style to stimulate consumption. This decision completely ignored the serious consequences of large quantities of material energy consumption, especially after the 1970s energy crisis it had been criticized widely.

In 1970s, capitalist world suffered a recession and energy crisis, while the blind pursuit of efficiency has brought serious ecological environmental problems. The manufacture of goods themselves is increasingly automated. Labor costs in the 1970s were a major consideration in factory location. That has declined as factory automation has expanded [4]. Since then, the protection of the ecological environment has become the important subject to all the mankind.

In 1970s, there is a clear shift from philosophy of technology to technoethics in the European and American philosophical area [16]. Most Western philosophers argued that the cause of the crisis was caused by technological progress and the expansion of technology to the social sphere of life that did not belong to it. The most notable feature of product manufacturing in this period was the busy expansion of production to meet the growing social needs. These problems have gone beyond the scope of traditional technical philosophy and entered the moral time category. What is morality in this era of technical decision? Hu Ning argued that technology and technical creation should serve the culture as a living condition of man and the peace in the world.

On the other hand, Wanda Orlikowski has studied the duality of technology: objective and human aspect of technology. The identifies prior views of technology-as either objective force or as socially constructed product-as a false dichotomy [17]. He thought technology is the product of human action, while it also assumes structural properties. So facing the technology issues, we need to do a strategic choice, that is we should consider the human aspect of technology not limited in the technology determinist.

The whole society should be careful to consider saving energy, maintaining ecological balance and control of economic issues, and industrial design made the most direct response to this. In this era of design theory began to become more diverse. Emerging design areas began to appear, like green design, universal design, sustainable design and high touch design.

The design of this era had been adapted to the new technology processing context. The focus of the design from the fashionable style of the shell wrapping new technology developed to pay more attention to the use of rationality and comfort issues. As technology philosophy turned to technical ethics, industrial design began to consider ethical issues. The most representative is the book 'Design for the real world', published by Victor Papanek, which marked the design ethics appeared. He criticized the reality of design was a commercialist vassal and called for the design required a sense of social morality and needed A Return from Form to Content.

At the same time, design for the human needs also began to specialization, systematic direction. Design for the old people, for the children, for the disabled and the universal design had been carried out in-depth research. Universal Design was coined by Ron Mace (1998) to reflect a type of design allowing all users to be accomodated [18].

Some of which are difficult to promote in the business community because of the special needs of the clients, but they must exist in our society. Erik Herlow mentioned in the 1970 Swiss Environment Discussion and Action Foundation that designers needed an ethical attitude and designer transformed the industrial design into a medium for mass consumption, mass production, and mass consumption. Design would be either a means of self-destruction or a tool for survival in a reasonable world.

3.4 In the 21st century

Mankind enters the 21st century, we have entered the digital and intelligent era. Over the last ten years there have been major digital advances in technology that have the potential to significantly change practice in Industrial Design. These include but are not limited to: Improvements in access to the internet worldwide – the expansion of the network, increased band width, improved communication tools supporting sophisticated interactions between designers, production companies and consumers. Advances in scanning technologies and the software to manipulate point cloud data. Developments in the use of RFID chips and the creation of digital tools to support an internet of things approach. Advances in additive manufacturing to include the use of metals, functional polymers and ceramics [10].

The lines between Industrial Design and digital design are blurring, just as is happening in other disciplines such as Graphic Design, Architecture, Forensics and a myriad of others. Digital technologies are disrupting the boundaries and academics and designers can respond positively or work against it [10].

3.5 Results

We use the timeline to sort the trivial information below as followed in table. 1

Table. 1 Historical Timeline

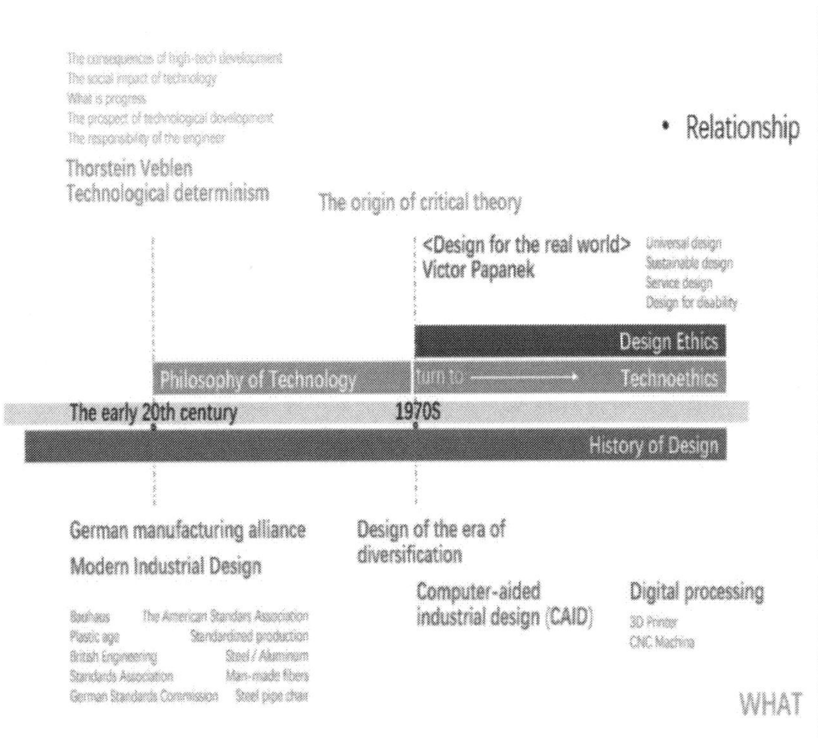

It is difficult to predict where technological innovation will take designers, how it will impact users' lives [10]. But from the human history process, the essence of the design is to reflect the good people to make technology and other material forces to achieve the value of improving human life [8].

4 Discussion and Conclusion

Based on the findings stated above, the following issues are worthy of further discussion. Machines are proficient, useful, timesaving, and labor-saving, but they are monotonous and lacking imagination. The ability to recognize two-in-one is the substratum of the human imagination, a thread running between art and design and everyday life. As a body extension of the machine, replace the work of people step by step. It underscores a more general proposition: the work of the future will have to be more imaginative, or else it will be turned over to machines [4]. Ethics has always been associated with human-to-human relations. Products, artefacts, built environments, communications, have only entered the ethical domain as tangential, and therefore neutral means, used by humans in their relations to other humans. To date, things, designed things, have not assumed a central or at least symmetrical, role with humans when it comes to ethics [12].

Technoethics and industrial design have similar developing processes. In this review, we can find this conclusion. In the period of blind worship technology, technical theory was one-sided, and industrial design was also not aware of ethical issues and recognized their own value in the process of technological development. Until 1970s, the ethical consciousness began to awaken. Philosophy of technology turned to technoethics and design ethics appeared to be independent from industrial process. Since then, the role of people has become increasingly heavy, and the two theories have begun to become three-dimensional pluralism.

Technology is the critical factor in the development of technoethics and industrial design. By convention, we judge a design by its functionalist and technoethics has developed based on negative issues from the progress of technology. Besides, technology revolution is the key power in the change of design.

The duality of technology: Objective reality (Deterministic Role) and Socially constructed product (Strategic Choice). In this process, the role of industrial design is the human who chooses what to do and how to realise it. After 1970, the beginning of the theory of technical criticism, the duality of technology, technoethics and to the ethics in the design field, they all due to put the human factors into the original reckless development considerations. In the process of industrial design, we need to make a strategic choice of materials, manufactories, new technologies.

Industrial design is a bridge between people and technology in modern society, and the material basis of industrial design is modern science and technology [8]. Industrial design will truly transform the new technology into human production practice and material products in our daily life, and it naturally becomes an intermediate link between the needs of human and manufacturing as the product of the division of labor. Technology represents only the possibility of substance, so that we need the power from design to make it really become part of the field of life. Design is the process of trying to make the world friendlier to us clumsy humans; it is the effort to make the world more caring toward us, more accepting of us and so more morally acceptable to us [12]. On one hand, the purpose of design is to guide the new technology to provide service for people, on the other hand the design continues to put forward new developing needs to technology.

Reference

1. Bras B. Incorporating environmental issues in product design and realization[J]. Industry and environment, 1997, 20(1): 7-13.
2. Fleming J R. Bethany R. Knorr of Colby College." History of the Clean Air Act". American Meteorological Society[J].
3. Russell S. Take a stand on AI weapons[J]. Nature, 2015, 521(7553): 415-416.
4. Murphy P. Design capitalism: Design, economics and innovation in the auto-industrial age[J]. She Ji: The Journal of Design, Economics, and Innovation, 2015, 1(2): 140-149.
5. Veloso M. Embrace a robot-human world[J]. Nature, 2015, 521(7553):418-418.
6. Galvan J M. On technoethics[J]. IEEE-RAS Magazine, 2003, 10(4): 58-63.
7. Zhiye Guo, Chengwei Wen. The three inherent ethical dimensions of the technique [J]. A Study of Dialectics of Nature, 2011(5):41-46.
8. Yongyi Lu, Xiaowei Luo. Product, Design, Modern life: The development of industrial design [M]. China Construction Industry Press, 1995.
9. Loy J. eLearning and eMaking: 3D Printing Blurring the Digital and the Physical[J]. Education Sciences, 2014, 4(1): 108-121.
10. Loy J, Canning S, Little C. Industrial design digital technology[J]. Procedia Technology, 2015, 20: 32-38.
11. Candi M. The role of design in the development of technology-based services[J]. Design Studies, 2007, 28(6): 559-583.
12. Tonkinwise C. Ethics by design, or the ethos of things[J]. Design philosophy papers, 2004, 2(2): 129-144.
13. Jonas H. Toward a philosophy of technology[J]. Hastings Center Report, 1979, 9(1): 34-43.
14. Beard C A. Time, Technology, and the creative spirit in political science[J]. American Political Science Review, 1927, 21(1): 1-11.
15. Dearstyne H, Spaeth D. Inside the Bauhaus[M]. 1986.
16. Guoyu Wang. Philosophy of technology turns to ethics in German [J]. Philosophy Research, 2005(5):94-100.
17. Orlikowski W J. The duality of technology: Rethinking the concept of technology in organizations[J]. Organization science, 1992, 3(3): 398-427.
18. Crews D E, Zavotka S. Aging, disability, and frailty: implications for universal design[J]. Journal of physiological anthropology, 2006, 25(1): 113-118.

MATEC Web of Conferences **167**, 02001 (2018) https://doi.org/10.1051/matecconf/201816702001
IC4M & ICDES 2018

Distributed Cooperative Control for Stepper Motor Synchronization

Sauro J Yague[1,a], Guillermo Reyes Carmenaty[1], Alejandro Rolán Blanco[1] and Aurelio García Cerrada[2]

[1] *IQS School of Engineering. Ramon LLull University. Via Augusta 390. 08017 Barcelona (Spain).*
[2] *ICAI School of Engineering. Comillas Pontifical University. Alberto Aguilera 23. 28015 Madrid (Spain).*

Abstract. This paper describes the design and simulation of a distributed cooperative control algorithm based on multi-agents to synchronize a group of stepper motors. Modeling of the two-phase hybrid stepper motor in closed loop is derived in $\{d-q\}$ rotary reference frame, based on field-oriented control techniques to provide torque control. The simulation obtained by MATLAB-Simulink shows that the distributed cooperative control effectiveness depends on the network topology defined by the graph.

1 Introduction

Distributed and decentralized synchronization of large groups of dynamic systems is an area of intensive research [1]. In this article, a cooperative control algorithm will be applied to synchronize five stepper motors in a topology network defined by its graph.

A cooperative control law has been implemented based on a multi-agent dynamical system where agents are interconnected by a "fixed" communication network, and each agent or node is mathematically modeled by a dynamical linear time-invariant system, with communication delays being neglected.

Multi-agent systems consist of autonomous agents which are able to interact with each other and/or with their environment. During the last 20 years, the cooperative control of multi-agent system has been attracting great attention in biology to understand animal group behavior, in situations such as the flocking of birds, the schooling of fishes, and the swarming of bacteria. Results have been translated into engineering disciplines, like formation control of aircraft and robots.

The stepper motor model implemented in the simulation is firstly discussed in Section 2, and a detailed review on field-oriented control to regulate flux and torque separately is given in Section 3. The closed-loop position control system based on $\{d-q\}$ rotary reference frame is justified in Section 4. A review on algebraic graph theory is presented in Section 5, together with the cooperative control law designed to run the simulations. The cooperative control law will be evaluated under different network topologies, with simulation results shown in Section 6. A closing chapter in Section 7 will address final conclusions of the simulation runs.

2 Stepper Motor Model

[a] Corresponding author : sauro.yague@iqs.url.edu

The first stepper motor model available was manufactured with a six-salient-pole stator, a four magnetless salient-pole rotor, it had three phases, and rotated in thirty-degree steps. Since then, stepper motors have evolved significantly over the years, and modern hybrid two-phase stepper motors incorporate toothed stator and rotor, as well as a permanent magnet in the rotor. The two-phase windings in the stator are physically oriented on axes that are 90° apart from each other. The current flowing in each stator winding produces a magnetic field vector, and by controlling the currents in the two stator windings, a magnetic field of arbitrary direction and magnitude can be produced by the stator.

Hybrid steppper motors have smaller size, higher stepping rate and higher torque than variable reluctance stepper motors. The dynamic model for a two-phase hybrid stepper motor modeled by [2], consists of (1), (2), (3), and (4).

$$v_\alpha = Ri_\alpha + L\frac{di_\alpha}{dt} - K_m w\sin(N_r\theta) \tag{1}$$

$$v_\beta = Ri_\beta + L\frac{di_\beta}{dt} + K_m w\cos(N_r\theta) \tag{2}$$

$$T_L + Bw = J\frac{dw}{dt} - K_m i_\alpha \sin(N_r\theta) - K_m i_\beta \cos(N_r\theta) \tag{3}$$

$$\frac{d\theta}{dt} = w \tag{4}$$

where v_α and v_β are the phase voltages $[V]$, i_α and i_β are the currents $[A]$, w is the rotor speed $[rad/s]$, θ is the rotor angle $[rad]$, and T_L is the load torque $[Nm]$. The parameters of the motor are phase resistance $R = 10\Omega$, phase inductance $L = 6mH$, torque constant $K_m = 2Nm/A$, number of rotor teeth per phase $N_r = 25$, inertia of motor $J = 1.9 \times 10^{-2} kg/m^2$, and coefficient of viscous friction $B = 1 \times 10^{-3} Nms/rad$.

3 Field Oriented Control

For the last two decades, servomotors have evolved from largely brush types to brushless, providing higher reliability and lower maintenance. Most high performance servo systems employ an inner control loop to regulate torque, which is enclosed by outer control loops to control rotor angular velocity and rotor angle position. While the design of the outer loop is considerably independent of the motor type, the design of the torque loop is inherently specific to the motor being controlled.

Torque control in a brush motor is rather simple, since the motor commutates itself. A Proportional-Integral (PI) controller which adjusts the voltage applied to the stator windings is used, in the effort to minimize the error between the requested and measured motor current. However, brushless motors, like stepper motors, are not self-commutating, resulting in the necessity to control currents/voltages applied to the stator windings, to produce a useful torque.

The two stator windings in the stepper motor are physically oriented on axes that are 90° apart from each other, while brushless synchronous DC (BLDC) motors includes three stator windings physically separated 120° instead. In the stepper motor, the current flowing in one of the stator winding produces a magnetic field vector, which sums with the field from the second winding. By controlling currents in the two windings, a magnetic field of arbitrary direction and magnitude can be produced by the stator. Consequently, torque is produced by the attraction or repulsion between the net stator field and the magnetic field in the rotor. For any rotor position, when the stator field is orthogonal to the field produced by the rotor, the torque will be maximum, and when the stator field is in line with the rotor field, they both superpose to each other, and no torque will be produced.

A generic stator field can be decomposed into two components: one parallel to the rotor field (direct component), and the other orthogonal to the rotor field (quadrature component), where only the quadrature-component produces torque.

By means of modeling the field produced by the current stator windings, a conceptual current space vector is used. The current space vector for a given winding has the direction of the field it produces, and a magnitude proportional to the current going through it. This makes it possible to represent the total stator field as a current space vector which is the vector sum of the current phasor components, one for each of the stator windings. The stator current space vector is a fictitious current that would flow in a single winding rotating so as to produce the same stator field direction and magnitude as the combination of the real currents through the real stator windings. The current space vector for two stator windings in a stepper motor and three stator windings in a brushless synchronous DC motor are defined in (5) and (6), respectively.

$$\bar{i}_{sv} = \frac{2}{3}\left[i_\alpha(t) + i_\beta(t) \cdot e^{+j\frac{\pi}{2}}\right] \tag{5}$$

$$\bar{i}_{sv} = \frac{2}{3}\left[i_a(t) + i_b(t) \cdot e^{+j\frac{2\pi}{3}} + i_c(t) \cdot e^{-j\frac{2\pi}{3}}\right] \tag{6}$$

In order to efficiently produce constant smooth torque, the stator current space vector should, ideally, be constant in magnitude and should turn with the rotor so as to always be in quadrature with it, regardless of the rotor position or speed. While the stator current space vector may be constant in magnitude and direction if viewed from the rotating reference frame of the rotor, from a fixed frame viewpoint the current space vector describes a circle as the rotor turns. Since the current space vector is produced by the vector sum of components from each of the motor windings, and due to the fact that the two windings in the stepper motor are physically orientated on axes that are 90° apart from each other, the motor currents should ideally be two sinusoidal waveforms, with a 90° phase shift.

In the effort to generate a smooth sinusoidal modulation of the motor currents as the stepper motor turns, an accurate measurement of rotor position is required (e.g. using high resolution encoders). These two sinusoidal current command signals are provided as inputs to a pair of PI controllers that regulate the current in the two stator windings. Tracking a sinusoidal waveform by a PI controller has some limitations at high rotor speeds. High rotor speed results in large errors which perturb the direction of the current space vector relative to the rotor, causing it to shift away from the quadrature direction, reaching a situation where no torque is delivered.

The limited bandwidth of PI controllers is tackled by using field oriented control [3]. The current space vector is controlled directly in the $\{d-q\}$ rotary reference frame of the rotor. In the ideal case, the current space vector is fixed in magnitude and direction (in quadrature) with respect to the rotor. Since the current space vector in the $\{d-q\}$ rotary reference frame is static, the PI controllers operate on DC values, rather than in time varying sinusoidal signals. This isolates the controller from the time varying winding currents and voltages, and therefore eliminates the limitation of the controller frequency response, proving an efficient torque control. For this reason, the measured stator currents in the stepper motor must be mathematically transformed from the two phase $\{\alpha-\beta\}$ static reference frame, to a $\{d-q\}$ rotary reference frame, before being processed by the PI controllers.

For brushless DC synchronous motors (BLDC), with three-phase stator windings, with phases physically separated 120°, the reference frame transformation can be performed in a single step: from three phase $\{a,b,c\}$ stationary frame to $\{d-q\}$ rotary frame, using Park's Transformation [4]. However, that transformation is best described as two step process, where the motor currents are first translated from the 120° physical frame of the motor stator windings to a fixed orthogonal reference frame. This transformation is known as Clark Transformation. Two PI controllers will be required to control the $\{d-q\}$ currents, and since only the quadrature current iq produces useful torque, the iq controller will track the requested torque, and the id controller will be configured with a zero input reference.

Figures 1 and 2 summarize stator winding current waveforms in a BLDC motor, as seen in three different reference frames.

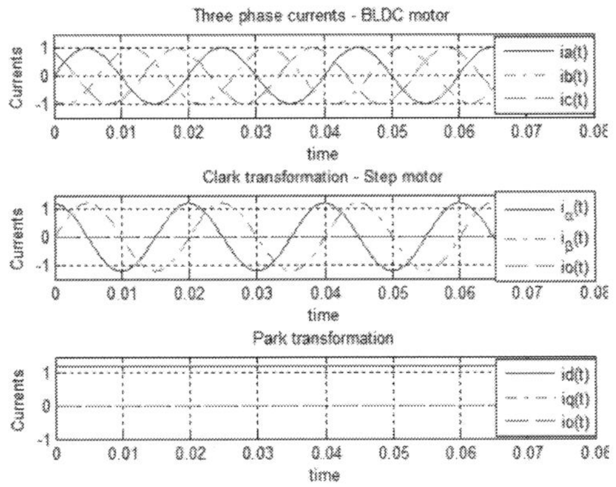

Figure 1. Clark and Park Transformations.

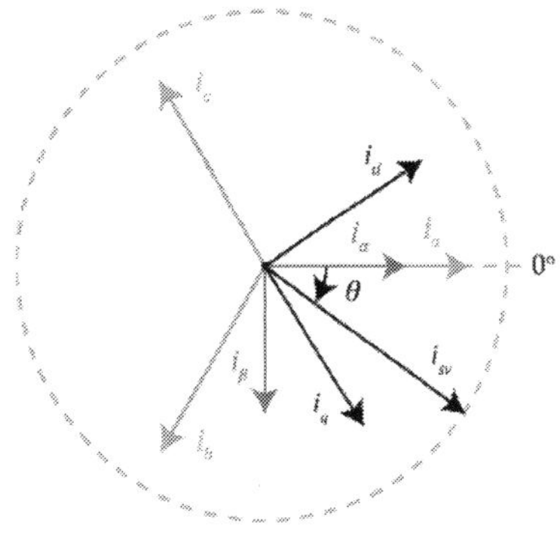

$\{i_a, i_b, i_c\}$ and $\{i_\alpha, i_\beta\}$ are stationary reference frames, while $\{i_d, i_q\}$ is a rotary reference frame

Figure 2. Stationary and rotary reference frame.

4 Position Control for a Stepper Motor

Two-phase hybrid stepper motors require a dual H bridge driver, and work reasonably well in open loop by half-stepping, full-stepping or micro-stepping.

However, most high performance servo systems employ an inner control loop to regulate torque, which it is enclosed by an outer control loop to handle rotor angular velocity, and followed by a second outer control loop for rotor angle positioning. While the design of the outer loops are considerably independent of motor type, the design of the torque loop is inherently specific to the motor being controlled. The block diagram in Figure 3, shows the three control loops employed at each stepper motor included in the group.

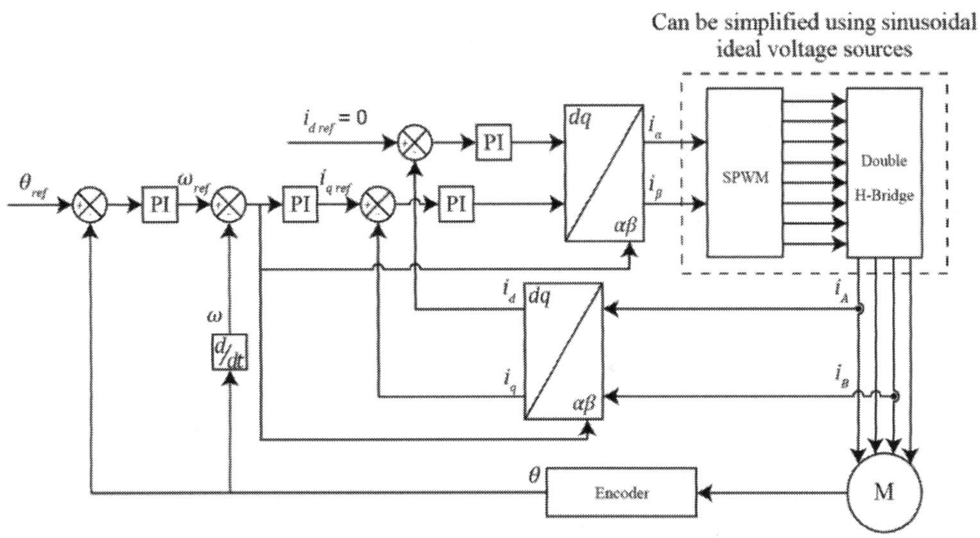

Figure 3. Block diagram for stepper motor position control

5 Distributed Cooperative Control

Cooperative control studies the dynamics of multi-agent dynamical systems linked to each other by a communication graph. The graph represents the allowed information flow between the agents, where each agent is confined to depend only on information about that agent and its neighbors in the graph [5].

Decentralized control systems refer to the case in which the agents are not following a centralized master/slave configuration. Instead, they are distributed within the system in a way that each agent is connected into a network of agents. Some of the advantages of distributed control systems are that a multi-agent system can perform tasks more efficiently than a single agent, increasing tolerance to possible agent failure. To coordinate with other agents in a network, every agent needs to share information with its neighbor peers, so that all can agree on a common goal of interest, like for instance, keeping the heading of aircraft in formation, or a target trajectory for a team of robots.

It was Reynolds [6] the first to propose a computer animation model to simulate collective behavior of multiple-agents, and after him several researches have provided some new in-sites of systematic framework of consensus problems. While most literature concentrates on studying consensus under fixed communication topology, in some applications the communication topology might change due to communication range limitations, mainly when agents are moving in space, like in the case of flight formation or robots synchronization.

5.1 Algebraic Graph Theory

The communication network is modeled by a graph and establishes the links and interconnections between the agents. The graph consists of nodes representing the agents, and directed edges corresponding to the allowed flow of information between the agents. Some basic graph theory concepts [7] are essential in the study of multi-agent dynamical systems.

A weighted order n graph is defined as a pair $G = (V, E)$, with a finite nonempty set of n nodes $V = \{v_1, \dots, v_n\}$, a set of edges or arcs $E \subseteq V \times V$, and a weighted adjacency matrix $A = (a_{ij})_{nxn}$. Each node v_i corresponds to an agent i, and each edge denoted as $(v_i, v_j) \in E$, is represented as an arrow from v_i (the tail) to v_j (the head). We assume the graph is simple, i.e., $(v_i, v_i) \notin E$, meaning that at any agent i there will be no "self-loops", and there will be no multiple edges between the same pairs of nodes.

Each edge $(v_i, v_j) \in E$ in a weighted direct graph (or digraph) represents the allowed flow of information from agent i to agent j, which means that agent j can received information from agent i. In contrast, the pairs of nodes in undirected graphs (also known as bidirectional graphs) are unordered, where an edge $(v_i, v_j) \in E$ denotes that agents i and j can receive information from each other.

The weighted adjacency matrix $A \in R^{nxn}$ of a digraph G is defined by: a_{ji} weights of the link (v_i, v_j) and $a_{jj} = 0$ for any $v_j \in V$, $a_{ji} > 0$ if $(v_i, v_j) \in E$, and $a_{ji} = 0$ otherwise.

The weighted adjacency matrix A of a weighted undirected graph is defined analogously except that $a_{ij} = a_{ji}$, $\forall i \neq j$, since $(v_i, v_j) \in E$ implies $(v_j, v_i) \in E$, resulting in a symmetric adjacency matrix A.

The weighted in-degree of a node v_i is the number of edges having v_i as a head, so it will be equal to the i-th row sum of A:

$$d_{in} = \sum_{j=1}^{n} a_{ij} \tag{7}$$

and the weighted out-degree of node v_i is the number of edges having v_i as a tail, so it will be equal to the i-th column sum of A:

$$d_{out} = \sum_{i=1}^{n} a_{ij} \tag{8}$$

The in-degree and out-degree are local properties of the graph. A (directed) tree is a connected digraph where every node except one, called the root, has in-degree equal to one. A spanning tree of a digraph is a directed tree formed by graph edges that connects all the nodes of the graph. A graph is said to have a spanning tree, if a subset of the edges forms a directed tree meaning that there is a node (called root node) with a direct path from that node to every other nodes in the graph with no cycles (a cycle is a simple path that starts and ends at the same node). The root set or leader set of a graph is defined as the set of nodes that are the roots of all spanning trees. A graph may have multiple trees; however, if it contains at least one spanning tree, the graph is declared as strongly connected.

The adjacency matrix A of an undirected graph is symmetric, $A = A^T$. A graph is said to be weight balaced if the weighted in-degree equals the out-degree for all i. If all the nonzero edge weights are equal to 1, this is the same as the definition of balanced graph. An undirected graph is weight balanced, since if $A = A^T$, then the i-th row sum equals the ith column sum.

The Graph Laplacian matrix is defined as follows:

$$L = D - A \tag{9}$$

where $D = diag\{d_{in}\}$ is the in-degree matrix, and A the adjacency matrix. Note that L has all row sums equal to zero.

Many properties of a graph may be studied in terms of its graph Laplacian matrix, providing a key role in the analysis of dynamical multi-agent systems on graphs. The eigenvalues of the Laplacian matrix explains properties of the underlying graph topology.

Any undirected graph has $L = L^T$, so all its eigenvalues are real and can be ordered as $\lambda_1 \leq \lambda_2 \leq \ldots \leq \lambda_n$.

A weighted undirected graph is connected, if and only if, its Laplacian matrix L has an eigenvalue zero with multiplicity one, and all other eigenvalues have positive real parts [8].

A weighted directed graph has a directed spanning tree, if and only if, its Laplacian matrix L has an eigenvalue zero with multiplicity one, and all other eigenvalues have positive real parts [9].

5.2 Cooperative Control Algorithm

When multiple agents agree on the value of a variable of interest, they are said to have reached consensus. Consensus algorithms are designed to be distributed, assuming only neighbor-to-neighbor interaction between agents, so agents will update the value of their information based on the information of their neighbors. Hence, the goal will be to design an updated law, so that the information of all agents in the network converge to a common value.

Consensus algorithms have been studied extensively in the context of cooperative control of multi-agents. The most general continuous time consensus algorithm was formulated by [10], and it is given by the first order single integrator dynamics:

$$\dot{x}_i = \sum_{j=1}^{n} a_{ij}\left[x_j - x_i\right] \tag{10}$$

where a_{ij} is the (i, j) entry of the adjacency matrix $A \in R^{nxn}$ associated with the graph G, and x_i is the information of the ith agent. Setting $a_{ij} = 0$ denotes the fact that agent i cannot receive information from agent j. A consequence of equation (10) is that the information of agent i is driven toward the information of its neighbors.

Equation (10) can be written in matrix form, considering the global dynamics of the state vector $x = [x_1 \ldots x_n]^T \in R^n$ as:

$$\dot{x} = -(D - A)x = -Lx \tag{11}$$

According to [10], the cooperative control law in (11) guarantees consensus, if and only if, the graph has a spanning tree, meaning that the dynamics given by (11) has a system matrix $-L$ with eigenvalues in the left-hand complex plane, except for only one of the eigenvalues, which will be located at the origin (stability condition).

In this research paper, distributed cooperative control of multi-agent systems has been applied for rotor synchronization of five stepper motors, with a communication network topology defined by their graphs. The cooperative control law applied is the following:

$$\partial\dot{\theta}_i = -k\left[\sum_{j=1}^{n} a_{ij}[\theta_i - \theta_j] + g_i[\theta_i - \theta_{ref,i}]\right] \quad (12)$$

where k is the control gain, a_{ij} is the adjacency matrix defined by the graph network topology, g_i is the pinning gain, with a non-zero value for only one of the stepper motors that has the reference rotor position angle (leader agent), and $\partial\dot{\theta}_i$ is the rotor angle tracker error.

The cooperative control law in (12), can be written in matrix form as shown in (13), and its implementation in block diagram will be as shown in the Figure 4.

$$\begin{bmatrix} \partial\dot{\theta}_1 \\ \partial\dot{\theta}_2 \\ \partial\dot{\theta}_3 \\ \partial\dot{\theta}_4 \\ \partial\dot{\theta}_5 \end{bmatrix} = -K\left[(L+G)\begin{bmatrix} \theta_1 - \theta_{ref,1} \\ \theta_2 - \theta_{ref,2} \\ \theta_3 - \theta_{ref,3} \\ \theta_4 - \theta_{ref,4} \\ \theta_5 - \theta_{ref,5} \end{bmatrix}\right] \quad (13)$$

where $K = diag(k)$ is the control gain matrix, $G = diag(g_i)$ is the pinning matrix, $L = D - A$ is the Laplacian matrix, A is the adjacency matrix, and D is the in-degree matrix of A.

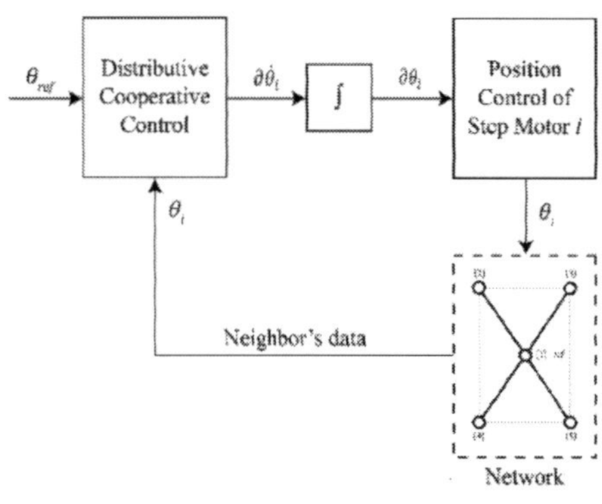

Figure 4. Cooperative control algorithm.

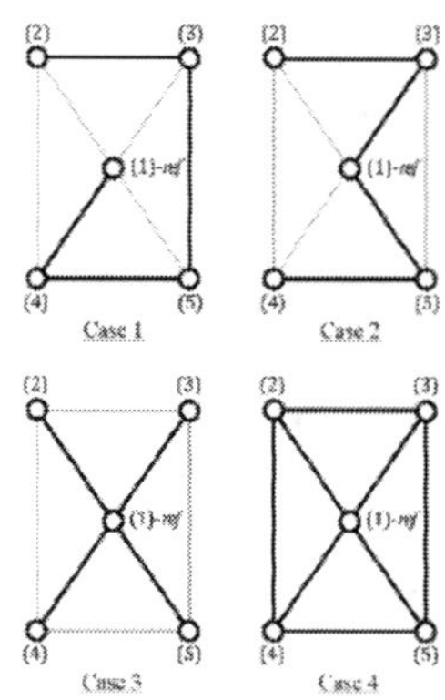

Figure 5. Network topology-graphs.

6 Simulation Results

A distributed cooperative control law based on multi-agents has been used to synchronize five stepper motors, where each agent will communicate with its neighbors according to a predefined network topology given by the graphs shown in Figure 5.

Leadership is appointed to one of the stepper motors by the pinning matrix $\boldsymbol{G}$, to which reference signal is applied, for the other agents to follow according to the cooperative control law defined at (12). Simulations are analyzed for a square, and triangular reference signal, with results shown in Figures 6-to-10.

Even though redundant communication links were assigned in the network of case 4, same results as in case 3 are obtained. The topology defined in case 1 is the most unfavorable, because each agent delays the information to the next agent, as confirmed by the simulation response.

7 Conclusions

The cooperative control algorithm implemented to synchronize rotor position of five stepper motors, with the internal closed-loop rotor position control derived based on field-oriented control techniques to provide torque control, gives different results depending on the topology of the selected network.

From the experiments carried out, it could be concluded that the network topology plays an important role when cooperative control based in multi-agents is considered. Hence, an optimization tool to find the optimal communication graph, attending to minimize a predefined cost function, could be an interesting future research topic.

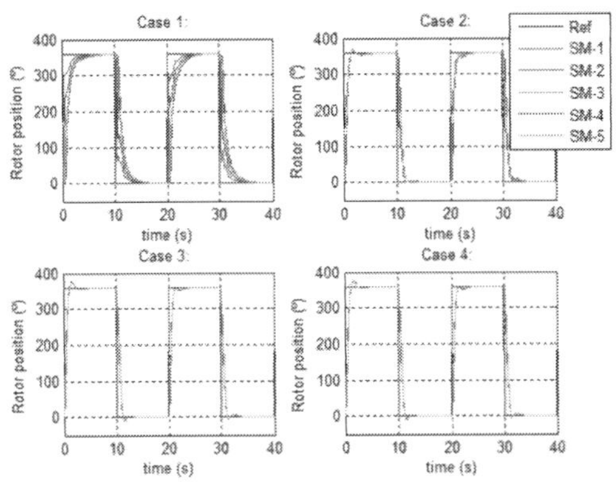

Figure 6. Square reference signal.

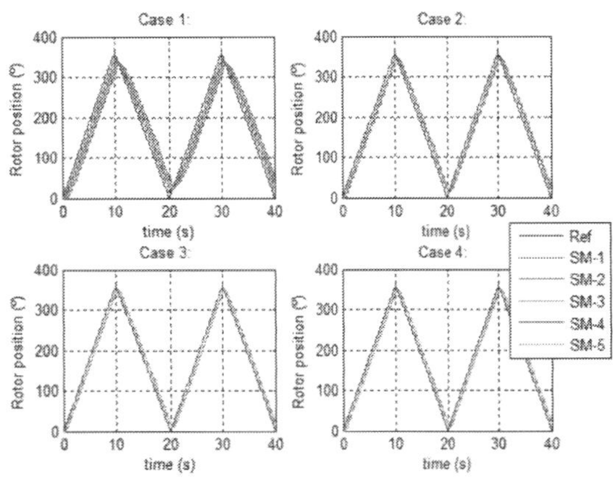

Figure 7. Triangle reference signal.

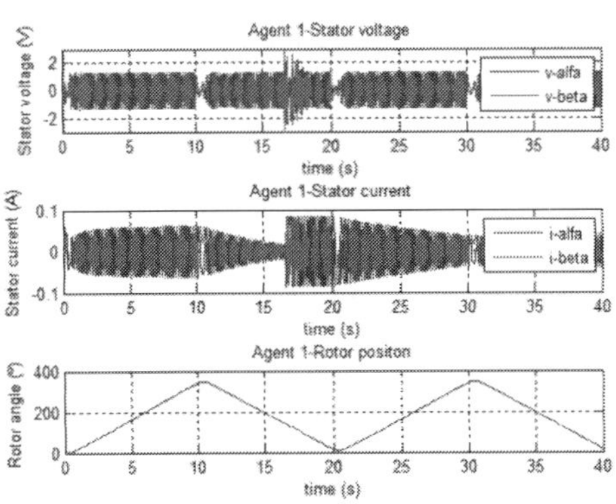

Figure 8. Stepper motor voltage and current.

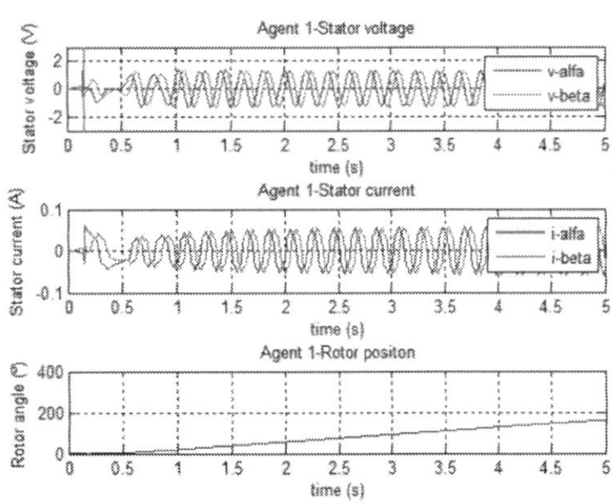

Figure 9. Stepper motor voltage and current.

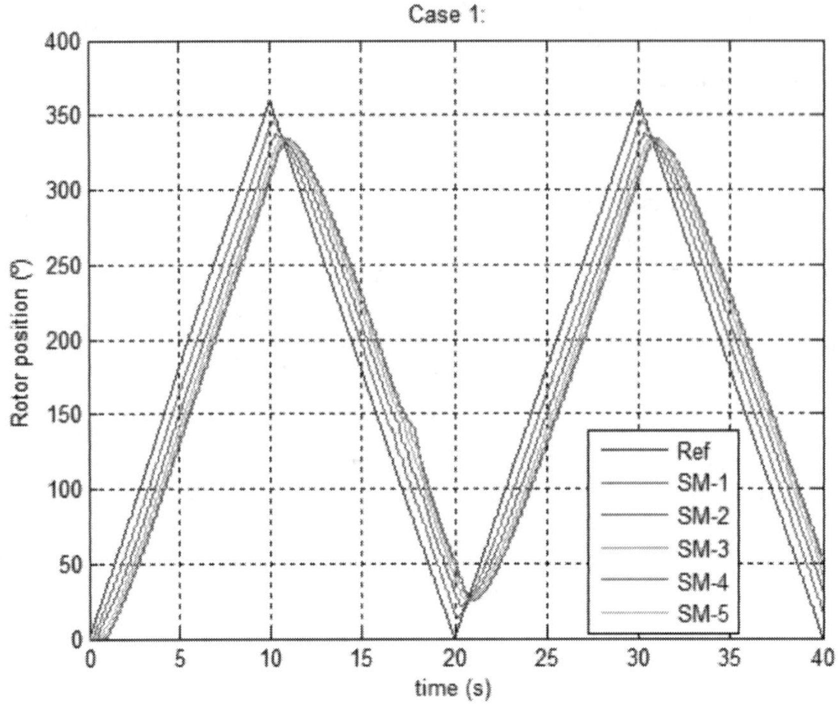

Figure 10. Delay due to the communication among agents.

Acknowledgement

"S. Yagüe and A. García-Cerrada collaborate with the support of the project PRICAM-CM (ref. S2013/IEC2933) partially financed by Madrid's Government."

References

1. Richard M Murray. Recent research in cooperative control of multi-vehicle systems. *International Conference on Advances in Control and Optimization of Dynamical Systems*, pages 1–27, 2007.
2. Takashi Kenjo and Akira Sugawara. *Stepping Motors and Their Microprocessor Controls*. Oxford University Press, 1995.
3. G. Brandenburg J. Heinzl H. Kellermann, P. Hildinger. Field oriented position control of a hybrid stepper motor. *In Proceedings of the EPE Conference*, pages 908–913, 1995.
4. R.H. Park. Two-Reaction Theory of Synchronous Machines, Generalized Method of Analysis, Part I. *Winter Convention of the A.I.E.E., New York*, 1929.
5. Kristian Hengster-Movric Frank L. Lewis, Hongwei Zhang and Abhijit Das. *Cooperative Control of Multi-Agent Systems*. Springer, 2014.
6. Craig W. Reynolds. Flocks, herds and schools: A distributed behavioral model. *ACM SIGGRAPH Computer Graphics*, (4):25–34, 1987.
7. Jorge Cort. *Distributed Control of Robotic Networks A Mathematical Approach to Motion Coordination Algorithms*. 2009.
8. F. R. K Chung. *Spectral Graph Theory*. American Mathematical Society, 1997.
9. Lixin Gao, Daizhan Cheng, and Jinhuan Wang. On consensus of multiagent systems under dynamically changing interaction topologies. *2007 IEEE International Conference on Control and Automation, ICCA*, pages 7–11, 2008.

[10] Ali Bidram, Frank L. Lewis, Ali Davoudi, and Zhihua Qu. Frequency control of electric power microgrids using distributed cooperative control of multi-agent systems. *2013 IEEE International Conference on Cyber Technology in Automation, Control and Intelligent Systems*, pages 223–228, 2013.

MATEC Web of Conferences **167**, 02002 (2018) https://doi.org/10.1051/matecconf/201816702002
IC4M & ICDES 2018

MEASUREMENT OF RESIDUAL STRESSES AND TOUGHNESS IN Ti-6Al-4V MATERIAL PRODUCED BY LASER BEAM MANUFACTURING

Olivier Quénard[1,a], Olivier Dorival[1,2], Philippe Guy[1] and Kevin Brethome[3]

[1]*Icam, 75 av. de Grande Bretagne 31076 Toulouse, France*
[2]*Université de Toulouse, Institut Clément Ader UMR CNRS 5312, INSA/UPS/ISAE/Mines Albi, 3 rue Caroline Aigle F-31400 Toulouse, France*
[3]*Centre National d'Études Spatiales, 18 av. Edouard Belin 31401 Toulouse, France*

Abstract. This study deals with the measurement of residual stress by the hole drilling method in Ti-6Al-4V samples obtained by Laser Beam Manufacturing after stress relieving. Indeed, it was of high interest to check the efficiency of such a thermal post-treatment. The influence of the orientation of both the sample and the notch is evaluated from a toughness point of view, and the obtained values are compared to that of similar parts submitted to Hot Isostatic Pressing after stress relief treatment. A more isotropic behaviour were detected on the HIPed samples (with respect to that of the stress relieved ones) with a higher average value. This development is discussed from both residual stress and microstructural points of view.

1 Introduction

For a few decades, additive layer manufacturing (ALM) appeared as a way of revolutionizing both the design possibilities and manufacturing standards. Now it has started to be used in space, aeronautical and automotive applications. Nevertheless, for each alloy, various parameters can modify the mechanical properties, such as the technique used (Laser Beam Manufacturing (LBM), Electron Beam Manufacturing (EBM), etc.), the manufacturing parameters, the heat post-treatment, powder characteristics, the loading orientation, etc., and very little data, furthermore with large scatter, can be detected in open literature.

This paper is related to the Laser Beam Manufacturing technique dedicated to Ti-6Al-4V which is increasingly being used in aerospace industries. A lot of work for that has been done to investigate the relationship between the process, the microstructure and mechanical properties for LBM shed light on the influence of thermal treatment [1-3] or the building orientation [4-6]. Several other papers dealt with the fatigue properties [7-8]. However, to the authors' knowledge, only a few papers have focused on the fracture toughness of ALM Ti-6Al-4V [7-12], although this is a key point in the damage tolerance approach.

In a previous paper [13], we presented tensile characteristics (Young modulus, Yield Stress, Ultimate Tensile Stress and Elongation) obtained for both stress relieved and post-Hot Isostatic

[a] Corresponding author : olivier.quenard@icam.fr

Pressed Ti-6Al-4V LBM samples, versus the loading orientation. Microstructural and surface fracture observations allowed the changes in the measured properties to be explained. Nevertheless, as outlined in this paper, residual stresses due to thermal conditions during manufacturing in some samples led to cracking in the as-built samples (before any thermal relief post treatment), inducing the inability to perform any mechanical test. The present study focuses on the measurement of residual stresses in the relieved Ti-6Al-4V LBM samples. The aim of this project is to determine if this soft heat treatment wholly releases samples free from thermal stresses. The toughness of the samples are presented for all cases of orientation and post-heat treatment. All these data were requested by the CNES in order to better design and size parts which would then be interesting to produce by LBM for weight saving in spatial applications.

2 Sample preparation and testing methods

2.1 Materials and manufacturing parameters

All the Ti-6Al-4V specimens were manufactured during a single batch in order to avoid any problem of reproducibility of the manufacturing parameters. By using a Ti-6Al-4V powder from Advanced Powders and Coatings, with 15-45 µm particle size, the samples were manufactured in an EOS M280 machine using a 1060-1100 µm wavelength laser. Successive layers were 60 µm thick. After a stress relief treatment on the base platform (670°C / 5h) to release residual stresses, half the specimens were submitted to an additional Hot Isostatic Pressing (HIP) process (920°C / 2h / 1020 bars).

2.2 Preparation for residual stress and toughness tests

For residual stress measurements, in order not to change the massiveness of the obtained samples (potentially leading to change in heat transfer and thus in the cooling rate after laser scanning) we chose to manufacture samples with dimensions (cubes of 20x20x20 mm^3) similar to those for toughness test specimens (at least from a width point of view). For the toughness tests, the size of produced samples was 45x45x20 mm^3.

After the ALM process and the corresponding thermal treatments, toughness specimens were machined to get the corresponding dimensions (32 mm width, according to ASTM E399 [14] norm (W32-B16)) and to create the notch along the expected direction (see Figure 1). Five samples were produced for each combination of orientation/thermal treatment.

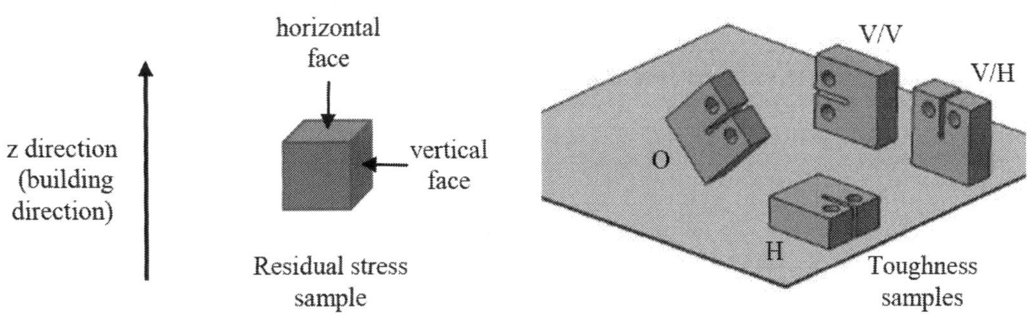

Figure 1. Orientations of the samples for both toughness (left) and residual stress (right) tests

For residual stress preparation, samples were gently manually polished in order to eliminate the surface oxide layer, and to allow good bonding with the strain gauge rosettes. Measurements were performed on both the horizontal and vertical surfaces (see Figure 1).

2.3 Residual stress and toughness tests set-up

Residual stresses were measured by the Hole Drilling Method (HDM). After bonding a 3-grid strain gauge rosette (type A) on the surface of the sample, this test consisted in measuring the resulting strain while gradually drilling a hole of 1.6 m in diameter, confirming with ASTM E387 [15]. The drilling was performed by steps of 50 μm, until 1 mm in depth was reached.

The three measured strain curves (see Figure 2), using Young Modulus and Poisson ratio measured (119 GPa and 0.33 respectively), lead to the determination of both the in-plane stress tensor (σ_x, σ_y and τ_{xy}) and the equivalent Von Mises stress.

Toughness tests were performed on an Instron electromechanical testing machine of UTM-EM type, using Bluehill 3.35 software. Apparatus involved a 2525-801 type 100 kN Instron load cell and a 2670-130 Crack Opening Displacement Gauge from Instron. Tests were performed following the ASTM E399 recommendation. Load rate of 0.33 kN/s was used up to the global failure of the specimen.

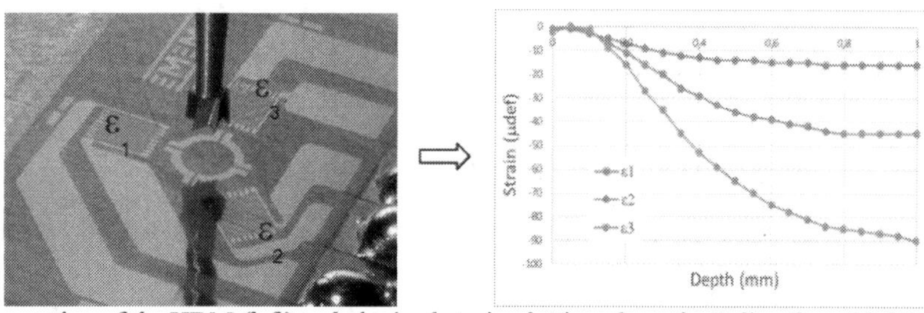

Figure 2. Presentation of the HDM (left) and obtained strain plotting along the 3 directions (right) versus the hole depth

3 Results

3.1 Residual stress

The raw data obtained on the horizontal surface of the relieved Ti-6Al-4V sample are presented in Figure 3-a. The calculated results reveal, on the one hand, that the stress along the X and Y directions presents very similar distributions (see Figure 3-b), while it seems that there is no shear stress. As a first conclusion, one can say that the tested sample has an isotropic horizontal in-plane behaviour, which can easily be explained by the fact that the laser manufacturing is performed by changing the scanning angle from layer to layer. Moreover, the positive values for the X and Y stresses correspond to the tensile stresses. On the other hand, one can note through the Von Mises stress curve (see Figure 3-c) that the residual stress in the material is superficially concentrated in the first 400 μm, with a maximum value centred around 150-200 μm. This maximum residual stress level is lower than 40 MPa, meaning that it is almost negligible. It seems that there is no residual stress after 250 μm depth.

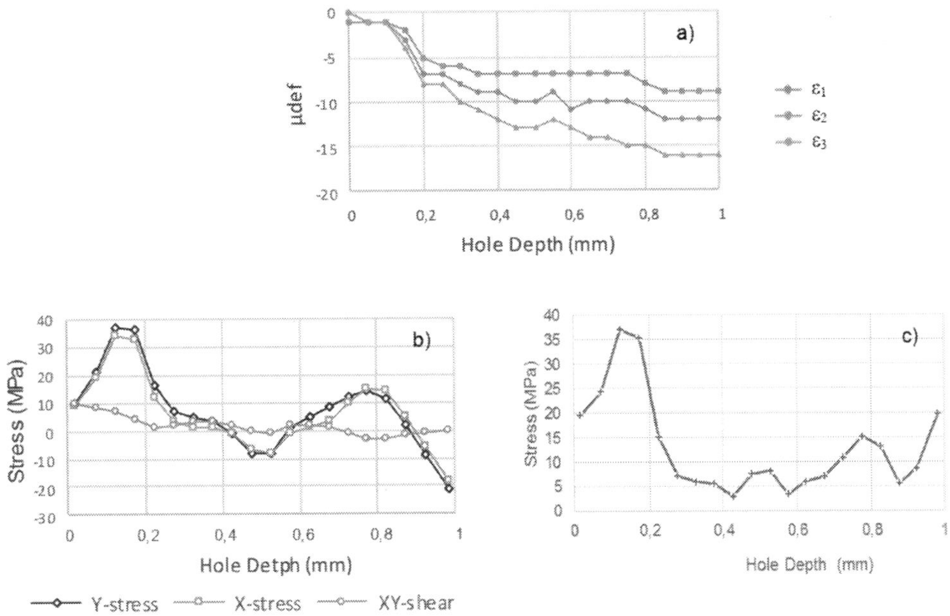

Figure 3. Strain curves obtained on the horizontal face by the HDM (a), and calculated stress tensor (b) and Von Mises Stress (c)

The results obtained on the vertical surface of the sample show similar curves concerning the X and Y stresses (see Figure 4-a) with reference to that obtained for the horizontal surface. The negative values measured at the extreme surface indicate a compressive stress, while these values quickly change for positive values representing tensile stresses. The similar stress distributions along the X and Y directions also reveal an isotropic vertical in-plane behaviour. The only difference is that a slight sub-superficial shear stress seems to be detected at around 200 μm depth, with a value lower than 20 MPa.

The calculated Von Mises stress curve presents (see Figure 4-b) a similar shape with reference to that of the horizontal surface. Indeed, only a slight sub-superficial residual stress is detected, the maximum value of which is close to 45 MPa, at around 200 μm depth. Almost no residual stress is detected after 300μm depth.

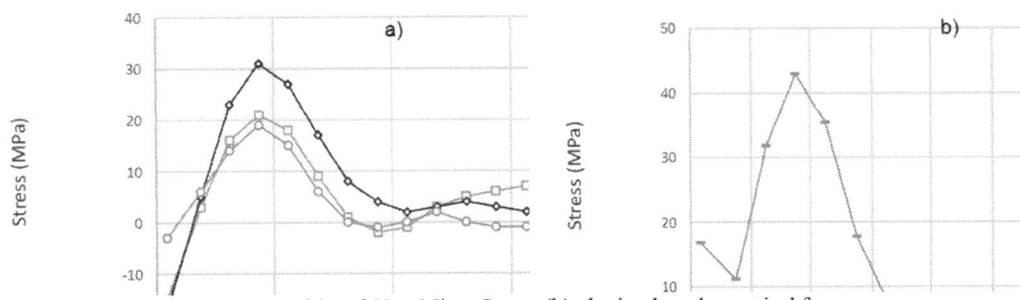

Figure 4. Calculated stress tensor (a) and Von Mises Stress (b) obtained on the vertical face

As a global conclusion for this part, one can say that the relieving heat treatment imposed on the LBM samples is efficient enough to erase almost all internal residual stress in the samples. Only a very slight sub-superficial stress, ranging from between 40 and 50 MPa, is detected at around 200 μm depth. In comparison, much higher values (up to 400 MPa) have been reported in literature on welded Ti-6Al-4V plates [16].

3.2 Toughness

3.2.1 Stress relieved Ti-6Al-4V

The results reveal that the orientation was a moderate influence on the measured toughness of the sample (see Figure 5). Indeed, only the V/H samples (see orientations in Figure 1) seem to present slightly higher average values (+17%) with reference to that of the three other orientations (the average values range between 38 and 46 $MPa.m^{1/2}$).

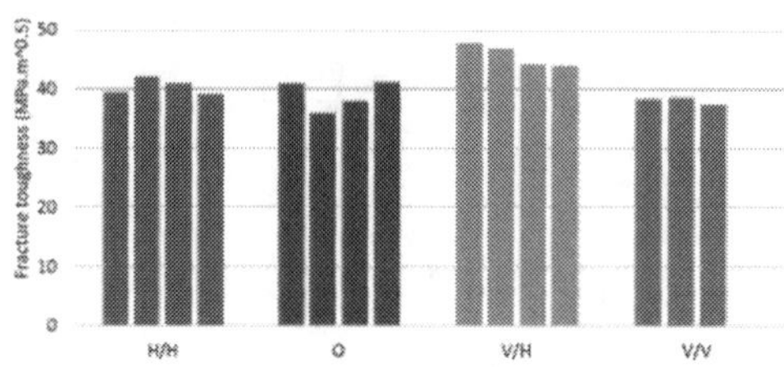

Figure 5. Fracture toughness vs orientation measured on the stress relieved Ti-6Al-4V samples

3.2.2 HIPed Ti-6Al-4V

The results for HIPed samples are presented in Figure 6. The fracture toughness of the samples strongly increased (+50-60% depending on the orientation) after Hot Isostatic Pressing, reaching an average value close to 65 $MPa.m^{1/2}$. Moreover, this post-treatment seemed give rise to a more isotropic material, as the slight anisotropy detected previously in the V/H direction had almost disappeared.

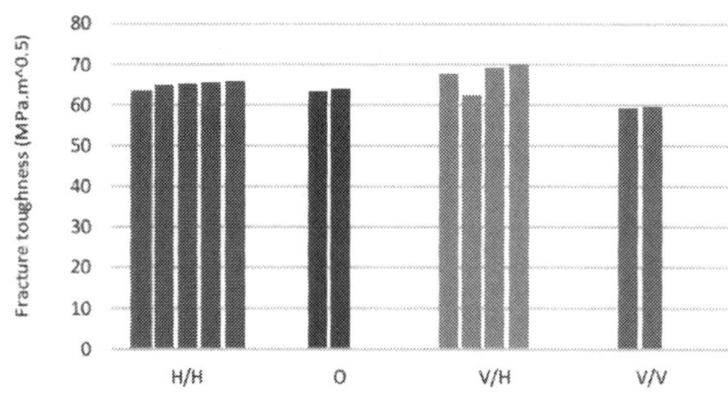

Figure 6. Fracture toughness vs orientation measured on the HIPed Ti-6Al-4V samples

4 Conclusions and prospects

As a first conclusion of this study, one can confirm that the stress relieving thermal post-treatment performed on the ALM Ti-6Al-4V samples is efficient enough to reduce internal residual stresses down to negligible values (lower than 40 MPa whatever the orientation of the measurement). Therefore, the difference of mechanical properties, and especially fracture toughness, before and after Hot Isostatic Pressing cannot be attributed, even partially, to residual stress. Only a change in the microstructure [13], changing from fully martensitic (for stress relieved samples) to α-β (for HIPed samples) seems to be responsible for the 50-60% increase in the obtained toughness.

Nevertheless, the HDM allows residual stress to be measured in only a narrow range of depth (lower than 1 mm), while our samples were 16 mm thick. A similar study could be performed on thinner samples, in order to check the influence of massiveness of the parts on the obtained microstructures and mechanical properties.

References

1. L. Thijs, F. Verhaeghe, T. Craeghs, J. Van Humbeeck, J.-P. Kruth, Acta Mater, **58**, 3303 (2010)
2. L. Facchini, E. Magalini, P. Robotti, A. Molinari, S. Höges, K. Wissenbach, Rapid Prototyping Journal, **16(6)**, 450 (2010)
3. B. Vrancken, L. Thijs, J.-P. Kruth, J. Van Humbeeck, Journal of Alloys and Compounds, **541**, 177 (2012)
4. C. Qiu, N.J.E. Adkins, M.M. Attallah, Materials Science and Engineering: A, **578**, 230 (2013)
5. M. Simonelli, Y.Y. Tse, C. Tuck, Materials Science and Engineering: A, **616**, 1 (2014)
6. S. Palanivel, A.K. Dutt, E.J. Faierson, R.S. Mishra, Materials Science and Engineering: A, **654**, 39 (2016)
7. B. Van Hooreweder, D. Moens, R. Boonen, J.-P. Kruth, P. Sas, Paul, Advanced Engineering Materials, **14(1-2)**, 92 (2012)
8. D. Greitemeier, F. Palm, F. Syassen, T. Melz, International Journal of Fatigue, **94(2)**, 211 (2017)
9. V. Cain, L. Thijs, J. Van Humbeeck, B. Van Hooreweder, R. Knutsen, Additive Manufacturing, **5**, 68 (2015)
10. P. Edwards, M. Ramulu, Fatigue & Fracture of Engineering Materials & Structures, **38(10)**, 1228-1236 (2015)
11. M. Seifi, M. Dahar, R. Aman, O. Harrysson, J. Beuth, JOM, **67(3)**, 597 (2015)
12. P. Edwards, A. O'Conner, M. Ramulu, J. Manuf. Sci. Eng, **135(6)**, 061016 (2013)
13. O. Quénard, O. Dorival, Ph. Guy, A. Votié, K. Brethome, ECSSMET, European Conference on Spacecraft Structures Materials and Environmental Testing (2016)
14. ASTM E8 / E8M-15a, ASTM International (2015)
15. ASTM E837, ASTM International (2001)
16. C. Casavola, C. Pappalettere, extras.springer.com (2007)

MATEC Web of Conferences **167**, 02003 (2018) https://doi.org/10.1051/matecconf/201816702003
IC4M & ICDES 2018

Structural Performance Evaluation for Vibration-based Energy Harvester utilized in Railway Vehicle

Ki-Weon Kang[1,a] and Ji-Won Jin[2]

[1] *Department of Mechanical Engineering, Kunsan National University, Kunsan, Jeonbuk, Republic of Korea*
[2] *R&D Group, Jeonbuk Institute of Automotive Convergence Technology, Kunsan, Jeonbuk, Republic of Korea*

Abstract. This study aims to assess the structural performance and structural integrity of vibration energy harvester (VEH). For this, the structural performance test were conducted to identify the natural frequency and structural response against frequency. And then, static structural analysis was performed using finite element analysis to investigate the failure critical locations (FCLs). Finally, we conducted the frequency response analysis in frequency domain to obtain the structural response with frequency and investigate the structural integrity of VEH. Using the above results, we assessed the structural performance and structural integrity of two types of VEHs.

1 Introduction

The vibration-based energy harvester (VEH) [1-3] uses vibrations generated from various structures or components, and in particular, maximizes the power generation using the resonance between the excitation frequency of the vibration load, which comes from structures or components, and the natural frequency of the VEH itself, thereby enhancing the efficiency of power production [4]. And the VEH is subjected to very high levels of acceleration and load because of the aforementioned resonance when the same level of vibration is applied. Hence, the VEH is very likely to be damaged structurally. It is, therefore, essential to assess the evaluation of the structural performance and integrity.

We assessed the structural performance and integrity of the VEH as shown in Fig. 1, which is utilized in high speed railway vehicle [5]. For this, a static performance test was conducted using the incremental sinusoidal sweep condition. Next, we conducted a static structural analysis to identify the natural frequency and failure critical location (FCL). We then conducted a frequency response analysis to identify the stress at the FCL against the frequency.

2 Results and Discussion

We investigated the structural performance and integrity for two types of VEHs (Type I and Type II). The VEH normally comprises a shaft carrying a vibration load, a coil case, a main body, a neodymium magnet and leaf-spring-shaped moving component [3]. In Type I VEH, the material of the spring was

[a] Corresponding author : kwkang68@kunsan.ac.kr

STS 410, and the thickness was 0.65 mm. This VEH was designed to have a natural frequency of $f_{n,1}$ Hz with the displacement limit of 4mm. Type II VEH had a rubber stopper that could restrict the displacement of 2mm at both ends of the spring. The material of the spring was SKD61 with thickness of 0.7 mm and the natural frequency of $f_{n,2}$ Hz.

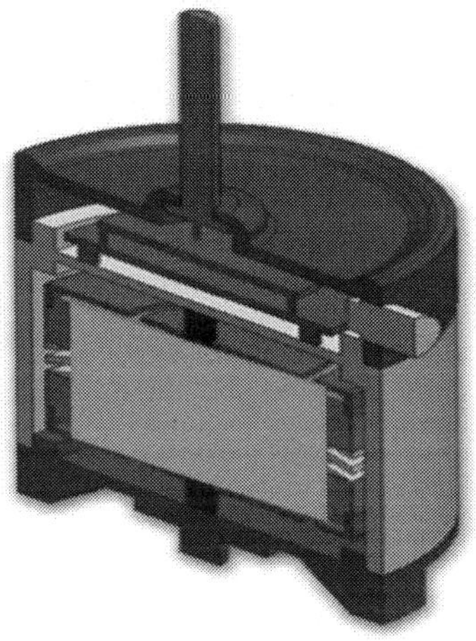

Figure 1. Vibration-based energy harvester utilized in railway vehcile

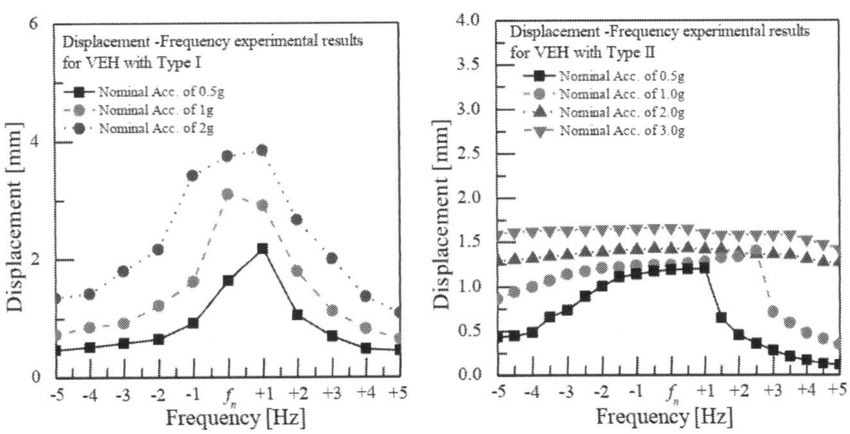

(a) Type I (b) Type II

Figure 2. Measured displacement–frequency curve

2.1 Static performance test

The static performance of the VEH was evaluated based on the incremental sinusoidal sweep tests, which their frequency ranges are natural frequencies of VEHs ± 5 Hz with an increment of 1 Hz and 0.5Hz for Type I and II VEHs, respectively, and their accelerations levels are 0.5, 1, 2, and 3g, respectively. Detail explanation for test procedure is in reference [6]. Fig. 2 shows the maximum displacement measured at outer rim of leaf-spring shaped moving component using the laser displacement sensor against the frequency of excitation. The first VEH shows quite different behaviour from the Type II VEH; this is attributed to that the displacement limitation of Type II VEH is just 2mm but 4mm for Type I VEH.

2.2 Static analysis

We conducted the static analysis using FE model in Fig. 3 [5]. The finite-element model was constructed using solid (CTETRA) and shell (CTRIA3) elements; the number of elements and nodes were 49,064 and 14,507, respectively. And, the multi-point constraints (MPC) were used between the solid and shell elements to avoid the hinge connections that could result from a mismatch in degrees of freedom of the solid and shell elements.

Fig. 4 shows the stress distribution obtained from the static FE analysis. The maximum stresses occur at the outer rim of the moving component for both VEHs and their values are 31MPa and 16MPa for Type I and Type II VEHs, respectively. These stresses are quite much lower than the experimental results and this behaviour comes from that the static analysis could not take into resonance between the excitation and natural frequency of VEHs.

Figure 3. FE model for VEH

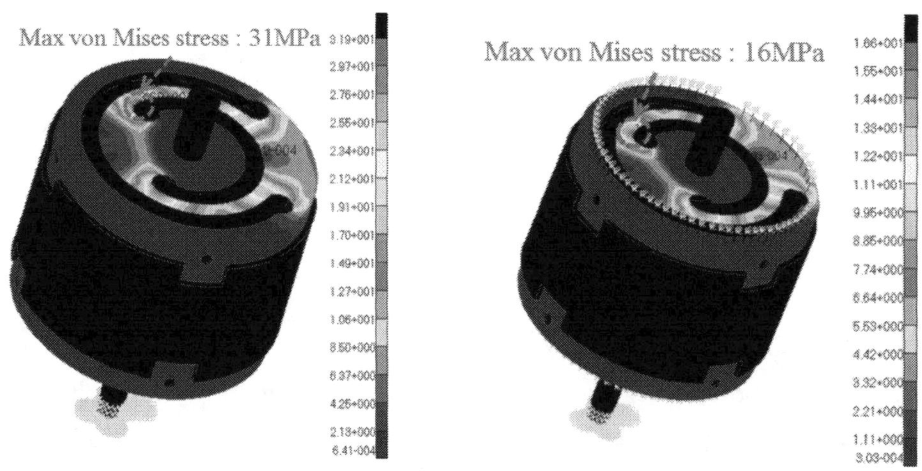

(a) Type I (b) Type II

Figure 4. Stress contour of VEH through static FE analysis

2.3 Frequency response analysis

The performance and structural integrity of vibration-based energy harvesters (VEHs) should be investigated considering the resonance between the excitation forces and natural frequency. The excitation forces are defined in the frequency domain and the structural response considering the resonance is expressed as the frequency response. Through the frequency response analysis, we could obtain the stress, displacement or acceleration response considering the resonance at specific locations in frequency domain.

Fig. 5 exhibits the stress contour on the leaf-spring shape moving component for Type I and Type II VEHs, respectively under the acceleration of 0.5g. The stress may quite differ from the static FE results in Fig. 4 and this behaviour may come from the resonance between the excitation and natural frequency in frequency response analysis. For further understanding these behaviours, the displacement - frequency response from the analysis are compared with the experimental results at the same points measured in the static performance test in Fig. 6 [5]. The figure states that the displacement-frequency results agree well with the experimental results for both the VEHs. It can be, therefore, seen that the frequency response analysis could well describe the response of the VEH with frequency.

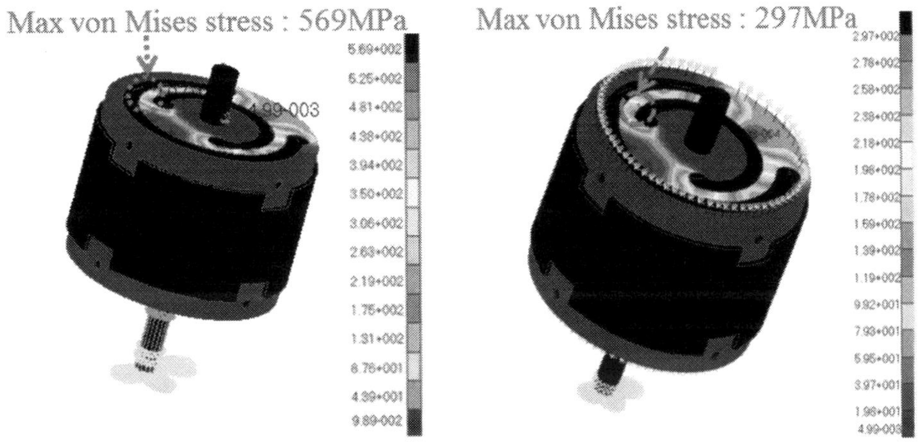

(a) Type I (b) Type II

Figure 5. Stress contour from frequency response analysis for VEH

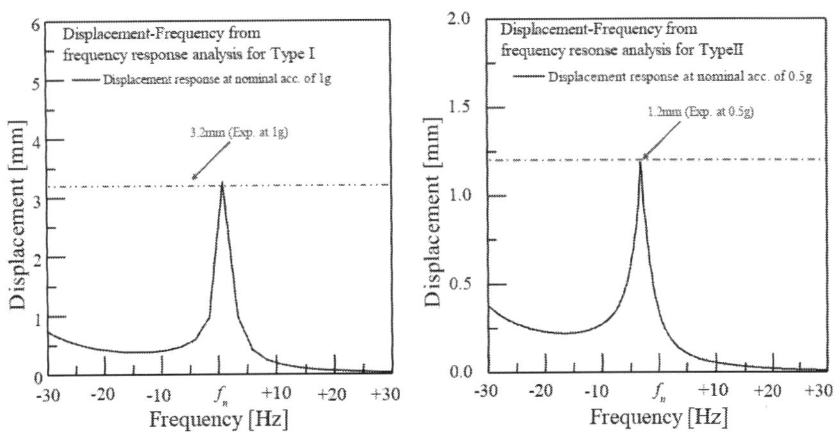

(a)Type I (b) Type II

Figure 6. Displacement response from frequency response analysis for VEH

3 Conclusions

We assessed the static performance and structural integrity of vibration-based energy harvesters. For this, we performed the static performance test, static structural analysis and frequency response analysis for two types of VEHs.

Acknowledgment

This research was supported by the Korea Institute of Energy Technology Evaluation and Planning (KETEP) and the Ministry of Trade, Industry & Energy (MOTIE) of the Republic of Korea(No.20174030201670) and by Basic Science Research Program through the National Research Foundation of Korea(NRF) funded by the Ministry of Education(NRF-2016R1D1A1A09918310).

References

1. S. Roundy, J. of Intelligent Material Systems and Structures, **16** (2005)
2. J.E. Kim, H. Kim, H. Yoon, Y.Y. Kim, B.D. Youn, Int'l J of Precision Engineering and Manufacturing-Green Technology, **2**, (2015)
3. R. Andosca, T.G. McDonald, V. Genova, S. Rosenberg, J. Keating, C. Benedixen, J. Wu., Sensors and Actuators A: Physical, **178** (2012)
4. G. Frank, W. Peter, J. of Micromechanics and Microengineering, **18** (2008)
5. J.H. Kim, J.W. Jin, J.H. Lee, K.W. Kang, Engineering Failure Analysis, **73** (2017)
6. J.W. Jin, K.W. Kang, J.H. Kim, Int'l J. of Precision Engineering and Manufacturing – Green Technology, **2** (2015)

Transverse and Longitudinal Stress Analysis in Reinforced Welded Oil Branched Pipes

H Safari[a] , R Pourrajab[a] and AYaghootian

National Iranian South Oil Company (NISOC), Ahvaz, Iran
Department of Mechanical Engineering, Shohadaye Hoveizeh University of Technology, Khuzestan, Iran
Department of Mechanical Engineering, Shahid Chamran University of Ahvaz, Ahvaz, Iran

Abstract. Welding process results in the non-uniform distribution of heat. Combined with thermal strain, plastic deformation and welding-induced distortion, it causes important problems such as failure. This study models welding process in a 12×16 split tee junction using finite element. This junction is a reinforcing branch used in the hot taping of oil pipe line. The model is studied in 3D state. This zone, i.e. heat affected zone (HAZ) is of high importance due to the existence of a severe temperature gradient and stress in the vicinity of HAZ. This study analyses single pass butt welding with a cooling time of 2700 seconds. In order to apply boundary conditions to the model, the studied structure should be in Free State according to practical procedure. The main aim of this simulation is to analyse Longitudinal and transverse residual stresses originated from co-directional and opposite-directional welding. This joint is formed of two semicircles welded to each other as linear single pass welding. Two welders simultaneously welded both sides of the pipe.

1 Introduction

In industrial factories, oil and gas refineries, petrochemical industries, thermal power plants, oil product pipelines and gas pipelines, there may be immediate need for cutting and welding operations on pipelines carrying hot, high pressure, flammable and toxic fluids. In such conditions, the operations should inevitably be done while the pipes are in service. The so-called hot tap technique is used for this purpose. According to studies, the majority of defects and failures in such junctions occur in the HAZ of carbon steels. The residual stress of welding serves as the main cause of the defects in the junctions [1]. Having an accurate prospect of the size and distribution of stresses as well as the effect of different welding parameters on residual stresses plays a vital role in controlling and minimizing their problems. Jiang and Yahiaoui [2] studied the effect of welding sequence on residual stress distribution in a multi-pass welded piping branch junction. High residual stress was formed in the vicinity of the weld region, irrespective of the sequence of welding. Deng et al [3] analysed the influence of deposition sequence on welding residual stress and deformation in an austenitic stainless-steel J-groove welded joint. Their results showed that welding sequence not only has a significant effect on results but it results in a significant change in the shape of deformation modes induced by welding sequence. Yaghi et al [4] evaluated residual stresses induced by dissimilar welding of a P92 steel pipe with weld metal IN625 using finite element simulation. Two subroutines were developed, one of which prescribed two phases of P92 steel while the other assumed three phases. Skouras et al

[a] Hossein.safari86@gmail.com; r.pourrajab@gmail.com

[5] measured welding-induced residual stresses in a P92 steel-In625 supperalloy metal in as welded and post welded heat-treated conditions. The results were discussed with respect to the residual stress magnitude and location as well as with respect to the effectiveness of the post welded heat-treatment procedure. Fu et al [6] analyzed the effect of welding sequence on residual stressed in a thin-walled octagonal pipe-plate structure. Nine different simulation sequences were carried out by single-pass TIG welding of an octagonal pipe-plate joint and the distribution of longitudinal and transverse residual stresses both on the outer and inner surfaces of the pipe were analyzed. Bajpei et al [7] evaluated the mitigation of residual stresses and distortions in thin aluminium alloy GMAW plates using different heat sink models. Their results show a good conformity between experimental and numerical results. This study analyzes the effect of welding-induced residual stresses on split tee reinforcing branch in a pipe-shaped joint of carbon steel (ASTM A36). The review of previous studies showed that there is no study on this subject. Co-directional and opposite-directional welding techniques were first modelled using ANSYS finite element. Then, the analysis results of Longitudinal and transverse residual stresses were studied in both welding techniques.

2 Modeling

2.1 Mechanical analysis

To simplify the welding simulation, the thermal and mechanical analyses were performed successively. Thermal analysis first determined heat distribution model during welding process. Then, the temperature history was applied to mechanical model as thermal load and stress distribution was obtained by solving the mechanical model. All items of mechanical model, except element type and boundary conditions, are similar to thermal model. Thermal model was obtained using the 3D element SOLID 70 with eight quadrilateral nodes with a single degree of freedom having temperature at each node while mechanical model was analyzed using the 3D element SOLID 185 with eight quadrilateral nodes; each have three degrees of freedom. In order to apply boundary conditions to the model, the structure should be in Free State in accordance with practical procedure. The studied junction is a 12×16-inch junction. Figures 1 and 2, and table 1, show the geometry and dimensions of the model, respectively. This model is studied in 3D state. Figure 3 shows the model meshing. There is a strong thermal gradient and stress in the vicinity of HAZ and this zone is of high importance [8, 9]. Therefore, finer meshes were used in this zone. It is impossible to generate or remove elements during solution process. Therefore, all elements already be created. To this ends, element birth and death capability was used [10].

Figure 1. View of the model.

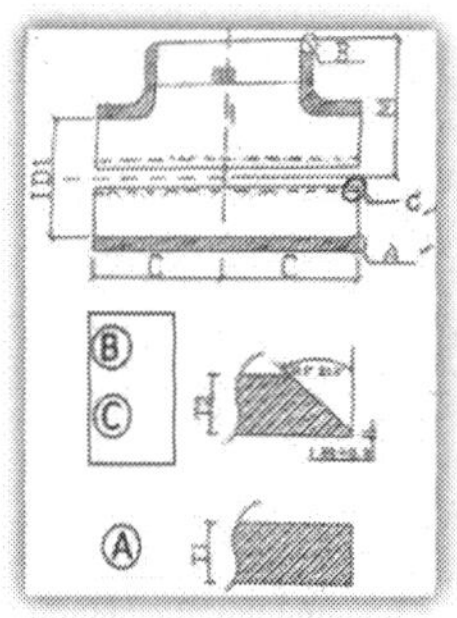

Figure 2. Dimension of split tee.

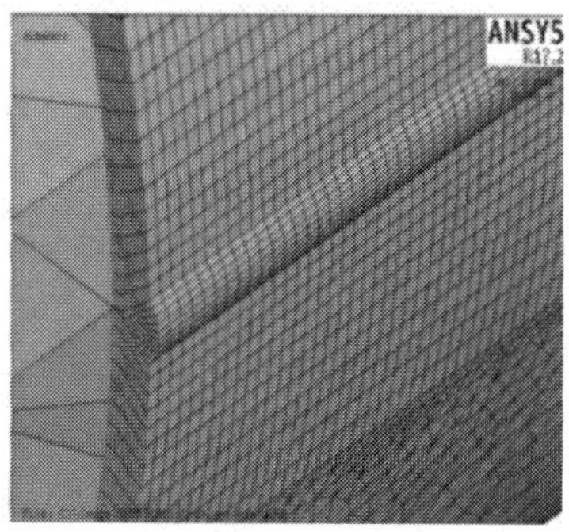

Figure 3. The model meshing.

Table 1. Dimension of split tee (Figure 2).

Size (inch)	WT (mm)	ID1 (mm)	OD2 (mm)	C (mm)	M (mm)	T1 (mm)	T2 (mm)
16×12	9.5	406.4	232.9	304.8	295.3	9.5	9.5

2.2 Thermal analysis

There are two main approaches to modelling input heat source in modelling welding process. The first one is to apply heat to a given volume of fusion zone in order to form an acceptable volume of melted material. The second one deals with the application of thermal flux or input heat source modelling [11]. This study required a large amount of calculations demanding a very specific system. Therefore, it adopted melting temperature method [9, 12]. Convection heat transfer coefficient in central line was zero due to the symmetry of the line and was considered to be 15 (W.m^2)/c in other surfaces. Ambient temperature was assumed 18°C [8, 9, 13, 14]. In this study, a constant heat was applied in each pass which was proportional to welding time and the position of welded elements.

2.3 Material specifications

With a ferrite and perlite microstructure, carbon steel (ASTM A36) is widely used in structures and equipment in dry lands. It shows acceptable welding property with no specific behaviour during welding. This simulation considers the same properties for the base metal, HAZ and welding metal [15, 16].

3 Results and discussion

3.1 Analysis of residual stress induced by co-directional and opposite directional welding

Figure 4 shows transverse residual stress of a selected path, which is perpendicular to welding line. According to this figure, in both co-directional and opposite-directional welding methods the type of stress is tensile stress from the highest to the lowest part of the studied model. By moving away from the welded zone, tensile stress decreases and it is converted to compressive stress. Finally, form a given distance the type of stress changes to tensile stress again. Figure 5 shows transverse residual stress distribution.

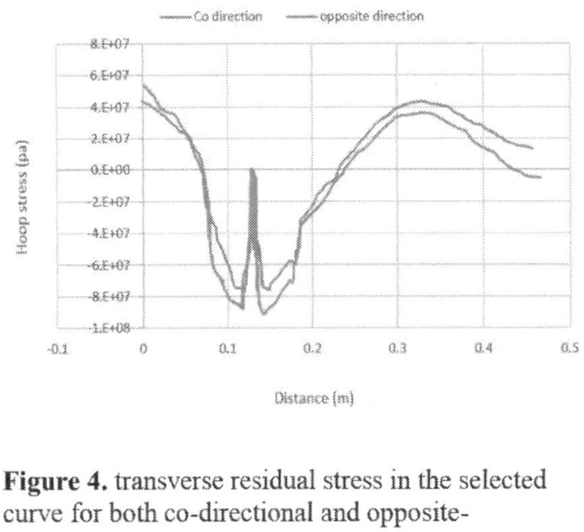

Figure 4. transverse residual stress in the selected curve for both co-directional and opposite-directional welding methods

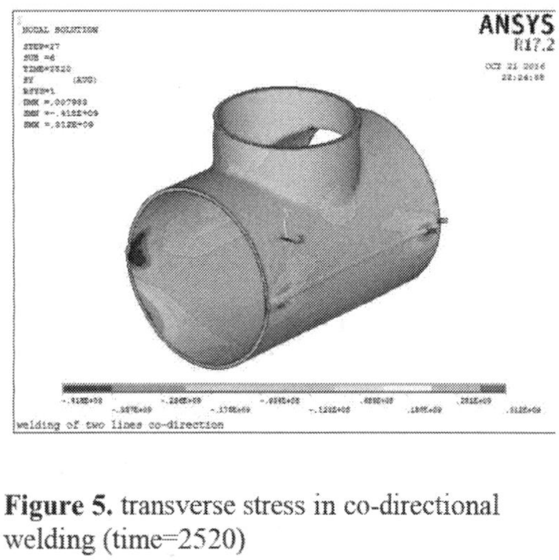

Figure 5. transverse stress in co-directional welding (time=2520)

https://doi.org/10.1051/matecconf/201816702004

Figure 6 compares co-directional welding with opposite-directional one. According to the figure, the value of residual stress has a tensile peak in the selected path. This point, on which the tensile stress becomes the maximum, is the point in which the two parts are welded to each other. After this point, the type of stress becomes compressive with a steep slope and finally the stress approaches a constant value. Figure 7 shows longitudinal residual stress distribution in different times.

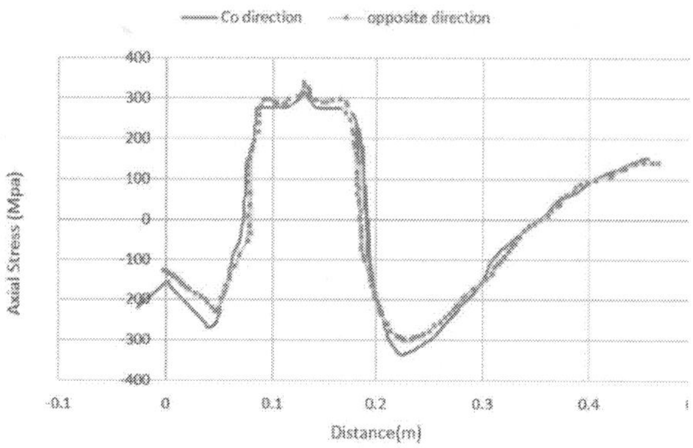

Figure 6. longitudinal residual stress in the selected curve in both co-directional and opposite-directional welding methods

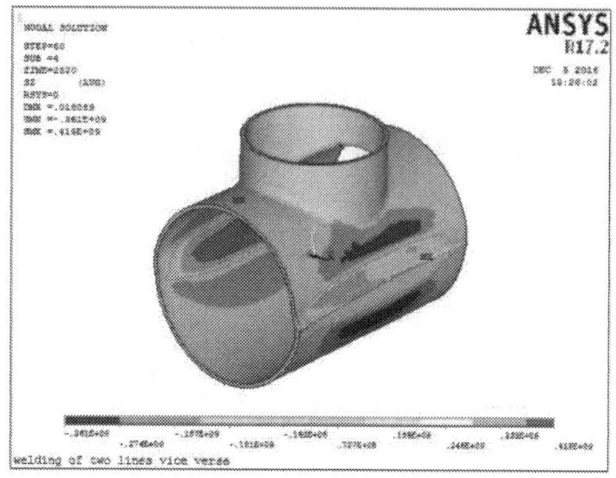

Figure 7. Longitudinal residual stress in opposite-directional welding (time=2520)

4 Conclusions

Analyse Longitudinal and transverse residual stresses originated from co-directional and opposite-directional welding were is studied in 3D state by using ANSYS finite element. Following conclusions are presented based on the Finite Element results:

- A large longitudinal tensile residual stress is created around HAZ and this value decreases away from HAZ.
- The maximum transverse residual stress is created in fusion zone. In far zones this stress almost tends to zero.
- The value of longitudinal residual stress in opposite-directional welding is slightly higher than that of co-directional welding.
- By moving electric arc and the deposition of welding beginning and ending effects in HAZ, transverse stress distribution was experienced considerable change.

5 Acknowledgement

Authors thank National Iranian oil company (NIOC), National Iranian South oil company (NISOC) and Marun Oil and Gas Producing Company (MOGPC) for their help and financial support.

References

1. T. Teng, P. Chang, W. Tseng , "Effect of welding sequences on residual stresses," Computers and Structures, (81),(2003).

2. W. Jiang, K. Yahiaoui, " Effect of welding sequence on residual stress distribution in a multipass welded piping branch junction", International Journal of Pressure Vessels and Piping,(95), (2012).
3. D. Deng, "Influence of deposition sequence on welding residual stress and deformation in an austenitic stainless steel J-groove welded joint",Materials and Design,(49),(2013).
4. A.H. Yaghi, T.H. Hyde, A.A. Becker, W. Sunb, " Finite element simulation of residual stresses induced by the dissimilar welding of a P92 steel pipe with weld metal IN625",International Journal of Pressure Vessels and Piping,(2013).
5. A. Skouras, P.E.J. Flewitt, M. Peel, M.J. Pavier, "Residual stress measurements in a P92 steel-In625 superalloy metal weldment in the as-welded and after post weld heat treated," International Journal of Pressure Vessels and Piping, (123-124), (2014).
6. D.f. Fu, Ch. Zhou, C. LI, G. Wang, L.LI,"Effect of welding sequence on residual stress in thin-walled octagonal pipe–plate structure," Transactions of Nonferrous Metals Society of China,3(24),(2014).
7. T. Bajpei, H. Chelladurai, M. Zahidansari, "Mitigation of residual stresses and distortions in thin aluminium alloy GMAW plates using different heat sink models," Journal of Manufacturing Processes,(22), (2016)
8. S.h. Cho, J.w. Kim, "analysis of residual stress in carbon steel weldment incorporating phase transformation ," science and technology of welding and joining,4(7), (2002).
9. L. Yajiang, W. Lura, C. Maoi , S. Xiaoqin , "finite element analysis of residual stress in the welded zone of high strength steel," Bulletin of Materials Science, 2(27),(2004).
10. Ansys, Rev 17.0,Ansys,Inc., Canonburg,PA,(2000).
11. D. Naeem, "Analysis of weld-induced residual stresses and distortions in thin-walled cylinders," ,Journal of Mechanical Science and Technology ,(23), (2009).
12. B. Taljat, B. Radhakrishnan, T. Zacharia ,"numerical analysis of GTA welding process with emphasis on post solidification phase transformation effects on residual stresses," materials science and engineering, 1-2(246) , ,(1998)
13. B.A.B. Andersson," thermal stresses in submerged-arc welded joint considering phase transformations," Journal of Engineering Materials and Technology, (100), (1978).
14. J. Goldak , A. Chakrorati, M.G. Bibby, " A new finite element model for welding heat sources," metallurgical tranctions ,2(15), (1984).
15. P.H. Chang, T.L. Teng, "numerical and experimental investigations on the residual stresses of butt-welded joints," computational materials science,4 (29), (2004).
16. S. Wen, P. Hilton, D. C Farragia, "finite element modeling of submerged arc welding process," journal of materials processing technology,1(119), (2001).

Cyber Physical Autonomous Mobile Robot (CPAMR) Framework in the Context of Industry 4.0

Yoon Ket Lee[1,a], Yeh Huann Goh[1] and Yiqi Tew[2]

[1]*Faculty of Engineering, Tunku Abdul Rahman University College, Kuala Lumpur, Malaysia*
[2]*Faculty of Computing and Information Technology, Tunku Abdul Rahman University College, Kuala Lumpur, Malaysia*

Abstract. Industry 4.0 or Smart Manufacturing creates intelligent object networking and independent process management through Internet of thing and data services. Cyber-Physical System (CPS) communicates among humans, machines and products through Internet of Things (IoT). In this paper a cyber physical autonomous mobile robot (CPAMR) IoT infrastructure system has been proposed. It is capable of performing human-machine interact by allowing users to place and manage orders using cloud platform. The Enterprise Resource Planning (ERP) system processes the data and send the product's data to Radio-frequency identification (RFID) tag system for storage and printout. Through Remote Telemetry Unit (RTU), the status of the product, CPAMR system and workstations or machineries are linked to the cloud platform. Initially, system identify the locations of CPAMR, product and the desired workstation. After gathering all the required information, Artificial Intelligence Algorithms (AIA) performs real time route map planning according to the shortest distance between CPAMR and the destination. This route map planning will then be sent to the CPAMR's micro-controller for operation. While the CPAMR is moving, it moves according to the planned route map with the assistance of the Obstacle Avoidance System until it reaches the destination and notifies cloud platform. Three individual projects representing three main functions of the proposed CPAMR have been carried out. Results show that the framework of the project is viable.

1 Introduction

Industry 4.0 or Smart Manufacturing is one of the hottest topics in the industry and become the world trend in the industry evolution [1-2]. Industry 4.0 concept is driven by the German whereas smart manufacturing leads by America. Both industry concepts are relatively similar. Industry 4.0 represents the coming fourth industrial revolution on the way to an Internet of thing, data services. Decentralized intelligence helps create intelligent object networking and independent process management, with the interaction of the real and virtual worlds representing a crucial new aspect of the manufacturing and production process [3]. One of the key factors to the success of implementing Industry 4.0 is through the Cyber-Physical System (CPS). CPS communicates among humans, machines and products with the help of Internet of Things (IoT). CPS system is able to acquire and process data, this

[a] Corresponding author: leeyk@tarc.edu.my

feature allows the CPS system to perform self-control in completing certain task and interact with humans via interfaces [4].

In an ideal production world, mass production with single product range and long product cycle life will definitely achieve the lowest product cost per unit. However in reality this is not feasible due to mass production, if the company continuously produce products with regardless of the sale demand from the market then it will end up keeping the stock in the warehouse even though the cost per production is low but the holding cost and deprecation cost of the products will end up a disaster to the company. Furthermore it almost impossible that has a good demand from market with only single produce range and long product cycle life. So the optimum solution to this is to increase the flexibility in production line but yet in the mass production environment. Flexibility in an automatic manufacturing system is always a challenge to the users or researchers. In order to achieve the desire flexibility, one of the solutions is through the mobile material handling system. It has to move independently to transport goods or process workpieces. It also need to movable lightweight robots to fully autonomously operating systems. In production shops, mobile robots have to be using modern navigation technique to move entirely without the need for floor markings, induction loops or magnets. It has to be real time application to deal with the any changes in production line. If production line has the mix product schedule, this Artificial Intelligence Algorithms (AIA) for the CPAMR can arrange the route map of the product according to the virtual world requirement, current product, machinery and also environment status.

In this research proposal framework, we investigate the usage of a novel CPS system in a smart material handling system. The proposed CPS system is equipped with the features of IoT, intelligent vision system and Autonomous Mobile Robot (AMR). The proposed Cyber-Physical AMR (CPAMR) system is embedded with AIA which allows the CPARM system to do decision making on production schedule planning, machinery, transportation requirement from production environment, work order for a part from virtual and part status based on the virtual and physical world interaction.

2 Related Work

An extensive literature reviews on several papers have been studied as categorizes below:

2.1 IoT and CPS Architecture

Lee et al. proposed 5-level CPS structure 5C, which consist of Connection, Conversion, Cyber, Cognition and Configure [4]. This 5C provides a viable and practical guideline for manufacturing industry to implement CPS for better product quality and system reliability with more intelligent and resilient manufacturing equipment. Gigli et al. make a comment on the IoT architecture that there still do not have the standard to put together different types of technologies such as sensors, actuators, network communications and computing technologies [5]. There are different IoT architecture models that have been proposed by the researchers. The architecture recommended by the International Telecommunication Union (ITU) has 5 layers which are Sensing Layer, Access Layer, Network Layer, Middle-ware Layer and Application Layers. This proposed IoT layer is similar to the Open Systems Interconnection (OSI) model for network and data communication. Qian et al. stated that traditional IoT structure is formed by three layers Perception Layer (responsible for recognizing and collecting information), Transportation Layer (consists of data communication networks) and Application Layer (applications which rely on the data from perception layer) [6]. Xu et al. suggested a collective summary of recommended architectures by different researchers which is similar to the structure proposed by the ITU above [7]. This proposed solution takes a service-oriented architecture (SOA) approach which is normally deployed in business software systems. The four-layer architecture is Sensing Layer, Networking Layer, Service Layer and Interface Layer. Trappey et al. proposed Perception layer, Transmission layer, Computation layer, Application layer in the IoT architecture [8].

The authors found that the above proposed IoT architecture even though there same to be different and every researcher try to have their own standard but the basic functionality and concept does not differ very much.

2.2 Autonomous Mobile Robot Device

A Propose AMR is a material handling device that consists of two major components. The moving component is an AGV and the material handling components is a robotic arm. The robotic arm is mounted onto the AGV. The type of robotic arm is flexible, it depends on the production line requirements however the mobility is fixed where it require the AMR to transport the robotic arm from one destination to another destination. Therefore the authors decide to review AGV research only. There are extensive researches have been conducted by the researchers on the topic of AGV. These papers can be further classified into the short summary below:

2.2.1 Localization Techniques

Jones suggested ideas on line following AGV, where the robot are able to localize itself once the AGV reaches a node or intersection of lines [9]. Tsumura and Markosov et al. found that Unable to localize AGVs along the lines, only at nodes [10-11] AGVs unable to travel without line, and unable to travel back to the line once its off-course. Require smooth surface for implementations. Lawrence proposed ideas of localization through odometry and Inertial Navigation System. However this method suffers from: Accumulative errors due to wheel slip. Unable to use the system for long period of time [12]. Highly dependent on modelling of system where many parameters needed to be considered such as slip, coefficient of friction, power draw from drivetrain, etc. Bosien et al. proposes the usage of RFID tags that are planted around the environment in the form of an array [13]. Thus allowing robots to localize depending on which RFID it is able to scan. The paper did not suggest a method to improve its accuracy, because by detecting RFID, there is a range that will produce inconsistency of results that lead to the pose of the robots cannot be accurately determined. Farkas et al. used triangulation method in their paper and propose the usage of Wi-Fi access points and the signal strength of the access points to determine the position of the AGV [14].

2.2.2 Docking Techniques

Lee et al. AGV maneuver and docking system based on bird's eye view image extraction [15]. The system used Hough line transform the estimate the distance between markers and the AGV itself. The testing data shows high efficiency and accuracy while traveling. This does not depend of fix line to travel to destinations. However does not include object avoidance system. Andersson et al. presents use of dead reckoning along with computer vision [16]. The AGV maneuver's along a pre-programmed path by using dead reckoning and odometry localization techniques. Computer vision is responsible for picking up images of markers located on the floor. However AGV suffers from accumulative errors due to the usage of odometry localization. Kim et al. published a paper regarding the implementation of object tracking with visual docking for performing tasks with multiple AGVs joining together [17]. The AGV's docking was done with 2 steps. Approaching and docking, where the first phase uses computer vision to track another AGV and determine the trajectory towards the AGV. The second phase starts when the AGVs are required to dock on another AGV using vision as well as with the help of compass sensors for heading and range finders to determine trajectory path. The testing of AGV docking system was done on surface of water and require high power processing unit due to requirement of high level of control of AGV. Vaz et al. presented the usage of infra-red sensing with known reflective panels strategically placed to allow localization of AGV [18]. This paper demonstrates high accuracy for the AGV to know where the docking area is but AGV may only be able to localize with the presence of the infra-red lasers only.

From the literature review we found that the paper presented by [15] research is very close to one of the research objectives or the function that we looking for [15] approach is a real time and flexible route map planning to the need of the requirement. However there is a major drawback of the system that the system does not include object avoidance systems. The feature is extremely important, during the traveling of the AGV, obstacle may appear randomly this will disturb the AGV performance or even end the process therefore in this proposal we should include the obstacle avoidance system.

Apart from this, the proposed system also need to include the AIA for object recognition to identify the destiny and to perform CPAMR route map planning.

2.3 Obstacle Avoidance System for (AGV)

Though out the literature review, we find that there are three types of obstacle avoidance system for an AGV, they are non-vision based, vision-based and sensor fusion based obstacle avoidance system.

2.3.1 Non-vision based

Arboleda et al. use simple obstacle avoidance system which only using ultrasonic or sonar sensor to detect obstacle is insufficient because sonar sensor has limitation in detecting obstacle surface which is not smooth or rigid [19]. Not effective when there are slopes, slants, and rough surfaces in front of AGV where the Sonar sensors might result in false readings Requires. Freitas et al. proposed use laser scanner for obstacle avoidance system [20]. Effective for rough surfaces, slants and slopes but only limited to certain range of coverage.

2.3.2 Vision based

Polanczyk et al proposed using stereo camera with the algorithm consists of three stages which are the camera motion estimation, building a map of obstacles and path planning [21]. Preliminary obstacle detection is carried based on the stereo-vision depth image. The ambient obstacles are searched in parallel with camera motion estimation. Points that are located at distances larger than 10 meters from the camera are discarded as disparity for a closer point can be estimated more accurately. On the other hand, the points within the range of 0.8 meters from the camera are considered to be an obstacle and taken into consideration for further construction of the map. This method is effective for any kind of surface but limited Region of Interest (ROI) depending of the Field of View (FOV) of the camera.

2.3.3 Sensor Fusion based

Monteiro et al. proposed system applied both Laser Range Finder and the camera are mounted in a common base where both sensors axes are arranged in parallel and aligned ideally [22]. The calibration is necessary to obtain a mapping expression to transform the points in the laser reference system to the camera reference system and then to image plane. Heavy algorithms yet integration of sensors system is required after sensors calibration.

A convincing conclusion can be drawn from literature review that this proposed CPAMR is feasible. There are no existing work have been conducted by researchers in this area. But yet there are feasible research has been proposed by the existing researcher individually. The authors integrating the existing research work by proposing adopting [7]. IoT Architecture in the project since this model is collective summary of different researchers and normally deployed in business system. Beside that, model is adopted for Real time and Flexible route map planning system and for the vision-based model will be selected for the obstacle avoidance system.

3 Proposed CPAMR System

The authors propose this CPAMR framework has 3 Basic core functions as shown in the Figure 1, 2 and 3. Whereby the communication between the real and virtual by applying IoT framework through the RFID sensing, RTU, Cloud, MQTT protocol and ERP platform. Besides, through the RFID sensing, the autonomous system or AMR can identify the process required by the cloud platform then the CPAMR is capable to perform the real time route map planning and obstacle avoidance system for the independent process management.

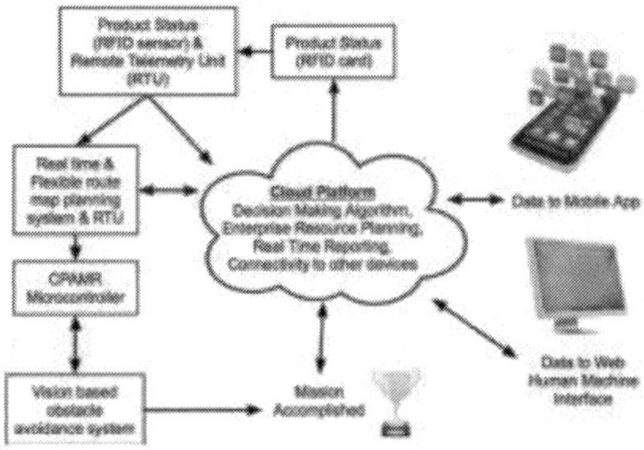

Fig. 1. Proposed CPAMR's IoT infrastructure framework

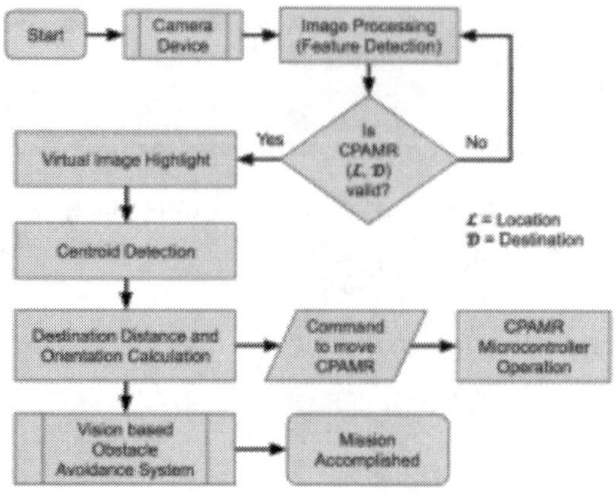

Fig. 2. Route Map Planning Flow Chart

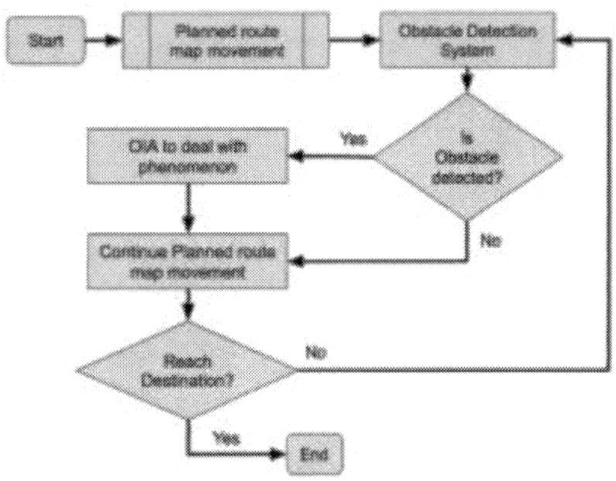

Fig. 3. Vision based Obstacle Avoidance System Flow Chart

3.1 System Platform

A CPAMR's IoT infrastructure system is developed in this study. Fig. 1 shows the proposed CPAMR's IoT infrastructure framework. Once the clients post the order and the requirements from the mobile or WEB based HMI to the cloud, the ERP system will process the data and send the product's data to RFID tag system for storage and printout. The RFID tag adhere to the product and send to the production line. Through Remote Telemetry Unit (RTU), the status by the product, CPAMR system and workstations or machineries are linked to the cloud platform. When the product is released to the production line, the RFID sensor will detect the product through the RFID tag. It will request the CPAMR to pick the product and transport to the corresponding workstation that required by the product. Before the CPAMR move, the cloud will notify the real time and Flexible route map planning system to start the route map planning as show in the Fig. 2. When the system starts, it will identify the CPAMR location, product location and the desired workstation. Once the location image is identified, the virtual image will be highlighted to extract the figure for centroid detection. AIA will calculate the shortest distance and orientation route map planning for the CPAMR. This route map planning will then sent to the CPAMR's micro-controller for operation. CPAMR mobile moves toward the destination according to the received road map. While the CPAMR is moving, the obstacle avoidance system will turn on as shown in Fig. 3. When it detects the obstacle, the AIA will trigger and deal with the phenomenon. CPAMR has to continue the planed route map until it reaches the destination and notify cloud platform.

3.2 System Testbed

This proposed CPAMR is a massive research work which requires different area of expertise for instances IoT, real time planned AMR and image processing for obstacle avoidance system. The authors divided the whole project into 3 project groups which each of the FYP students will conduct the initial study in the similar area of this research work. For the IOT project, Amazon Web Service platform has been employed as a base for center hub, the module like Amazon S3, Amazon DynamoDB and Amazon SNS is being employed for the IOT processing part. In term of connectivity and data control, protocol Message Queue Telemetry Transport (MQTT), RFID sensor and tags, Arduino, Esspressolite V 2.0 (RTU) have been used to build prototype. Real Time Route Map Planned AMR, a web camera, mechanical structure, motor, Raspberry Pi, Arduino, motor driver, Esspressolite V 2.0 and MQTT protocol have been engaged to construct the AMR. For the obstacle avoidance system, a series of imaging processing technique has been applied to develop AIA to avoid the obstacle during the moving. Imaging processing technique like Raw RGB Image captured, Converted Gray-scale and Blurred Image, Canny Edge Detection, Binary Image with Segmented Regions with Noise, Filtered and Eroded Binary Image with Segmented Region, Binary Image with Center of Obstacle Detected have been applied to develop the AIA [23].

4 Results and Discussion

This primary research project has been conducted under the Faculty of Engineering, Tunku Abdul Rahman University College. The 3 primary functions of CPAMR feasibility study has been investigated by 3 different undergraduate FYP students. The primary results of the investigations are positive. A student has achieved the fundamental function of IOT framework by using Amazon Web Service package and MQTT protocol to communicate between the modules to acquire data from the field and send the signal to the field, a student demonstrated using the web camera to identify the location of AGV and destination, by applying image processing algorithms the student is able to plan the route map for the AGV and pass over the command to micro controller (Arduino) through MQTT protocol to control the AMR to move to the destination on real time random route map planning basis. Meanwhile another student also successfully demonstrated the control of the AGV to avoid the obstacles during the moving. These primary results are very encouraging and also proven the CPAMR framework concept is viable. Further research and development challenge is on how to integrate these 3 different projects into a master project. Parameters and performance measures of the CPAMR need to be defined to suit the Industry 4.0 research.

5 Conclusion

On the whole, the initial study of the CPAMR demonstrate the framework of the project is viable, the three main functionality of the proposed CPAMR are being achieved at the basic level from different project group. Although it is beginning stage of the research project, it shows a good start to a final massive project which requires expertise from various area such as IoT, Big data analytic, autonomous system, manufacturing planning and control, data security and etc.

References

1. McKinsey Digital, "Industry 4.0 after initial hype", https://www.mckinsey. de/ files/mckinsey industry 40 2016.pdf, 2016, [Accessed on 31 July 2017].
2. R. Geissbauer, S. Schrauf, V. Koch, S. Kuge, "Industry 4.0 - Opportunities and Challenges of the Industrial Internet," PricewaterhouseCoopers Aktiengesellschaft Wirtschaftsprufungsgesellschaft (PwC) ", pp. 1-52, December 2014.
3. W. MacDougall, "Industrie 4.0 Smart Manufacturing for the future," Germany Trade and Invest, pp. 1-40, July 2014.

4. J. Lee, B. Bagheri, H-A. Kao, "A Cyber-Physical Systems architecture for Industry 4.0-based manufacturing systems," Manufacturing letter, vol.3, pp.18-23, January 2015.
5. M. Gigli, S. Koo, "Internet of Things: Services and Application Categorization," Advances in Internet of Things, vol.1, pp.27-31, July 2011.
6. X. Qian, J. Zhang, "Study on the structure of Internet of Things (IOT) business operation support platform," IEEE International Conference on Communication Technology (ICCT), pp.1068-1071, January 2011.
7. L. D. Xu, W. He, S. Li, "Internet of Things in Industries: A Survey," IEEE Transactions on Industrial Informatics, vol.4, pp.2233-2243, January 2014.
8. A. J. C. Trappey, C. V. Trappey, U. H. Govindarajan, A. C. Chuang, J. J. Sun, "A review of essential standards and patent landscapes for the Internet of Things: A key enabler for Industry 4.0," Advanced Engineering Informatics, IN PRESS, December 2016.
9. J. L. Jones, A. M. Flynn, B. A. Seiger, "Mobile Robots: Inspiration to Implementation, Second Edition" CRC Press, April 1993.
10. T. Tsumura, "Survey of autonomous guided vehicles in japanese factory", IEEE International Conference on Robotics and Automation, pp.1329-1334, April 1986.
11. D. Markosov, H. Durrant-Whyte, "Mobile Vehicle Navigation in Unknown Environments: Multiple Hypothesis Approach", IEE Proceedings of Control Theory nad Applications, vol.142(4). pp.385-400, Jul 1995.
12. A. Lawrence, "Modern Inertial Technology: Navigation, Guidance, and Control," Springer Science & Business Media, September 2001.
13. A. Bosien, M. Venzke, V. Turau, "A rewritable RFID environment for AGV navigation", International Workshop on Intelligent Transportation, March 2008.
14. K. Farkas, A. Huszak, G. G ′ odor, "Optimization of Wi-Fi Access Point ′ Placement for Indoor Localization". Journal Informatics & Information Technology Today, vol.1(1), pp.28-33, July 2013.
15. J. Lee, C-H. Hyun, M. Park, "A Vision-Based Automated Guided Vehicle System With Marker Recognition for Indoor Use," Sensors, vol.13(8), pp.10052-10073, August 2013.
16. D. Andersson, E. Mansson, "Positioning and Docking of an AGV in a Clinical Environment" Master's thesis in Biomedical Engineering, Chalmers University of Technology, 2012.
17. Y-H. Kim, S-W. Lee, H. S. Yang and D. A. Shell, "Toward Autonomous Robotic Containment Booms: Visual Servoing for Robust inter-Vehicle Docking for Surface Vehicles," Intelligent Service Robotics, vol.5(1). pp.1-18, January 2012.
18. P. M. Vaz, R. Ferreira, V. Grossman, M. I. Ribeiro, "Docking of Mobile Platform based on Infra-red Sensors," IEEE International Symposium on Industrial Electronics, pp.735-740, July 1997.
19. E. R. Arboleda, M. C. T. Alegre, K. F. Idica, "Development of a Low Cost Electronic Wheelchair with obstacle Avoidance Feature," Journal of Mechatronics, Electrical Power and Vehicular Technology, vol.06, pp.89- 96, December 2015.
20. G. Freitas, B. Hamner, M. Bergerman, S. Singh, "A Practical Obstacle Detection System for Autonomous Orchard Vehicles," International Conference on Intelligent Robots and Systems, pp.3391-3398, October 2012.
21. M. Polanczyk, M. Strzelecki, K. ′ Slot, "Obstacle Avoidance Procedure ′ and Lee Algorithm Based Path Replanner for Autonomous Mobile Platform," International Journal of Electronics and Telecommunications, vol.59(1), pp.85-91, March 2013.
22. G. Monteiro, C. Premebida, P. Peixoto, U. Nunes, "Tracking and Classification of Dynamic Obstacles Using Laser Range Finder and Vision," International Conference on Intelligent Robots and Systems, pp.1-7, 2006.
23. YK Lee, JM Lim, KS Eu, YH Goh, Y Tew," Real Time Image Processing Based Obstacle Avoidance and Navigation System for Autonomous Wheelchair Application," 2017 Asia-Pacific Signal and Information Processing Association Annual Summit and Conference (APSIPA ASC),accepted, December 2017

MATEC Web of Conferences **167**, 02006 (2018) https://doi.org/10.1051/matecconf/201816702006
IC4M & ICDES 2018

Direct Adaptive Compensation Control of Mechanical Systems With Unknown Actuator Failures and Dead-Zone Nonlinearities

Xiaohang Su and Zhi Liu[a]

School of Automation, Guangdong University of Technology, Guangzhou, China

Abstract. In this paper, a new tuning function backstepping control scheme is proposed for a class of parametric strict feedback nonlinear systems to accommodate actuator failures/faults and dead-zone constraints, where the failures/faults are uncertain in time, pattern, and values, and the dead-zone parameters are not available. Roughly speaking, such a scheme is developed in two steps below. First, by using an adaptive smooth inverse function to compensate for the dead-zone nonlinearity, we separate the coupling actuator dynamics into two parts, i.e., the dead-zone compensation errors and the nominal failure dynamics. Afterward, we further handle these two parts based on the techniques of robust adaptive approach and parametrization method. With our scheme, the global boundedness of the signals in the closed-loop system are ensured, and the tracking error is steered to zero asymptotically. These results have also been verified through simulation studies.

1 Introduction

In a practical control system, dead-zone characteristic ubiquitously exists in gear friction, hydraulic valves, DC motors and mechanical connection, which usually degrades the performance of the system. In recent years, adaptive compensation control schemes for dead-zone nonlinearity in the presence of parametric uncertainties have been developed. In [1], the dead-zone nonlinearity is regarded as an external bounded disturbance which can be mitigated through robust control design by using a direct decomposition method (DDM). Apart from DDM, the inverse compensation method (ICM) is another effective strategy to eliminate the effects of nonsmooth nonlinearities. Compared to the compensation method DDM, ICM can completely cancel actuator dead-zone by constructing an inverse compensator through parameter identification; see [2]. However, it is usually costly for offline parameter identification. Thus it is of significance to develop an adaptive inverse compensation method (AICM) by using the adaptive technique to estimate unknown dead-zone parameters online; see [3].

Besides the inherent actuator nonlinearities mentioned above, the failures/faults of suddenly getting stuck and losing partial effectiveness may also occur in practical actuation mechanisms. Recently, studies on actuator failures started by handling the linear system failure in [4], and the

[a]Corresponding author : lz@gdut.edu.cn

results were further extended to nonlinear system failure in [5] by the utilizing backstepping technique. With backstepping recursive design, a new prescribed performance bounds (PPB) based controller is proposed in [6] to guarantee the tracking error within the prescribed bounds. Moreover, in the recent work [7], a robust adaptive fault-tolerant control scheme is further proposed for compensating failures in dead-zone actuators. However, such a scheme mainly treats the actuator failures and dead-zone nonlinearity as bounded disturbance-like effects, and thus the perfect asymptotic tracking performance cannot be obtained even for the case of finite number of actuator failures.

Motivated by observations above, in this paper, we study the problem of direct adaptive failure compensation control for a class of uncertain multiple inputs and single output (MISO) nonlinear systems with actuator failures and dead-zone constraints. Such problem is successfully addressed with our newly proposed control methodology with the following contributions.

1) Note that we take actuator failures and dead-zone nonlinearities into account simultaneously in control design, which thus is more general than the existing work [8] that only considered dead-zone effect and [5] that solely focused on actuator failure compensation. Moreover, perfect asymptotic tracking control performance is achieved, irrespective of the existence of such both adversaries.

2) The smooth adaptive inverse is firstly utilized to compensate for dead-zone nonlinearity, such that the compensating errors containing failure parameters between actual function value of dead-zone nonlinearity and its designed dead-zone output, which is then eliminated in the subsequent design.

3) Recently, the authors in [7] proposed a new and novel robust adaptive control scheme, which has been shown to be applicable to compensate for actuator dead-zone and failures/faults. However, such a scheme cannot recover the perfect asymptotic tracking performance when the number of actuator failures become finite. This issue is also successfully addressed with our proposed scheme.

The outline of the paper is organized below. In Section 2, the control problem is formulated, and the related assumptions are given. In Section 3, a desired controller for addressing control problems is constructed and the stability analysis is provided at the end of this section. Finally, simulations on mechanical system with dead-zone and failures/faults are shown in Section 4.

2 System model and problem statement

In this paper, the following uncertain nonlinear system for the adaptive actuator failure compensation problem is considered.

$$\begin{aligned} \dot{x}_i &= x_{i+1} + \varphi_i^T(x_1, x_2, \ldots, x_i)\vartheta, i = 1, 2, \ldots, \lambda - 1 \\ \dot{x}_\lambda &= \varphi_0(x,\eta) + \varphi_\lambda^T(x,\eta)\vartheta + \sum_{j=1}^{q} b_j \gamma_j(x,\eta) u_j \\ y &= x_1 \end{aligned} \tag{1}$$

where $x_i \in \Re^i$ are state variables, $y \in \Re$ is the system output, $u_j \in \Re$ for $j = 1, 2, \ldots, q$ denotes the j th control input to the plant, $\varphi_0 \in \Re, \varphi_i(x) \in \Re^p (i = 1, 2, \ldots, \lambda)$ and $\gamma_j(x,\eta) \neq 0 (j = 1, 2, \ldots, q)$ are known smooth nonlinear functions, $\vartheta \in \Re^p$ and $b_j (j = 1, 2, \ldots, q)$ are unknown parameters and control coefficients, respectively. Denote τ_j as the input of the j th actuator ($j = 1, 2, \ldots, q$), u_j as the output of the j th actuator ($j = 1, 2, \ldots, q$). An actuator with its input equals to its output, i.e., $\tau_j = u_j$, namely the failure-free actuators.

As previous analysis, dead-zone is inevitable in most mechanical devices, which can be denoted as

$$v_j = D(\tau_j) = \begin{cases} m_r(\tau_j - d_+) & \tau_j \ge d_+ \\ 0 & d_- < \tau_j < d_+ \\ m_l(\tau_j - d_-) & \tau_j \le d_- \end{cases} \quad (2)$$

where v_j for $j = 1,2,...,q$ is the j th output of the dead-zone actuator. $d_- \le 0, d_+ \ge 0$ and $m_r > 0, m_l > 0$ are unknown constants.

Considering the dead-zone actuator with possible unknown failures/faults, the effect of failure nonlinearity can not be simply taken as an external disturbance-like term without considering its structural information, thus the mathematic model for failure is needed below:

$$u_j = \rho_j v_j + v_{sj}, \forall t \ge t_{jF} \quad (3)$$

$$\rho_j v_{sj} = 0, j = 1,2,...q \quad (4)$$

where $\rho_j \in [0,1)$, v_{sj} and t_{jF} are all unknown constants.

The control objectives of this paper are to ensure all the signals of the nonlinear system bounded and the system output y asymptotically tracks the reference signal y_r . To achieve the objectives, the following assumptions are needed.

Assumption 1. The dead-zone parameters satisfy $m_r \ge m_{r0}, m_l \ge m_{l0}$, where m_{r0}, m_{l0} are two small positive constants, as described in [8].

Assumption 2. The reference signal y_r and its first λ th order derivatives $y_r^{(i)} (i = 1,2,...,\lambda)$ are known, bounded and piecewise continuous. $\gamma_i(x,\eta) \ne 0$, and the signs of b_i , i.e., $sgn(b_i)$ are known.

3 Objective controller design for solving the problem

3.1 Dead-zone Inverse Model

In this part, to prevent the system from potential chattering phenomenon in the backstepping recursion, v_j needs to be designed as

$$v_{dj} = -\hat{\theta}_j^T \hat{w}_j, j = 1,2,...,q \quad (5)$$

where $\hat{\theta}_j^T = [m_r, m_r d_+, m_l, m_l d_-]$ is an estimate of θ_j , $\hat{w}_j^T = [-\Phi_1(\tau)\tau_j, \Phi_1(\tau), -\Phi_2(\tau)\tau_j, -\Phi_2(\tau)]$ is an identity matrix. $\Phi_1(v_j) = \dfrac{e^{\frac{v_j}{e_0}}}{e^{\frac{-v_j}{e_0}} + e^{\frac{v_j}{e_0}}}$ and $\Phi_2(v_j) = \dfrac{e^{\frac{-v_j}{e_0}}}{e^{\frac{-v_j}{e_0}} + e^{\frac{v_j}{e_0}}}$ are designed as smooth continuous functions, where $e_0 > 0$ is a user-defined parameter.

Then the corresponding actuator smooth dead-zone inverse model is obtained as:

$$\tau_j = \frac{v_{dj} + m_r d_+}{\hat{m}_r}\Phi_1(v_{dj}) + \frac{v_{dj} + m_l d_-}{\hat{m}_l}\Phi_2(v_{dj}) \tag{6}$$

The compensating errors between v_j and v_{dj} are obtained as

$$v_j - v_{dj} = -\tilde{\theta}_j^T \hat{w}_j + D_{Nj} \tag{7}$$

where $\tilde{\theta}_j = \theta_j - \hat{\theta}_j$, $D_{Nj} = \theta_j^T(\hat{w}_j - w_j)$. D_{Nj} can be guaranteed to be bounded for all $t \geq 0$, see, [8].

3.2 Controller Design

In this subsection, the novel adaptive compensation scheme for actuator failures and dead-zone constraints of nonlinear system is proposed. First of all, the following error variables are introduced as

$$z_1 = x_1 - y_r \tag{8}$$

$$z_i = x_i - \alpha_{i-1} - y_r^{(i-1)}, i = 2,3,\ldots,\lambda \tag{9}$$

Step i ($i = 1,2,\ldots,\lambda-1$). Based on the tuning function design scheme, the i th stabilizing function α_i and tuning function ς_i are chosen as

$$\alpha_1 = -c_1 z_1 - \varpi_1^T \hat{\vartheta} \tag{10}$$

$$\alpha_i = -z_{i-1} - c_i z_i - \varpi_i^T \hat{\vartheta} + \frac{\partial \alpha_{i-1}}{\partial \hat{\vartheta}}\Gamma\varsigma_i + \sum_{j=1}^{i-1}\left[\frac{\partial \alpha_{i-1}}{\partial x_j}x_{j+1} + \frac{\partial \alpha_{i-1}}{\partial y_r^{(j-1)}}y_r^{(j)}\right] + \sum_{j=2}^{i-1}\frac{\partial \alpha_{j-1}}{\partial \hat{\vartheta}}\Gamma\varpi_i z_j, i = 2,\ldots,\lambda-1 \tag{11}$$

$$\varsigma_i = \varsigma_{i-1} + \varpi_i z_i, \varpi_i = \varphi_i - \sum_{j=1}^{i-1}\frac{\partial \alpha_{i-1}}{\partial x_j}\varphi_j \tag{12}$$

Step λ. From the above definitions, $\dot{z}_\lambda$ is obtained as

$$\dot{z}_\lambda = \varphi_0 + \varphi_\lambda^T \vartheta - \dot{\alpha}_{\lambda-1} - y_r^{(\lambda)} - \sum_{i=1}^{q} b_i \gamma_i(x,\eta)\rho_i \tilde{\theta}_i^T \hat{w}_i + \sum_{i=1}^{q} b_i \gamma_i \rho_i D_{Ni} + \sum_{i=1}^{q} b_i \gamma_i(\rho_i v_{di} + v_{si}) \tag{13}$$

Remark 1: Based on the adaptive inverse compensation method (ACIM), the compensating errors between actual function value and its designed dead-zone output are obtained, as seen in (7). By combining the failure model (3)-(4), we can separate the coupling actuator dynamics into two parts: the dead-zone compensation errors containing failure parameters and the nominal failure dynamics, as shown in (13). These two parts can be compensated by synthesizing robust control approach, parametrization method and direct adaptive approach in following steps.

Supposing one or more of failures suddenly occur in the dead-zone actuators during time interval $[T_{K-1}, T_K)$, where T_{K-1}, T_k for $k = 1,2,\ldots,e,e+1(e \leq q)$ are a finite number of time instants. We define two sets such that $Q_{tot_k} = \{j_{1,1}, j_{1,2}, \ldots, j_{1,q_{tot_k}}\}$ and $Q_{par_k} = \{j_{1,1}, j_{1,2}, \ldots, j_{1,q_{par_k}}\}$. As previous analysis, the unknown parameters D_{Ni} in (13) can be computed as

$$|b_i D_{Ni}| \le |b_i| \bar{D}_{Ni} = |b_i| (\tilde{\bar{D}}_{Ni} + \hat{\bar{D}}_{Ni}) \quad (14)$$

Then the Lyapunov function V_λ is chosen as

$$V_\lambda = V_{\lambda-1} + \frac{1}{2} z_\lambda^2 + \sum_{i=1, i \notin Q_{tot_k}}^{q} \frac{\rho_i |b_i|}{2} \tilde{\kappa}^T \Gamma_k^{-1} \tilde{\kappa} + \sum_{i=1}^{q} \frac{\rho_i |b_i|}{2} \tilde{\theta}^T \Gamma_s^{-1} \tilde{\theta} + \sum_{i=1}^{q} \frac{\rho_i |b_i|}{2} \tilde{\bar{D}}_{Ni}^2 \quad (15)$$

where Γ_k and Γ_s are two positive-definite matrices. If b_i, ρ_i and v_{si} for $i = 1, 2, \ldots, q, h \in Q_{tot_k}$ are known, κ is a desired constant vector such that

$$\sum_{i=1, i \notin Q_{tot_k}}^{q} |b_i| \rho_i \kappa^T \varrho = \alpha_\lambda - \sum_{h \in Q_{tot_k}} b_h \gamma_h v_{sh}, \kappa_1 = \frac{1}{\sum_{i=1, i \notin Q_{tot_k}}^{q} |b_i| \rho_i}, \kappa_{2,h} = \frac{-b_h v_{sh}}{\sum_{i=1, i \notin Q_{tot_k}}^{q} |b_i| \rho_i} \quad (16)$$

where $h \in Q_{tot_k}$ and $h \in \{1, 2, \ldots, q\} \Rightarrow \kappa_{2,h} = 0$, $\hat{\kappa} = [\hat{\kappa}_1, \hat{\kappa}_2^T]^T$, $\hat{\kappa}_2 = [\hat{\kappa}_{2,1}, \hat{\kappa}_{2,2}, \ldots, \hat{\kappa}_{2,q}]^T$, $\varrho = [\alpha_\lambda, \gamma_1, \ldots, \gamma_q]^T$. Then $\dot{V}_\lambda$ can be computed as

$$\begin{aligned}
\dot{V}_\lambda \le & -\sum_{i=1}^{\lambda-1} c_i z_i^2 + z_\lambda \Big\{ z_{\lambda-1} + \alpha_\lambda + \varphi_0 + \varpi_\lambda^T \hat{\vartheta} - \sum_{i=1}^{\lambda-1} \Big[\frac{\partial \alpha_{\lambda-1}}{\partial x_i} x_{i+1} + \frac{\partial \alpha_{\lambda-1}}{\partial y_r^{(i-1)}} y_r^{(i)} \Big] n - \frac{\partial \alpha_{\lambda-1}}{\partial \hat{\vartheta}} \dot{\hat{\vartheta}} \\
& + \sum_{i=1}^{q} b_i \gamma_i (x, \eta)(\rho_i v_{di} + v_{si}) + \sum_{i=1}^{q} |b_i| \rho_i \gamma_i \hat{\bar{D}}_{Ni} - y_r^{(\lambda)} \Big\} + \tilde{\vartheta}^T (\varsigma_\lambda - \Gamma^{-1} \dot{\hat{\vartheta}}) + \sum_{i=2}^{\lambda-1} z_i \frac{\partial \alpha_{i-1}}{\partial \hat{\vartheta}} (\Gamma \varsigma_{\lambda-1} - \dot{\hat{\vartheta}}) \\
& - \sum_{i=1, i \notin Q_{tot_k}}^{q} |b_i| \rho_i \widetilde{\kappa_i^T} (\varrho z_\lambda + \Gamma_k^{-1} \dot{\hat{\kappa}}) + \sum_{i=1}^{q} |b_i| \rho_i \tilde{\theta}_i^T \big[|\gamma_i \hat{w}_i z_\lambda| - \Gamma_s^{-1} \dot{\hat{\theta}}_i \big] + \sum_{i=1}^{q} |b_i| \rho_i \widetilde{\bar{D}_{Ni}} \big[|\gamma_i z_\lambda| - \dot{\hat{\bar{D}}}_{Ni} \big]
\end{aligned} \quad (17)$$

To make $\dot{V}_\lambda$ non-positive, the tuning function ς_λ and the stabilizing function α_λ are defined as

$$\varsigma_\lambda = \varsigma_{\lambda-1} + \varpi_\lambda z_\lambda, \ \varpi_\lambda = \varphi_\lambda - \sum_{j=1}^{\lambda-1} \frac{\partial \alpha_{\lambda-1}}{\partial x_j} \varphi_j \quad (18)$$

$$\begin{aligned}
\alpha_\lambda = & -z_{\lambda-1} - c_\lambda z_\lambda - \varphi_0 - \varpi_\lambda^T \hat{\vartheta} + \frac{\partial \alpha_{\lambda-1}}{\partial \hat{\vartheta}} \Gamma \varsigma_\lambda + \sum_{i=2}^{\lambda-1} \frac{\partial \alpha_{i-1}}{\partial \hat{\vartheta}} \Gamma \varpi_\lambda z_i \\
& + \sum_{i=1}^{\lambda-1} \Big(\frac{\partial \alpha_{\lambda-1}}{\partial x_i} x_{i+1} + \frac{\partial \alpha_{\lambda-1}}{\partial y_r^{(i-1)}} y_r^{(i)} \Big) - \sum_{i=1}^{q} |b_i| \rho_i \gamma_i \hat{\bar{D}}_{Ni} + y_r^{(\lambda)}
\end{aligned} \quad (19)$$

The final parameter update laws are chosen as

$$\dot{\hat{\vartheta}} = \Gamma \varsigma_\lambda, \dot{\hat{\kappa}} = -\Gamma_k \varrho z_\lambda, \dot{\hat{\theta}}_i = \Gamma_s |\gamma_i \hat{w}_i z_\lambda|, \dot{\hat{\bar{D}}}_{Ni} = |\gamma_i z_\lambda| \quad (20)$$

The control law v_{di} is finally determined as

$$v_{di} = sgn(b_i)\frac{1}{\gamma_i}\hat{\kappa}^T\varrho, for\ i = 1,2,...,q \tag{21}$$

3.3 Stability Analysis

Theorem: Considering the closed-loop adaptive system consists of the plant (1) with the control law (21) and the parameter update laws (20) in the presence of dead-zone nonlinearity (2) and actuator failures (3)-(4) and under Assumptions 1-2. The boundedness of all the signals are ensured and the tracking error approaches to zero asymptotically, i.e., $\lim_{t\to\infty}[y(t) - y_r(t)] = 0$.

Proof. From **Step i**, we conclude that

$$\dot{z}_1 = z_2 - c_1 z_1 + \varpi_1^T \tilde{\vartheta} \tag{22}$$

$$\dot{z}_i = z_{i+1} - c_i z_i - z_{i-1} + \varpi_i^T \tilde{\vartheta} + \frac{\partial \alpha_{i-1}}{\partial \hat{\vartheta}}(\Gamma \varsigma_i - \dot{\hat{\vartheta}}) + \sum_{j=2}^{i-1} \frac{\partial \alpha_{j-1}}{\partial \hat{\vartheta}} \varpi_i z_j \tag{23}$$

Among $[0, T_1)$, with the control law (20), supposing that b_i is known and $\kappa_{2,h} = 0$ for $k = 1,2,...,q$ is chosen, the derivative of z_λ then satisfies

$$\dot{z}_\lambda = -c_\lambda z_\lambda - z_{\lambda-1} + \varpi_\lambda^T \tilde{\vartheta} - \sum_{i=1}^{q} |b_i| \tilde{\kappa}^T \varrho + \sum_{j=2}^{i-1} \frac{\partial \alpha_{j-1}}{\partial \hat{\vartheta}} \varpi_i z_j - \sum_{i=1}^{q} b_i \gamma_i(x,\eta) \rho_i (\tilde{\theta}_i^T \hat{w}_i - D_{Ni}) \tag{24}$$

Assuming during the time interval $[T_{k-1}, T_k)$ for $k = 2,3,...,r$, the system with the controller (21) and update laws (20) undergoes dead-zone nonlinearity and PLOE type of failure, then we have

$$\dot{V}_{k-1} \le -\sum_{j=1}^{\lambda} c_j z_j^2 \tag{25}$$

where $V_{k-1}(T_k + \Delta(t)) \le V_{k-1}(T_k)$, which shows that $z(t), \tilde{\vartheta}, \tilde{\kappa}, \tilde{\theta}, \tilde{D}_N \in L_\infty$, then the boundedness of $x_1(t),...,x_\lambda(t)$ can be derived successively. Besides, by using Barbalat's lemma and combining with $\dot{z}(t) \in L_2$, the error vector $z(t)$ asymptotic convergence to zero can be proved, i.e., $\lim_{t\to\infty} z(t) = 0$. The proof of the theorem is completed.

4 Simulation studies

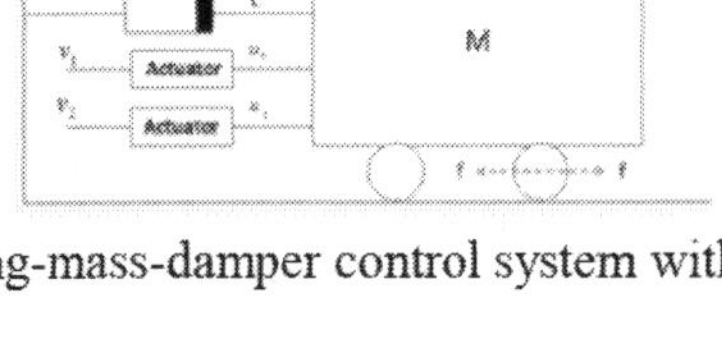

Figure 1. Spring-mass-damper control system with dual actuators.

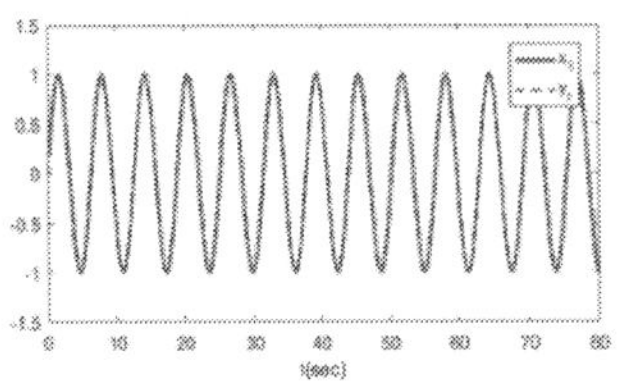

Figure 2

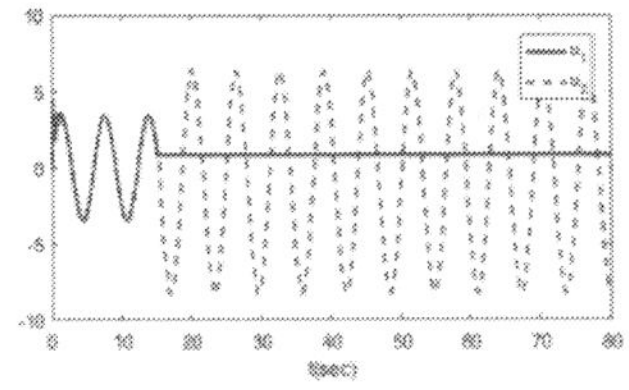

Figure 3

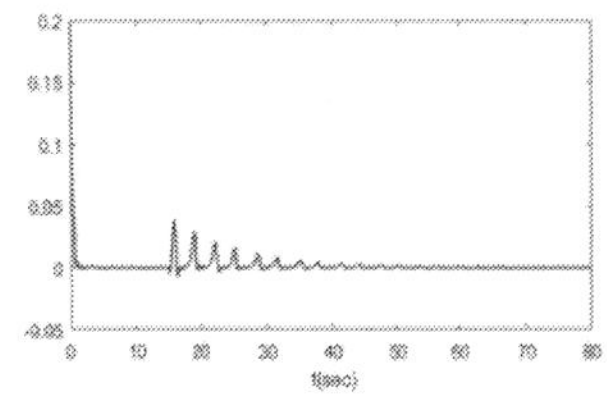

Figure 4

In this example, we apply our approach to the following physical system with dual actuators. For the, visualization, the corresponding physical model is designed in Fig. 1. In addition, the dynamic mathematic model is given as

$$\begin{aligned} \dot{x}_1 &= x_2 \\ \dot{x}_2 &= -\frac{k}{M}x_1 - \frac{c}{M}x_2 + \frac{1}{M}\sum_{i=1}^{2} u_i \end{aligned} \tag{26}$$

where $y = x_1$ is regarded as the output position of the car whose mass is denoted by $M = 1kg$. $k = 8N/s$ is the stiffness of the spring, and $c = 2N \cdot m/s$ represents the damping coefficient. For simulation, the initial values are given as $x_1(0) = 0.1m, x_2(0) = 0m/s, u_1(0) = u_2(0) = 0N$ and the reference signal is $y_r = sin(t)$. The dual actuator failure terms are modeled as: at 15th second, the actuator u_1 suffers partial loss of effectiveness (PLOE) as $u_1 = 70\%v_1$. Besides, the maximum static friction of this physical model is chosen as $1N$, which can be regarded as a dead-zone nonlinearity whose parameters are defined as $d_+ = 1N, d_- = -1N, m_r = 1, m_l = 1$.

By choosing the other simulation parameters as $c_1 = 3, c_2 = 2.8$, the results are shown in Fig. 2-Fig. 4. As presented in Fig. 3, the actuator u_1 suffers failure and actuator u_2 works in good condition. By using our proposed control scheme, it can be seen from Fig. 2 that the output signal x_1 tracks the reference signal y_r perfectly. Besides in Fig. 4, the tracking error asymptotically approaches to zero.

5 Acknowledgments

This work was supported in part by the National Natural Science Foundation of China under Grant 61573108, in part by National Natural Science Foundation of China (U1501251), in part by the Natural Science Foundation of Guangdong Province 2016A030313715, in part by the Natural Science Foundation of Guangdong Province through the Science Fund for Distinguished Young Scholars under Grant S20120011437, in part by the Ministry of Education of New Century Excellent Talent under Grant NCET-12-0637.

References

1. Z. Li and C. Y. Su. Neural-adaptive control of single-master-multiple-slaves teleoperation for coordinated multiple mobile manipulators with time-varying communication delays and input uncertainties. IEEE Transactions on Neural Networks and Learning Systems, 24(9):1400-1413, September 2013.

2. D. A. Recker, P. V. Kokotovic, D. Rhode, and J. Winkelman. Adaptive nonlinear control of systems containing a deadzone. Proceedings of the 30th IEEE Conference on Decision and Control, 1991, pages 2111 - 2115, December 1991.
3. G. Tao and P. V. Kokotovic. Adaptive sliding control of plants with unknown dead-zone. IEEE Trans. Autom. Control, 39:59 - 68, January 1994.
4. G. Tao, S. Chen, and S. M. Joshi. An adaptive actuator failure compensation controller using output feedback. IEEE Transactions on Automatic Control, 47(3):506 - 511, March 2002.
5. X. D Tang, G. Tao, and S. M. Joshi. Adaptive actuator failure compensation for nonlinear MIMO systems with an aircraft control application. Automatica, 43(11):1869 - 1883, November 2007.
6. W. Wang and C. Y Wen. Adaptive actuator failure compensation control of uncertain nonlinear systems with guaranteed transient performance. Automatica, 46(12):2082 - 2091, December 2010.
7. Y. X. Li and G. H. Yang. Adaptive fuzzy decentralized control for a class of large-scale nonlinear systems with actuator faults and unknown dead zones. IEEE Transactions on Systems, Man, and Cybernetics: Systems, PP(99):1 - 12, 2016.
8. J. Zhou, C. Wen, and Y. Zhang. Adaptive output control of nonlinear systems with uncertain dead-zone nonlinearity. IEEE Transactions on Automatic Control, 51(3):504 - 511, March 2006.

Transportation and Production Lot-size for Sugarcane under Uncertainty of Machine Capacity

Kanchala Sudtachat[1,a]

[1]*School of Manufacturing Engineering, Institute of Engineering, Suranaree University of Technology,111 University Avernue, Mueng, Nakhon Ratchasima, 30000,Thailand*

Abstract. The integrated transportation and production lot size problems is important effect to total cost of operation system for sugar factories. In this research, we formulate a mathematic model that combines these two problems as two stage stochastic programming model. In the first stage, we determine the lot size of transportation problem and allocate a fixed number of vehicles to transport sugarcane to the mill factory. Moreover, we consider an uncertainty of machine (mill) capacities. After machine (mill) capacities realized, in the second stage we determine the production lot size and make decision to hold units of sugarcane in front of mills based on discrete random variables of machine (mill) capacities. We investigate the model using a small size problem. The results show that the optimal solutions try to choose closest fields and lower holding cost per unit (at fields) to transport sugarcane to mill factory. We show the results of comparison of our model and the worst case model (full capacity). The results show that our model provides better efficiency than the results of the worst case model.

1 Introduction

The production system of sugar industry in Thailand obtains raw material from small fields that located around the mill factory. Currently, the farmers delivery units of sugarcane in a small lot size to mill factory by using farm trucks or six-wheeled trucks in which do not have schedule time accounting to production planning of the mill factory. Thus, large queues are waiting in front of machines (mills). Furthermore, the important problem of production system occurs due to breakdown time of the machines (mills) that affect to uncertainty of machine capacities. Therefore, one operation system is to schedule an arrival of sugarcane from fields to mill factory including considers a several possible way of production lot size planning based on random machine capacities occurring in real world problem. The operation system may encounter breakdown problems of machines (mills) that leading to situation of lower full machine capacities. In harvesting season, we employ a transportation lot size strategy. The decisions of transportation schedule are made before the harvesting season. In the production schedule of each period, we would determine the production lot size based on scenarios occurring of uncertainty of machine (mill) capacities. The objective is to minimize expect total cost. We develop a two stage stochastic programming model for integrated transportation and production lot size problem when the situation of plant is uncertainty of machine (mill) capacities. The model is investigated using a random input parameters in a small size problem.

[a] Corresponding author : kanchala@sut.ac.th

2 Literature Review

Essentially, the models for the sugarcane systems addressed a complex harvesting and transportation system. Guise and Ryland [1] presented an optimization model for the scheduling of production in sugar mills. Higgins et al. [2] introduced the model of the harvesting and transportation system. They evaluated the value chain and built a model for harvesting sugarcane and the transportation system, including the financial flow. In addition, Milan et al. [3] introduced mixed integer programming to formulate the model for sugar cutting and sugarcane transport problems. Grunow et al. [4] investigated the production of raw sugar in decisions related to cultivation, harvesting and dispatch planning. The objective was to minimize the associated costs of systems. Piewthongngam et al. [5] also presented a supply chain for the sugarcane operation system in terms of cultivation and harvesting using a mathematical model. The extension work addressing the supply chain for the sugarcane operation system was introduced by Khamjan et al. [6]. They proposed the location of sugarcane loading stations. The sugarcane fields were allocated to each loading stations. In this paper, we develop the mathematical formulation to analyze the transportation and production lot size models. Our model is presented by using the two stage programing model when the first decision is transportation lot size, after the capacity of machines (mills) realized that each machine (mill) can not operate in full capacity.

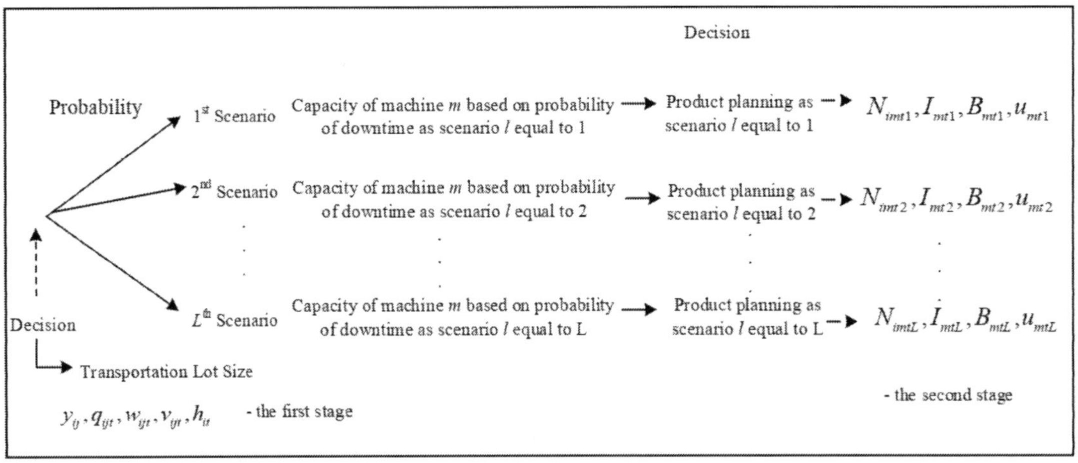

Figure 1. The two stage stochastic programming model of the transportation and production planning for sugarcane

3 Model Formulation of Transportation and Production Lot-size Model

3.1 Mathematical model

The proposed model can be used to determine the problem of how to make decisions of the transportation lot size of each field and assign trucks to transport sugarcane from fields to the mill factory for each period. We consider a single type of vehicle. At the mill factory, we consider a breakdown probability of machines (mill) that resulting in uncertainty capacity for each machine (mill) for each period after transporting sugarcane to front of mill occurs already. The problem is stated as how many of transporting lot size for each period given a number of vehicle, likelihood of machines capacity realized and determine the production lot size for each machine (mill) for each period. We formulated the model as the two stage stochastic programming model; the first stage of transportation lot size problem is to make decision before the transporting sugarcane for cane field to the mills occurs and evaluate the first state cost, while the second stage is to evaluate the cost of production lot size due to a realized of machine capacities. The second stage cost (recourse cost) is random variable dependent on the first stage transportation lot size decision for each field. The minimum overall expected cost is determined with respect to the random variables of probability of machine capacities. The extensive form of stochastic of transportation and production lot-size for sugarcane under uncertainty of machine (mill) capacity shows in Fig 1.

The parameters of transporting and production lot size models is shown below.

Notation		**Description**
Indices		
n		number of fields
J		number of trucks
T		number of time periods
L		number of scenarios
MC		number of machines (mills)
i		fields index as $i = 1, 2, \ldots., \mathrm{n}$
j		existing trucks index as $j = 1, 2, \ldots, J$
t		time periods as $t = 1, 2, \ldots, T$
m		machines (mills) index as $\mathrm{m} = 1, 2, 3, \ldots, MC$
l		scenarios as $l = 1, 2, \ldots, L$
$BigM$		a large integer number
Parameters		
d_{ij}		prepare cost of field i to factory by truck j
s_{ijt}		transportation cost per time (fixed cost) of truck j at time period t, where consist of load-unload cost, cutting cane cost and transportation cost
c_{it}		holding cost per ton of sugarcane leaving at field i at time period t
Q_{it}		yield (tons) of field i at time period t
$Setup_m$		setup cost of machine (mill) m
g_{imt}		production cost of sugarcane from field i produce on machine (mill) m at time period t
o_t		holding cost per ton of sugarcane at front of machine (mill) at time period t
b_t		lost production cost per ton of sugarcane at time period t
$prob_l$		probability of scenario l occurring for any time
p_{ml}		percent of capacity of machine (mill) m with scenarios l based on probability of breakdown of machine (mill) or stopping production line by any situations that make machine (mill) capacity being lower than full capacity
COS		number of trucks
COV		capacity of each truck
COM		capacity of machine (mill) at full capacity per time period
COF		requirement of factory for each time period
Decision variables		
y_{ij}	1	if we allocate truck j to transport sugarcane at field i
	0	otherwise
u_{mtl}	1	if we assign sugarcane at least a ton to machine (mill) m at time period t as scenario l
	0	otherwise
v_{ijt}	1	if we allocate truck j to transport sugarcane at field i at time period t
	0	otherwise
w_{ijt}		number of travels of truck j to transport sugarcane at field i at time t
q_{ijt}		quantity of sugarcane at field i, which transport by truck j at time t
h_{it}		holding units (tons) of field i, kept at its field at the end of time period t
N_{imtl}		quantity of sugarcane of field i at front of mill assign to machine (mill) m at time period t as scenario l
I_{mtl}		holding unit of sugarcane at front of mill for assigning to machine (mill) m at time period t as scenario l

B_{mtl}	lost production unit of sugarcane of machine (mill) m at time period t as scenario l
f	optimal value

Objective Function $\quad Minimize\ f = \sum_{i=1}^{n}\sum_{j=1}^{J} d_{ij} y_{ij} + \sum_{i=1}^{n}\sum_{j=1}^{J}\sum_{t=1}^{T} s_{ijt} w_{ijt} + \sum_{i=1}^{n}\sum_{t=1}^{T} c_{it} h_{it}$

$$+\sum_{l=1}^{L} prob_l \cdot \left(\sum_{m=1}^{MC}\sum_{t=1}^{T} Setup_m \cdot u_{mtl} + \sum_{i=1}^{n}\sum_{m=1}^{MC}\sum_{t=1}^{T} g_{imt} \cdot N_{imtl} + \sum_{m=1}^{MC}\sum_{t=1}^{T} o_t \cdot I_{mtl} + \sum_{m=1}^{MC}\sum_{t=1}^{T} b_t \cdot B_{mtl} \right) \quad (1)$$

Stage I:

$$v_{ijt} \le y_{ij} \qquad \forall i,j,t \qquad (2)$$

$$q_{ijt} \le Q_{it} \cdot v_{ijt} \qquad \forall i,j,t \qquad (3)$$

$$\sum_{j=1}^{J} q_{ijt} \le Q_{it} \qquad \forall i,t \qquad (4)$$

$$q_{ijt} \le COV(w_{ijt}) \qquad \forall i,j,t \qquad (5)$$

$$w_{ijt} \le BigM(v_{ijt}) \qquad \forall i,j,t \qquad (6)$$

$$\sum_{i=1}^{n}\sum_{j=1}^{J} q_{ijt} \ge COF \qquad \forall i,t \qquad (7)$$

$$h_{it} \ge h_{it-1} + Q_{it} - \sum_{j=1}^{J} q_{ijt} \qquad \forall i,t \qquad (8)$$

Stage II:

$$BigM \cdot u_{mtl} \ge \sum_{i=1}^{n} N_{imtl} \qquad \forall i,m,t,l \qquad (9)$$

$$I_{mtl} \ge I_{mt-1l} + \sum_{i=1}^{n} N_{imtl} - p_{ml} \cdot COM \qquad \forall i,m,t,l \qquad (10)$$

$$\sum_{m=1}^{MC} N_{imtl} = \sum_{j=1}^{J} q_{ijt} \qquad \forall i,t,l \qquad (11)$$

$$B_{mtl} \ge p_{ml} \cdot COM - \sum_{i=1}^{n} N_{imtl} - I_{mt-1l} \qquad \forall t,l \qquad (12)$$

$$y_{ij} \in (0,1) \qquad u_{mtl} \in (0,1) \qquad v_{ijt} \in (0,1)$$

$$w_{ijt} \in \text{int+} \qquad q_{ijt} \in \text{int+} \qquad h_{it} \in \text{int+}$$

$$N_{imtl} \in \text{int+} \qquad I_{mtl} \in \text{int+} \qquad B_{mtl} \in \text{int+}$$

Objective function (1) the first term is a cost of preparing road of travel from vehicle j to field i which we assign vehicle j to transport sugarcane of field i based on distance between field i and vehicle j. The second term, we compute the setup cost for load, unload and transportation cost per time per period that the load and unload cost are constant value. However, the transportation cost depends on distance between field i and vehicle j. The third term is the holding cost for leaving sugarcane at field i at period t because we do not harvesting on time. The forth term, we present the probability of scenarios l occurring for machine (mill) that resulting from uncertainty capacity of machine (mill) m. The probability of scenarios l occurring for each machine (mill) is multiply by the recourse cost that occurring in the second stage. The equation consists of sum of four terms, the setup cost of machine (mill) m if we assign at least a ton to produce on machine (mill) m, the production cost of sugarcane production, the holding cost for keeping sugarcane at front of mills and the lost production cost for all periods. Constraint in the *Stage I*, constraints (2) and (3) if we transport sugarcane at least a ton from field i by vehicle j ensure y_{ij} equal one. The constraint (4) ensure that we transport tons of sugarcane from field i less than yield of field i. Constraint (4) ensure the transported sugar cane from field i by all vehicles is less than yield of field i. The constraints (5) and (6) use to calculate the number of travels of vehicle j to transport sugarcane from field i to the mill factory for each period. Constraint (7) ensure that the transported sugarcane from all field i by all vehicles is larger than requirement of the mill factory. The constraint (8) calculate number of holding unit of field i that do not harvest on time period. Constraint in *Stage II,* constraint (9) ensure if we assign only a unit of sugarcane to produce on machine (mill) m, the set up cost of machine (mill) m would be computed. Constraint (10) use to calculate the holding units of sugarcane at front of machine (mill) m for each period for each scenario. Constraint (11) ensure that we assign units of sugarcane to all machines is less than available number of units at front of mills. Constraint (12) calculate number of lost production units of sugarcane of machine (mill) m for each period for each scenario.

4 Numerical Examples

This section, we generate random data set to test the two stage stochastic programming model. We present the comparison of results of our model and a model that do not consider the full capacity of all machines (mills) called a worst case model. The models coded by using the IBM ILOG CPLEX Optimization Studio Version 12.5. The input parameters of problem size 10 fields, 4 vehicles, 2 machines (mills), 2 time period and 2 scenarios, capacity of each vehicle 30 tons and full capacity of each machine (mill) 400 tons.

4.1 Optimal solutions

The mathematical model is investigated using a small problem 10 fields, 4 vehicles, 2 mills, 2 time periods. The results show that we increase the tons of requirement, the expected total cost would be increased and the optimal policy choose the fields 2, 6, 7, 9 to transport sugarcane to the mill. The results show in Table 1. Because of two reasons, the holding cost per unit of sugarcane yield of fields 2, 6, 7, 9 are larger than other fields if we do not harvest and leave them at fields, there are higher cost than other fields, and these fields are located nearby the mill factory that resulting in lower transportation cost if we would choose them to transport sugarcane to the mill factory. The results of comparison of the expected total cost of our two stage model (optimal solutions) at which assumptions approach to realistic world problem and the worst case model (optimal solutions) that we assume at full capacity for each machine (mill) for current operation. The results show that our model provides lower cost than the results of the worst case model at average 9%. Figure 2 show a sensitivity analysis of the expected total cost when we increase the percent of each type of cost. The results of sensitivity shows that increased holding cost at fields will affect to the expected total cost more than increased holding cost at mill and increased lost production cost at mill. The results imply that the expected total cost depends on the function of holding cost at fields in linear function and not depend on the function of holding cost and function of lost production cost at mill.

Table 1. The expected total cost of the two-stage model (optimal solutions) that comparing to the total cost of worst case model (optimal solution) shown in Thailand currency (Bahts)

ID	Fields	Requirement of factory (Tons)	Chosen cane fields to transport in full yield	Two-stage model (optimal solutions)	Worst case model (optimal solution)	% Deviation
1	10	500	2, 6, 7, 9	23476	26194.4	10
2	10	525	2, 6, 7, 9	23816	26494.1	10
3	10	550	2, 6, 7, 9	24026	26717.4	10
4	10	575	1, 2, 6, 7, 9	24572	27017.1	9
5	10	600	1, 2, 6, 7, 9	25201	27310.6	8

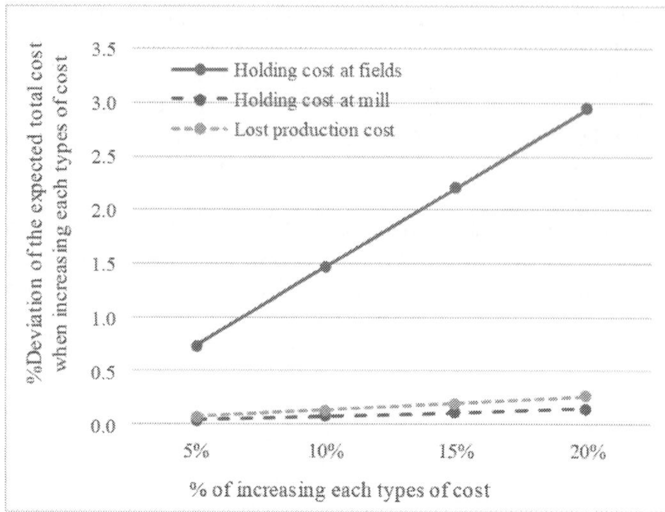

Figure 2. The sensitivity analysis of increasing percent for each types of costs; holding cost per ton at fields, holding cost per ton at the mill factory and lost production cost per ton at the mill factory

5 Conclusion

In this paper, we present the two stage stochastic programming model for the integrated transportation and production lot size problem that consider uncertainty of machine capacities into the model. We consider the probability of scenarios occurring as stage of system based on fluctuating of machine (mill) capacities. We showed the results of investigating the mathematical model for small size problem. The numerical results show that the optimal solution try to choose the closest fields and choose fields that are larger holding cost per unit of kept sugarcane at fields to transport sugarcane to mill factory. The comparison results of our model (optimal solution) and the worst case model (optimal solutions) showed that results of our model provides more efficient of system than the results of the worst case models.

References

1. J.W.B. Guise, G.J. Ryland, Production Scheduling and Profitability in Sugar Milling, *Economic Analysis and Policy*, **2**, 1, 18 (1971)
2. Higgins, G. Antony, G. Sandell, I. Davies, D. Prestwidge, B. Andrew, A framework for integrating a complex harvesting and transport system for sugar production, *Agricultural Systems*, **82,** 2, 99-115 (2004)
3. E. L. Milan, S. M. Fernandez, L. M. P. Aragones, Sugarcane transportation in Cuba, a case study, *European Journal of Operational Research* **174,** 1, 374-386. (2006)
4. M. Grunow, H. O. Günther, R. Westinner, Supply optimization for the production of raw sugar, *International Journal of Production Economics,* **110**, 1, 224-239. (2007)
5. K. Piewthongngam, S. Pathumnakul, K. Setthanan, Application of crop growth simulation and mathematical modeling to supply chain management in the Thai sugar industry. *Agricultural Systems* **102**, 1, 58-66 (2009)
6. W. Khamjan, S. Khamjan, S. Pathumnakul, Determination of the locations and capacities of sugarcane loading stations in Thailand, *Computers & Industrial Engineering* **66**, 4, 663-674 (2013)

MATEC Web of Conferences **167**, 02008 (2018) https://doi.org/10.1051/matecconf/201816702008
IC4M & ICDES 2018

The Use of Virtual Manufacturing Features in forward Manufacturing Process Planning

Janusz Pobozniak[a]

Faculty of Mechanical Engineering, Cracow University of Technology, Al. Jana Pawła II 37, 31-864 Cracow, Poland

Abstract. CAPP system is the missing link between CAD/CAM system. Two types of operation of CAPP systems can be distinguished. The first one is the backward process planning, where the part is transformed from its ready state to the stock state. The second method is the forward planning, where the stock is transformed into the finished part. Backward process planning allow for the greater degree of automation, but is difficult for the acceptation among the industrial users. One of the problem in the forward process planning is the necessity to use the manufacturing features which do not occur in the finished part, they occur only in the part intermediate states. The paper propose so called virtual manufacturing features to solve this problem.

1 CAPP system as the link between CAD/CAM

There are several phases in the product lifecycle like concept development, cost estimation, marketing research, preliminary and final product documentation, manufacturing process planning, CNC program preparation, production start-up, etc. Computer Aided systems (CAx) with different degree of automation are used on each of these stages. The degree of automation means the degree of human assistance in the decision taking. Computer Aided Design (CAD) systems are used to develop the design documentation. The modern systems, apart from geometrical model store also the non-geometrical data like dimension tolerances, location tolerances, surface roughness, etc. After the development of the design documentation, the manufacturing process planning is started. The sequence of operation is determined during this stage. In case of machining operations, the machine tools, tools, machining fixtures, tools and machining cycles are selected. The number of set-ups necessary to machine the part is also determined. On this stage, Computer Aided Process Planning (CAPP) systems are used. The next stage is CNC machine tool programming with Computer Aided Manufacturing (CAM) systems. Usually, all works at this stage are done manually based on the already developed manufacturing processes. Because of this, CAPP systems are treated as the link between CAD and CAM. Automatic generation of manufacturing process plans or significant degree of human assistance allow to shorten the technical production preparation time and increase the degree of automation during CNC machine tool programming. Automation of these activities with CAPP systems offers several advantages. The first of them is the possibility to store and reuse the manufacturing knowledge. The good quality knowledge coming from experienced process planes or confirmed by laboratory research can be stored. Then it can be reused by less experienced persons, so

[a] Corresponding author : pobozniak@mech.pk.edu.pl

the newly developed processes will have the same, high quality. Additionally, such solution promotes the standardization and full use of company know-how. This aspect is especially important in view of nowadays noticed, missing engineering resources and growing complexity of manufacturing processes. The typical process plans presented in academic books contains usually only a few operations whereas the manufacturing processes in the aerospace industry usually have several dozen of operations. Their correct selection, order and contents are the factors deciding about the quality of manufacturing process plans. The other reason promoting the use of CAPP systems is the time consumption of manufacturing process planning. The typical manufacturing processes developed in the aerospace industry have several hundred pages. The use of CAPP systems significantly shortens the time needed to develop the documentation. The next point presents the use of Knowledge Based Systems (KBS) in CAPP, and forward and backward process planning technique. The goal of this point is not to make the exhaustive review of CAPP systems, as it can be found for example in [1, 2]. but present the factors blocking the wider use of CAPP systems in industry. On this background, the virtual manufacturing features and the benefits of technology are presented.

2 Forward and backward machining process planning

CAPP systems are often based on expert system technology. The main elements of such solution are the knowledge database, reasoning mechanism, normative database and user interface. The knowledge databases store the appropriately represented knowledge. The most common form of knowledge representation are the production rules. The rules can test the geometrical model and PMI data. For example, they check the occurrence of particular manufacturing features like hole, pocket, dimensional and location tolerances, status of the part surfaces or part material. The values of variables used in the conditions rule can be read from the normative database, evaluated using other rules or submitted by external programs doing some calculations. The literature presents also other knowledge representations [1, 3] like for example frames or multi-level decision networks. Even if other Artificial Intelligence (AI) techniques are used like Artificial Neural Networks (ANN), Simulated Annealing (SA), Fire Fly algorithm or hybrid evolutionary approaches [1], the data about the features are still needed. The feature data are also used in heuristic procedure for process selection and operation sequencing [4]. The control mechanism is responsible for reasoning process. It check the rules (using forward or backward approach). If more than one rule is true, the rule which must be fired is selected. In more advanced solutions, the algorithm of manufacturing process planning is implemented in the control mechanism as in [3], where it is responsible for processing of manufacturing knowledge in the form of multi-level networks. For example, first the group of rules for stock selection is run. Then the group of rules for set-up determination is checked and next the group of rules for machining cycles selection is run. The user interface presents the results of reasoning and explains why the particular decisions were taken. In case of CAPP systems, this mechanism is also used to interactively execute the tasks which were not automated.

Two approaches to manufacturing process planning can be distinguished in literature [5]:

- backward process planning,
- forward process planning.

In the case of backward process planning, the part passes through a series of intermediate states, from the finished state to the stock state. The subsequent intermediate states are determined by adding the material. In order to illustrate, Figure 1 shows the rotational part along with the machining cycles. The first of these cycles is facing, the next is roughing, then groove turning, and finally hole drilling and reaming. In the backward process planning, the starting point is the finished part. The rules checking whether there are internal holes with higher accuracy are run. Since the hole meeting these requirements is found, the appropriate allowances (for final reaming and drilling) will be added to this hole. Next the rule adding the material to external rotational groove is run. Then the material corresponding to rough turning is added. This processing is continue until the shape of the stock is reached. It shall be noted that the rules refer primarily the state of the finished part. This is because all information about the design task is given in model submitted by design engineer.

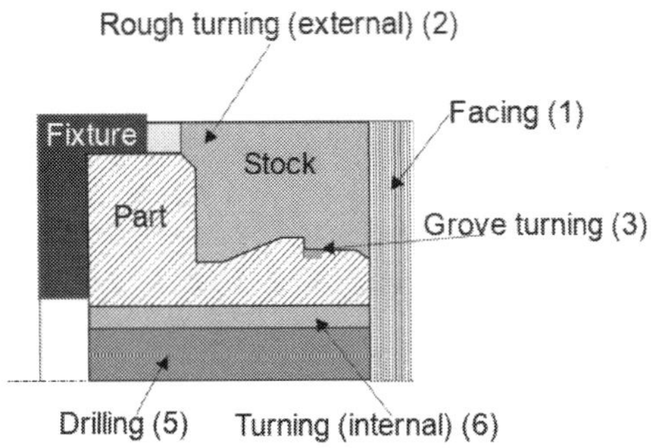

Figure 1. Machining cycles for rotational part

In the forward process planning, the subsequent machining cycles are selected, starting from the stock and ending with the finished product, which corresponds to the standard workflow of the manufacturing engineer. According to Author of this work, **the main disadvantage of this method is the fact that there are manufacturing features in part intermediate states, that are not present in the part finished state**. For example, the profile for external turning is not visible in the part presented on Figure 1. Such profile is the part of the profile used to create this rotational part, but it doesn't include the right planar face and profile of groove. In addition, the contour may be offset from the object when additional lathe finishing operations or grinding is used. When developing the systems for industrial users, it is not possible to use the forward design method without using such features. This is the main difficulty and the disadvantage of forward process planning method.

It shall be noted that the recognition of manufacturing features is currently one of the main obstacles in the development of CAPP systems. This is especially true in case of interacting manufacturing features [1, 6]. If the same geometric and topological model can be decomposed into different sets of manufacturing features, this leads to the occurrence of interacting manufacturing features. An example of such features is given in Figure 2. Figure 2a) presents another version of the profile for rough turning of external surfaces for analysed part. In this case, the roughing cycle is performed first (before the facing cycle), and as previously, the pocket is not processed during this cycle. The shape of this profile for roughing cycle is quite different from the one shown earlier. Another example of the decomposition into manufacturing features is given in Figure 2b. In this variant, it is assumed that first machining cycle is the facing, then the contour is turned but without the pocket and without the chamfers. The pocket and chamfers are machining in the next cycles. Such strategy is usually selected in case of manually controlled lathes.

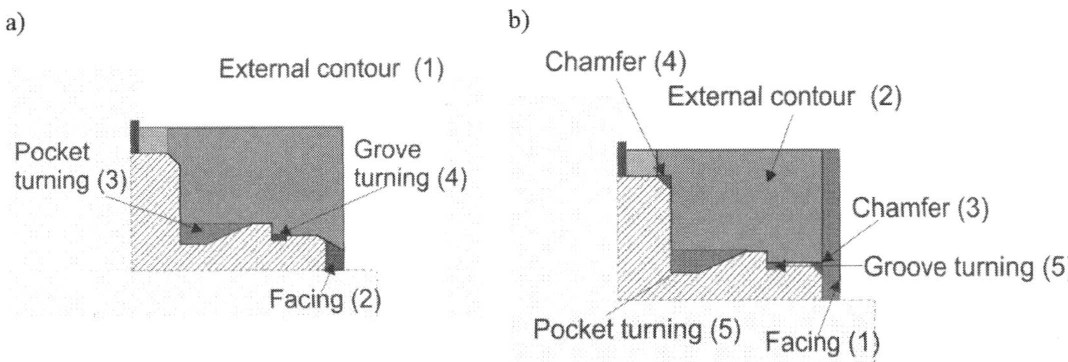

Figure 2. Interacting manufacturing features

To summary, despite many research projects on CAPP [1] and the unquestionable benefits of their usage, the provided level of automation is still unsatisfactory for commercial users. Many promising results have been obtained for CAPP systems based on backward process planning method [3]. This method could be accepted in an industrial environment in the case of complete automation. If the automation is partial, implementation of CAPP based on the backward method is not acceptable, especially in the case of complex manufacturing processes, consisting of several dozen operations. The paper presents the virtual manufacturing features. This solution that can be used during the development of CAPP systems based on forward process planning to use the manufacturing features which do not occur in the finished part, but occur only in the part intermediate states.

3 Virtual manufacturing features

Virtual manufacturing features are part of a manufacturing oriented part model.

The simplified, manufacturing oriented part model is a set of intermediate states:

$$MPM=\{IS_1, IS_2, IS_i, IS_{(i+1)},\ldots,IS_F\} \quad (1)$$

where:

MPR Manufacturing feature oriented part model

IS part intermediate state, i=1...F, for i=1 intermediate state IS_1 represents the stock, and for i=F intermediate state IS_F represents the finished state.

Each intermediate state IS_i is described by a set of manufacturing features, constraints between features and operations transforming the manufacturing features into the virtual manufacturing features.

$$IS_i=\langle F_i, CS_i\rangle \quad (2)$$

where:

F_i A set of manufacturing features in the intermediate state i, such as holes, grooves, pockets, etc. Each feature has a semantic representation in the form of attribute sets and a full B-Rep geometric representation.

CS_i Set of constraints describing the relations between features.

Manufacturing features are a basic component of manufacturing oriented part model. Manufacturing features is a specific fragment of geometric data, topological data, and non-geometric data, important from the point of view of process planning.

Virtual manufacturing feature is a result of geometric transformations on manufacturing feature using some operators. These operators can do the following changes:

- Elimination of the elements that do not meet some conditions. For example, this may involve the removal of concave elements from the milled profile, so the resultant profile is convex. This is the case for long-length profiles that include a small width concave element. The created virtual feature in the form of a convex profile is machined with a large diameter cutter, and the concave elements with the smaller diameter tool, which reduces, among others, the machining time.
- Changing the dimensions of a feature. One of the basic roughing cycles is facing. The surface for facing is always smaller than the corresponding surface of the stock. Creation of the virtual manufacturing feature extending this surface to the boundary of stock allows for proper machining of this surface.
- Changing the position of the feature relative to the geometry of the object. Such virtual manufacturing feature is always created for multi-step machining, for example divided into rough, final and super final steps. Virtual manufacturing feature is created by offsetting the profile to the distance equal to the amount of the allowance for further machining steps.
- Selection of feature elements meeting certain quality criteria. For example, operator can select surfaces with a roughness of 6.25 or less. These surfaces (elements of feature) can be subjected to final machining.

An important characteristic of a virtual manufacturing feature in comparison to normal manufacturing features is the possibility to test allowance reaming for machining. It is possible to check whether the allowance was removed at all (i.e. whether any machining cyclers related to feature were done), current dimension of the allowance and its accuracy (for example dimensional tolerance).

The use of virtual manufacturing features is presented on Figure 3. The part is machined on a 3-axis milling machine with vertical spindle. This figure shows the rules. These are typical production rules ***IF <Condition> Then <Actions>***. In order to reference the manufacturing features in a condition, it is necessary to use ***Exists*** keyword, followed by the round bracket and the feature class name. The ***Exists*** function is True when there is an instance of the given feature class in the part model i.e. when specified feature exists in the part model. After the feature name, the conditions for the feature must be given, linked by logical operators. It is acceptable to use parentheses to build more complex logical expressions. For example, the rule on Figure 3b has the condition ***I.z = 1***. This condition checks whether the hole feature Z axis is parallel to the to the Z axis of the global coordinate system. The next condition, ***Diameter <25*** checks if the hole diameter is less than 25 mm. If the diameter is larger, apart from drilling, also pre-drilling is required. To check if the feature was ever machined, the value of the ***IS.Status*** (Intermediate Status.Status) is checked. If this attribute is equal to ***NonMachined***, it means that the feature was not machined at all (the volume of material belonging to feature volume was not removed). This condition is checked by performing a geometric comparison operation.

Fig. 3c) shows the determination of a virtual manufacturing feature for surface facing. Operators acting on original manufacturing feature and creating the virtual manufacturing feature are introduced by the suffix ***Mod*** (like ***Modify***). For example, ***Mod.ExtendXY*** operator extends the dark coloured surface in the XY plane, so the surface for facing is created. Another operator, ***Mod.Offset=1*** shifts the top surface of the part by 2 mm, so the allowance for final facing will remain. Similarly, the virtual manufacturing feature for profile milling is created (Fig. 3d). The operator ***Mod.Convex*** transform the profile for milling into the convex one. This operator removes four non-through grooves from the profile. In these examples, both surface for facing as well as concave profile are not present in the final part. These virtual manufacturing features are created, because they are needed to create the correct machining process plan. The set of such operators is defined for each manufacturing feature.

The last example (Figure 3e) shows the conditions refer to the intermediate state of the virtual manufacturing feature. There is a hole after previous drilling operation. Due to the tolerance of this hole (***Hole.Tolerance.Class = H and Hole.Tolerance.Zone<10***), final reaming is required. This reaming is used if the amount of reaming allowance is relatively small. In order to check the amount of remaining allowance to remove, the ***IS*** prefix with the ***Allowance*** attribute is used. If the oversize is greater than 0 and less than 1.5 mm, this reaming is recommended. In case of a larger allowance, further drilling would be recommended. If there is a need to check the current tolerance for of the intermediate state of this hole, the condition of type ***IS.Tolerance.Class = H*** and / or ***IS.Tolerance.Zone > = 10*** can be used. In order to check the surface roughness of this hole in the current intermediate state against the threshold value, the condition ***IS.Roughness.Ra <= 6.25*** is used.

After executing THEN part of any of the rules, a subtraction Boolean operation is performed to update the condition of the object. The first parameter of this operation is the volume of the part in the current intermediate state and the second parameters is the volume of material removed by the cycle. As the result, the new intermediate state is generated. So CAPP system works in the cycle:

1. Checking the rule conditions and selection of the rule to be fired from the conflict set. Other advanced processing mechanisms may also be used at this moment.
2. Transformation of the part by making the appropriate geometric Boolean operation, which results in the generation of the next intermediate state. Transformation also includes modification of PMI information, such as surface roughness or dimensional tolerance of a transformed feature.
3. Jump to step No. 1 unless the state of the finished part was reached.

The significance of the generation of intermediate states shall be noticed. These states are necessary if many CNC machine tools are used during manufacturing process, which is common in the aerospace industry. Intermediate states are also required for the selection of workpiece fixtures. If there are interacting manufacturing features, the decisions made by the manufacturing engineer are reflected in the generated intermediate states, which significantly reduces the number of interacting features during the next decision problems.

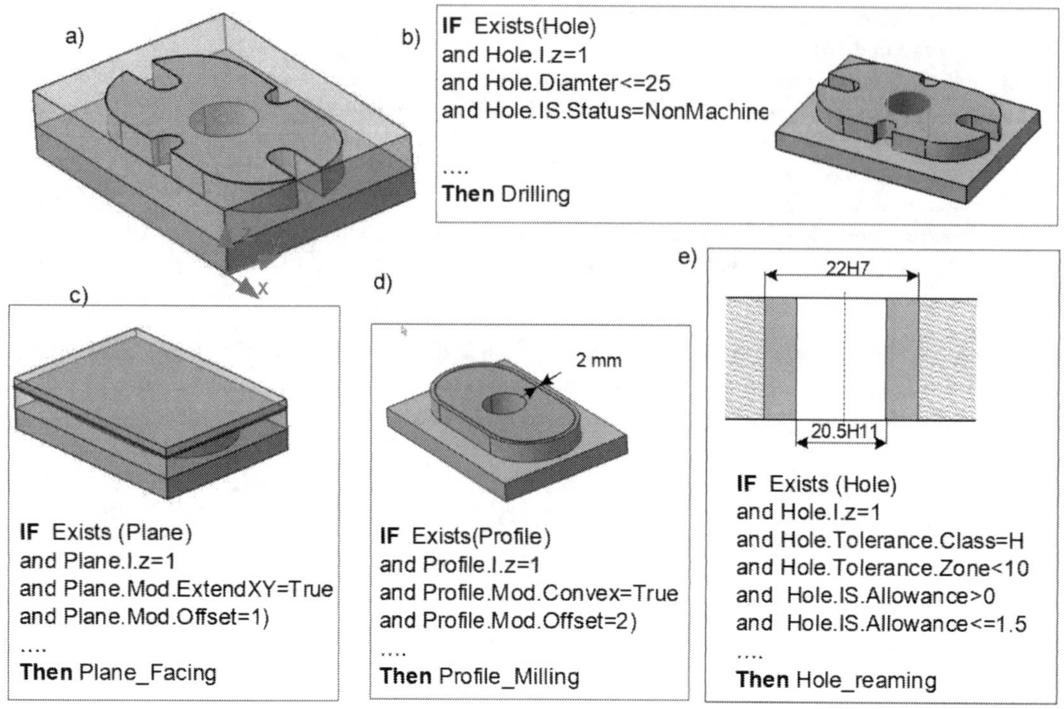

Figure 3. The use of virtual manufacturing features

4 Summary

The paper introduces so-called virtual technological features. These are manufacturing features that appear only in the intermediate states of the part, and are not present in the finished part. Their use allows the generation of manufacturing processes with forward planning method. The presented concept was partly verified by the development of software for recognition of manufacturing features defined in STEP-NC [2]. Further implementation work will be directed to the generation of part intermediate states based on the virtual manufacturing features, using one of the available commercial graphics kernels (OpenCascade, ACIS or C3D). The results of the work can be used during the development of CAPP systems.

References

1. Kumar Leo, S.P., State of The Art-Intense Review on Artificial Intelligence Systems Application in Process Planning and Manufacturing, In Engineering Applications of Artificial Intelligence, Volume 65, 2017
2. Chlebus E., Krot K., CAD 3D models decomposition in manufacturing processes, Archives of Civil and Mechanical Engineering, Vol. 16, 2016
3. Duda J., Pobozniak J., The Architecture of Intelligent System for CNC Machine Tool Programming, Procedia Manufacturing, Volume 11, 2017
4. Deja M., Siemiatkowski M., Feature-based generation of machining process plans for optimised parts manufacture, Journal of Intelligent Manufacturing, 2013
5. Chryssolouris, E.L.K.E, Manufacturing Systems: Theory and Practice, Springer-Verlag New York, 2006
6. Dipper T., Xu X., Klemm P., Defining, recognizing and representing feature interactions in a feature-based data model, Robotics and Computer-Integrated Manufacturing, Volume 27, Issue 1, 2011

Glass Gob Modeling and Experimental Validation Using a Drop Test

Adrià Biosca[1, 2], Salvador Borrós[1], Vicenç Pedret Clemente[2] and Andrés-Amador García Granada[1,a]

[1]*IQS School of Engineering, 08017 Barcelona, Spain*
[2]*Ramon Clemente, 08320 El Masnou, Spain*

Abstract. Glass blowing to create bottles with specific thickness distribution profiles requires several experimental iterations. Such iterations are expensive and increase the time to market. The use of simulation pretends to decrease the amount of prototypes by doing virtual validation of glass blowing molds. To feed simulations with realistic physical values, a gob drop test has been designed. This test provides valuable experience on the use of the software and validates heat transfer, viscosity and other physical parameters. Gob drop test was chosen for the possibility to record the test with infrared thermal cameras. Gob obtained similar shapes when dropped on a cast iron plate for both central and side sections with longer cooling of about 25°C. Such technique allowed the user to gain experience on the use of software and obtain valuable physical parameters for future glass blowing optimization.

1 Introduction

Glass containers have been used during ages. For packaging industry certain qualities are required, and glass has most of them: it is an innocuous and hermetic material that allows long-term preservation, it is a recyclable material and it adds a lot of value to the final product. The individual machine section system (IS) was implemented at the beginning of the 20th century, first as manual machines and later as automatic machines.

Ramon Clemente (RC) is a glass container manufacturing company for the perfume sector. It has about 75 years experience, but like many other glass companies, developing new containers and its mold equipment is still made in the old fashioned way. There is a lot of empirical knowledge on molds design, but at the end most of it is reduced to trial and error. This means that for every new container development, new molds are designed, machined and tested at the production line. The formed containers are analyzed and adjustments are made in the mold and in the process conditions until desired quality is reached.

Numerical simulations of glass forming process are complex because there is a conjugated heat transfer problem, glass viscosity has a very high dependence with temperature, glass domain suffers extreme free surface deformations, etc. Over the last decades, taking advantage of the development and improvement of numerical simulations, the narrow neck pressed blow (NNPB) forming process has been studied to optimize the weight of the containers [1-3].

[a] Corresponding author : andres.garcia@iqs.edu

To speed delivery times, reduce costs and plant downtime, a simulation tool capable to predict container glass distribution in function of blank mold cavities and production process properties it is being developed. The success of this project would help RC to be much more competitive.

2 Glass forming process description

Glass forming process begins with the melting of raw materials in the furnace, refining in the working end and conditioning the molten glass in the forehearth. The glass stream flows through the feeder and the orifice ring and it is cut by shear blades forming discrete gobs. Delivery equipment distribute the gobs to the machine sections where container forming occurs.

Many forming steps are needed to manufacture a container with an IS machine, the whole process can be seen in Figure 1: glass starts being a gob loaded inside the blank mold, after two blows in opposite directions the parison shape has been defined; an arm transfers the parison from the blank to the blow mold where glass parison is reheated, then it stretches and is blowed again. Final containers are sent through an annealing lehr, where its internal stresses are released. At the beginning glass behaves as a fluid and ends up being a solid due to the temperature dependence on viscosity.

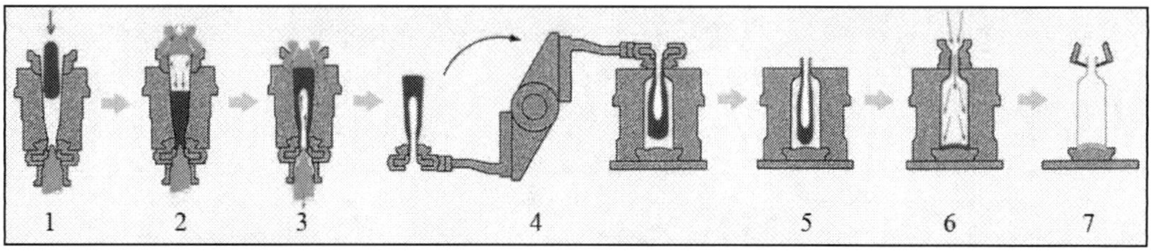

Figure 1. Container forming IS machine stages for blow and blow process. Described stages are: gob loading (1), settle blow (2), counter blow (3), transfer (4), reheat and stretch (5), final blow (6) and take out (7).

3 Gob drop validation for future blowing simulation

The development of a simulation tool capable of predicting the final container glass thickness distribution in function of the blank mold cavities and production process conditions is a very complex problem. Moreover, running the simulations requires a lot of sensitive information (i.e. material properties, initial values, boundary conditions, etc.) that needs to be previously obtained from the process itself.

Understanding, interpretation and validation of the results is also complex, as most of the time during the production process, glass remains inside the blank or blow mold. It is hard to predict initial and final glass geometries inside the molds, and define its transition during blowing because the lack of visibility on closed molds. Only the use of thermocouples installed on the mold cavity surface as a contact sensors [4] may show some clues about the flow of glass.

Due to all the modeling difficulties, it seemed interesting to work on a simpler case like studying glass deformation during a gob drop test. This case has been an initial step that has helped to solve many modeling and data acquisition issues. In addition, drop test simulations have been helpful to obtain a better understanding of the model simulations. All the improvements achieved with the drop test have been also used to improve the ongoing simulations of the blow and blow model.

The objective of this paper is to correlate numerical and experimental results of a drop test, and to apply the knowledge learned to the blow and blow forming simulations.

3.1 Gob drop numerical simulation

A 2D axisymmetric model has been built using ANSYS Polyflow. The simulation describes the impact of the free falling glass gob against a cast iron plate. Glass gob geometry has been simplified

as a cylinder with rounded edges. Glass volume is described by gob length and diameter, and it is in concordance with the container weight. Gob dimensions were obtained from experimental measurements.
The numerical model contains two domains:
• The glass domain is defined as a generalized Newtonian non-isothermal material. Material data is set, including temperature dependence of viscosity by the Vogel-Fulcher-Tammann viscosity equation [5]. Density, thermal conductivity and heat capacity per unit mass are also specified.

Gob velocity at impact has been obtained from experimental measurements.

Experimental measurements have been performed to determine glass temperature in the gob skin. During loading gob heat is released mainly by contact with the delivery equipment, temperature distributions in glass gobs at loading stage have been a subject of study in the past [6]. A temperature gradient has been defined from the skin to inside the gob. Such temperature distributions were obtained from transient cooling simulations during the time that the gob travels from the exterior of the feeder to the blank mold, through the delivery equipment.

A thermal contact conductance is defined between glass and mold at its contact interference, convection heat loss is defined too.
• The plate domain is also defined as non-isothermal. In order to describe a similar thermal contact between glass and mold than in the loading, material properties from blank mold cast iron were used to set the plate. Density, thermal conductivity and heat capacity per unit mass are also specified.

Due to the extreme high deformations of the glass domain, an adaptive remeshing has been configured, the quality of the mesh elements is checked after each time step. A finer mesh has also been defined near the interference of the glass with the mold to correctly define the thermal gradient. Volume conservation is verified so as not to modify the initial volume of the glass domain with all the mesh modifications.

3.2 Gob drop experimental testing

A plate has been machined using the same cast iron material as the blank mold and with the same surface finish as in the mold cavities. This plate has been fixed on the funnel. The funnel arm is blocked in loading position and all other individual section (IS) movements are also blocked. Discrete gobs are loaded through the delivery equipment to the plate to perform the drop test.

A FLIR A325sc infrared thermal camera has been used to measure the temperatures in a range from 300 to 2,000°C and also the glass deformation along the impact was recorded.

Tests were performed under different conditions described in the results and discussion section.

4 Results and discussion

Tests were performed on a six section IS machine with a single gob (SG) cooled distributor. Several gob drop tests and simulations have been performed under different working conditions.

During the gob drops different stages can be defined describing the glass gob mechanical behavior:
1. The drop test starts with the gob at a certain height and zero velocity.
2. The gob accelerates during free fall until the impact velocity is reached.
3. The gob strikes the surface of the cast iron plate.
4. During the impact the gob undergoes extreme deformation and flows over the plate surface, the initial shape of a cylinder ends up being a kind of amorphous medallion.
5. The glass stops flowing due to heat loss at mold contact interface and on the skin.

4.1 Gob drop test conditions: delivery equipment

Height between the shear blades and the top of the blank mold is about 2.45 m, this is the free fall gob height. In a free fall scenario, the total fall time would be 0.707 s and impact velocity would be

6.93 m/s. But at the IS machine, between the shear and the funnel and the blank mold there is the delivery equipment, which is formed by three different parts: scoop, trough and deflector.

The orifice ring and the shear are situated in the middle of the IS machine and they describe a symmetry axis when the IS machine is viewed from the front view, so on the left and right side there are three IS which are symmetric. IS1 delivery equipment is symmetric with IS6 (side sections), IS2 with IS5 and IS3 with IS4 (central sections).

There is one scoop for all six sections. There are six troughs which are symmetrical; central IS troughs are shorter than side ones. The slope of all the troughs is almost the same. There are six deflectors which also are symmetrical; all the deflectors have a similar curvature, but at the end a straight section is added for the central IS, so central IS troughs are longer than side ones (see Table 1).

Table 1. Delivery equipment (DE) length measurements (m) in function of IS machine.

	IS1 (side)	**IS2**	**IS3 (central)**
Scoop	0.62	0.62	0.62
Trough	1.22	0.85	0.61
Trough length increase	+100.0 %	+39.3 %	0 % (reference)
Deflector	0.56	0.63	0.72
Deflector length increase	-22.2 %	-12.5 %	0 % (reference)
Total DE length	2.40	2.10	1.95
Total DE length increase	+23.1 %	+7.7 %	0 % (reference)

The fact that there are six IS machines and the orifice ring and the gob distributor with the scoop are between IS3 and IS4, defines that: IS3 and IS4 (central sections) are closer and gobs travel at close range; On the other hand, S1 and IS6 (side sections) are further away and gobs travel farther. This delivery equipment differences between sections have consequences for the glass properties, comparing IS1 with IS3:

• The gob displacement distance from orifice ring to the blank mold is longer, and the gob needs more time to be delivered.

• Longer distance and longer delivery equipment mean longer contact time and more friction between glass and the delivery equipment.

• Friction with the delivery equipment affects the shape of the gob, during the falling the gob is lengthened.

Therefore, in the end, there is more heat loss on the gob surface due convection and contact conductance with the delivery equipment; consequences of that heat loss can also be seen on [6]. The gob impact velocity will be lower. There is no appreciable impact on gob elongation.

Gob drop tests and its simulations have been performed under IS1 and IS3 working conditions.

4.2 Gob drop experimental results

Gob drop tests were performed from the central and side sections of the IS machine.

The temperature of the glass gobs has been measured at the end of the three IS deflectors. Figure 2 shows the temperature values of 14 consecutive gobs loaded at each individual section. It can be appreciated how IS1 temperature values are lower than IS2 and IS3. Deviation from all three series is under ±5 °C when compared to the mean values. Side section gobs showed higher cooling with lower

 https://doi.org/10.1051/matecconf/201816702009

glass temperatures of around 25 °C when compared to central section gobs with a good statistical repeatability.

Total fall time and impact velocity are also consistent: side section fall times are longer than central section ones and impact velocities show slower values when compared to central section velocities. However, there is no appreciable impact on gob elongation.

Figure 3 shows four drop tested gob samples, IS1 on the left and IS3 on the right. Although gob velocity and surface temperature before impact were lower, its final dimensions are almost the same. There is no appreciable difference on the final glass drop tested samples when side and central sections are compared.

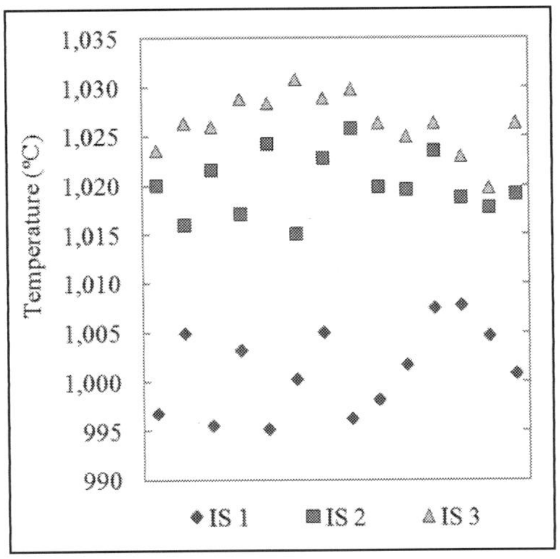

Figure 2. Gob temperature measurements.

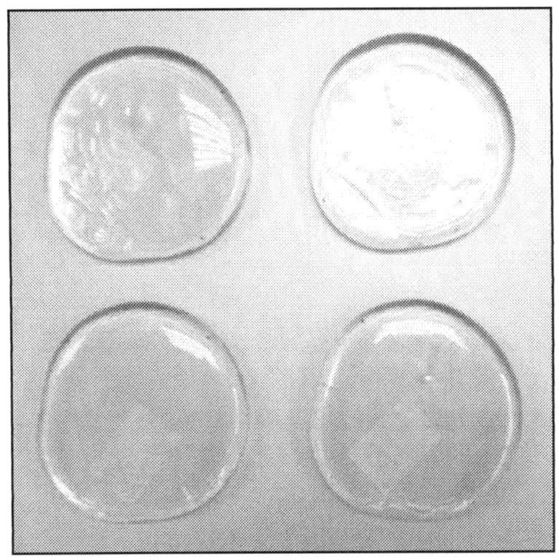

Figure 3. Drop tested gob samples.

Deflectors can be adjusted to guide the gob's path and perform a normal impact against the surface of the horizontal plate. Despite that, observing the amorphous deformed gob samples no circular symmetry can be found. During deformation, the gob is slightly offset from the vertical axis.

4.3 Gob drop simulation results

Gob drop test simulations have been performed under different working conditions. The glass temperature distributions and the impact velocities have been set to reproduce the drop test conditions at central and side sections of the IS machine.

Figure 4 shows representative deformation results from different simulation time steps. Each time step has local scale temperature contours in order to appreciate big temperature gradients in the final step (E), but also the small temperature gradients of (A).

(A) shows the temperature gradient from the skin to inside the gob before impact. (B) and (C) are intermediate simulation steps that describe gob deformation due to impact energy. At (D) most of the kinetic energy has been used to deform the initial gob shape.

But glass is at high temperature and its low viscosity allows it to flow over the plate surface, reason why the deformation of the glass continues at a slower rate. The contact conductance defined in the interface between the glass and the plate removes heat from the glass. The viscosity of the glass increases and after a long period of time, glass stops flowing (E). Without a non-isothermal model, the glass would have flowed indefinitely, obtaining very different results.

The shape of the gob and its skin temperature has been recorded with the IR thermal camera. Experimental and numerical results can be compared in Figures 4 and 5.

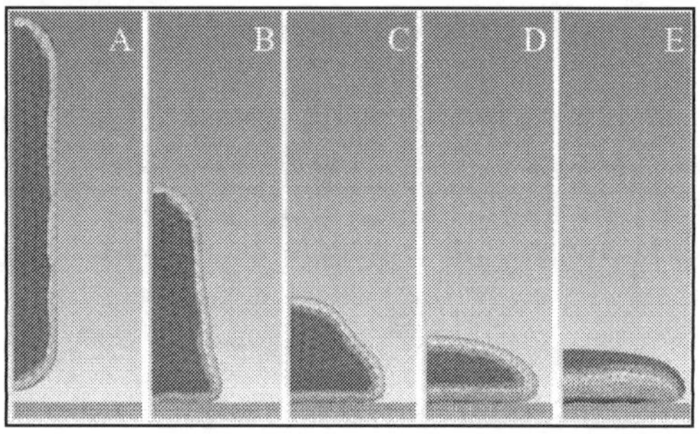

Figure 4. Representative simulation steps.

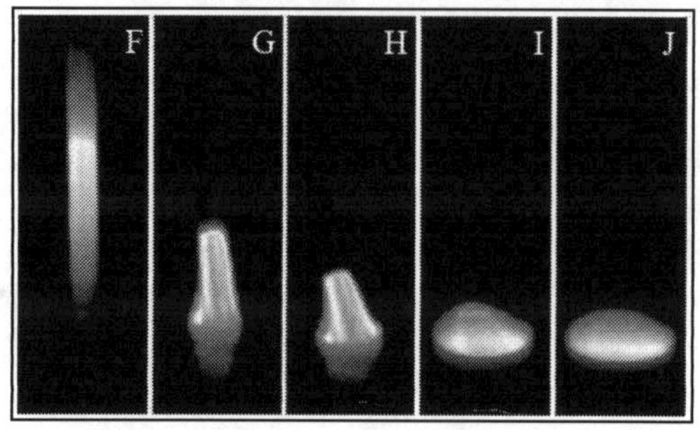

Figure 5. Representative IR captures.

As it can be seen, there is a good correlation between numerical and experimental results. Both results show a good agreement but in the future more drop tests will be performed in order to obtain stronger correlations.

There are no relevant differences in the results of the glass dropped geometries when comparing the simulation results at the side and central sections.

Simulations have been done with a 2D axisymmetric model, so the model cannot describe symmetry axis deviations of the experimental results.

5 Conclusions

An experimental gob drop test case has been designed and numerically modeled. This test provides valuable experience on the use of the software and validates heat transfer, viscosity and other physical parameters.

The temperature distribution in the glass domain defines its viscosity and flow. As seen in the numerical results, a non-isothermal model is required to describe the viscosity highly dependence on temperature to reproduce the glass flow of the experimental results. Apparently, an isothermal model will also not be suitable for describing the glass flow in blow and blow glass container forming simulations.

Gob loading conditions at the central and side IS machine sections have been studied and initial assumptions have been verified. Side section gobs showed higher cooling with lower glass temperatures of around 25 °C when compared to central section gobs with a good statistical repeatability. The total fall time and impact velocity are also consistent: side section fall times are longer and impact velocities are slower than central section ones.

Initial gob conditions are different depending on the section, however experimental and numerical results do not show relevant differences.

References

1. J.M.A. Cesar de Sa, Eng. Computation, **3** **[4]**, 266-275 (1986).
2. M. Hyre, J. Am. Ceram. Soc., **85** **[5]**, 1047-1056 (2002).
3. J.A.W.M. Groot, C.G. Giannopapa, R.M.M. Mattheij, J. Press. Vess.-T. Asme, **136**, 1-9 (2014).
4. D.M. Shetterly, N.T. Huff, J. Non-Cryst. Solids, **38** ~ **39**, 867-872 (1980).
5. G.S. Fulcher, J. Am. Ceram. Soc., **8** **[6]**, 339-355 (1925).
6. H. Müller-Simon, G. Bergmann, K. Kessler, *70th Conference on Glass Problems*, 151-158 (2010).

MATEC Web of Conferences **167**, 02010 (2018) https://doi.org/10.1051/matecconf/201816702010
IC4M & ICDES 2018

Effect of Gr Contents on Wear Properties of Al2024/MgO/Al2O3/Gr Hybrid Composites

İsmail Ovalı[1,a], Cemal Esen[2], Sevda Albayrak[3] and Halil Karakoç[4]

[1] *Pamukkale University, Department of Manufacturing Engineering, 20160 Denizli, Turkey*
[2] *Celal Bayar University,Department of Mechanical and Manufacturing Engineering,Manisa,Turkey*
[3] *Gazi University, Department of Metallurgy and Materials Eng., 06500 Ankara, Turkey*
[4] *Hacettepe University, Department of Mechanical Program, 06935 Ankara, Turkey*

Abstract In the present study, hybrid metal matrix composites, Al2024/10Al2O3, Al2024/10Al2O3/3MgO, Al2024/10Al2O3/6MgO, Al2024/10Al2O3/3MgO/1.5 Gr, Al2024/10Al2O3/3MgO/3Gr, and reinforcement samples (AA 2024) produced with powder metallurgy process. AA 2024 and reinforcement powders were determined mixture rations and separately mixed during 30 minutes in a three-dimensional Turbula mixer. The mixed compositions were pressed at 300 MPa and sintered at 550°C during 1 h. After that, three materials were extruded at the same temperature. Experimental results show that hybrid metal matrix composites (HMMCs) a better wear resistance than the reinforcement samples because of higher hardness. Gr behave as a lubricant during wear process. The wear resistance of HMMCs can be optimized with controlling of the reinforcement content and type.

1 Introduction

Metal matrix composites (MMCs) reinforced with harder particles have been extensively used in automotive and aerospace industries because of their superior mechanical properties. Gr also are used in the MMCs as lubricator [1-3].Adding of Gr significiantly improve wear performance of MMCs.

Şahin et al devoloped the regression models. The model show that with increasing load incrérase the wear rate. However, it decreases with increasing of sliding distance [4]. In a study by Al-Rubaie, increasing in reinforcement size of Al-SiC composites improved wear resistance [4]. Basavaraja et al. (2006) investigated the dry friction wear behavior of hybrid composites. They reported that the wear resistance increased with the increase of SiCp and graphite % wt and the wear rate decreased with increasing SiC % wt [5]. Alahelisten et al. studied that, the tribological performance of a composite depends on the microstructure properties of the material and the wear conditions [6]. Basavarajapa et al. fond that using of graphite particles in aluminum matrix improve wear resistance and form a lubricant layer between the friction surfaces [7].

Topcu et al. studied the mechanical properties of B4C reinforced Al composites the composite hardness increased with increasing of reinforcement % wt. [8]. Deuis et al. reported the effect of particle size and % wt. on the wear behavior of aluminum matrix composites and found that the most important parameters are sliding distance and load [7]. Suresha and Shridhara revealed that adding hard particle such as SiC and Gr improve the wear resistance of hybrid composites [7]. The effect of

[a] Corresponding author : iovali@pau.edu.tr

B4C on the wear performance and mechanical properties of Al 7075-B4C composites were investigated by Baradeswaran and Elaya Perumal. They observed that increasing B4C particles increase the wear performance of the composites when compared to the matrix alloy [9].

In the present work, the effects of reinforcement type and % wt. content, (Al6061), (%12B4C), (%12SiC), (%9B4C+%3SiC), (%6B4C+%6SiC) ve (%3B4C+%9SiC),on the abrasive wear performance of Al2024 hybrid matrix composites produced with hot pressing methods were investigated.

2 Experimental

2.1 Material and methods

The 6061 series aluminium matrix material powder size was taken less than 100 µm. The chemical composition of AA 2024 has shown in the Table 1. The powder properties also is given in Table 2. First of all, AA 2024 and reinforcement powders were determined mixture rations and separately mixed during 30 minutes in a three-dimensional Turbula mixer. After that, AA 2024 and hybrid composites were prepared in the same parameters. All of volume fractions were shown in Table 3. During experimental work, the reinforcement was added slowly for homogeneous distribution. The mixed compositions were pressed at 300 MPa and sintered at 550°C during 1 h. After that, three materials were extruded at the same temperature. It was used a pre-heated hydraulic extrusion mold and consist of shaped metal blocks. Flow diagram of process is shown in Fig. 1.

Table 1. The chemical composition of Al2024 alloy.

Component	Wt. %	Component	Wt. %
Aluminum (Al)	90.7 to 94.7	Zinc (Zn)	0 to 0.25
Copper (Cu)	3.8 to 4.9	Zirconium(Zr)	0 to 0.2
Magnesium (Mg)	1.2 to 1.8	Titanium (Ti)	0 to 0.15
Manganese (Mn)	0.3 to 0.9	Chromium (Cr)	0 to 0.1
Iron (Fe)	0 to 0.5	Residuals	0 to 0.15
Silicon (Si)	0 to 0.5		

Table 2. Properties of matrix and reinforcing powders

Material	Manufacturing Process	Density (g/cm3)	Melting Temperature (°C)	Grain size
Al2024	Gas atomization	2.78	502 - 638	<100 µm
Al_2O_3	---	3,97	2040	<8 µm
MgO	---	3.58	2852	10-30 nm
Gr	---			<10 µm

Table 3. Hiybrid Composite materials

Composite materials	**% wt.**			
	Al2024	**Al2O3**	**MgO**	**Gr**
Al2024/10Al_2O_3	90	10	-	-
Al2024/10Al_2O_3/3MgO	87	10	3	-
Al2024/10Al_2O_3/3MgO	84	10	6	-
Al2024/10Al_2O_3/3MgO	85,5	10	3	1.5
Al2024/10Al_2O_3/3MgO/1.3Gr	84	10	3	3

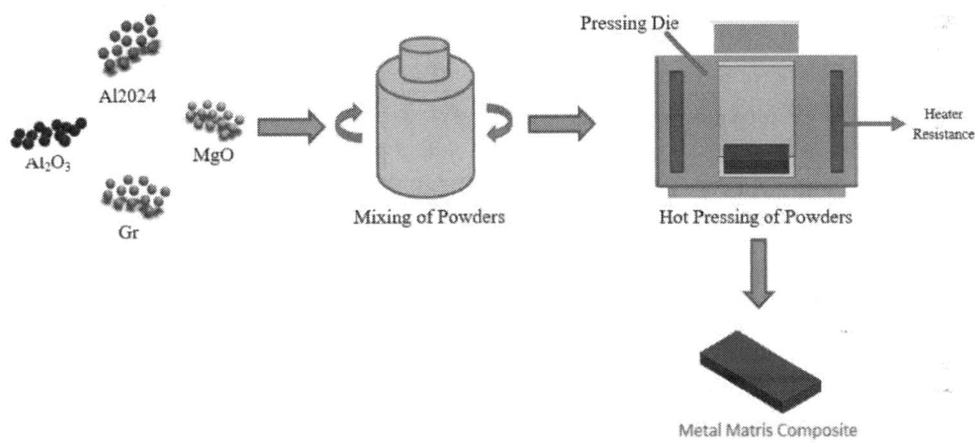

Figure 1. Schematic representation of AMC composite fabrication by hot pressing.

The wear test were carried out by a pin-on-disk wear-testing devices according to ASTM G132-96 (Figure 2). The wear tests were carried out by a pin-on-disk wear-testing device according to ASTM G132-96. Samples surfaces were grounded with 80 SiC paper for removing rough surface than the ACM test pin, with a diameter of 10 mm, was fixed and counterface abrasive disk (abrasive paper 150 grit) during the test .The wear test was performed in distance 100, 200 and 300 m, at speed of 1.2 m/s and with imposed load of 5, 10 and 15 N. Prior to measuring, samples were cleaned in acetone for cleaning of surface contaminants, dried and then weighed using an electronic balance having a resolution of 0.001 mg.

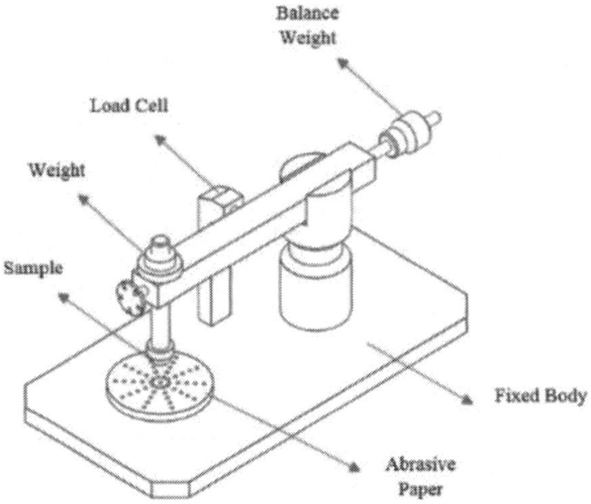

Figure 2. Pin-on-disk wear-testing set-up

3. Result and Discussion

3.1 Effect of the sliding distance and load on the weight loss of HMMCs

The variation in weight loss of HMMCs and unreinforced composite Al 2024 as a function of load are shown in Figure 3.

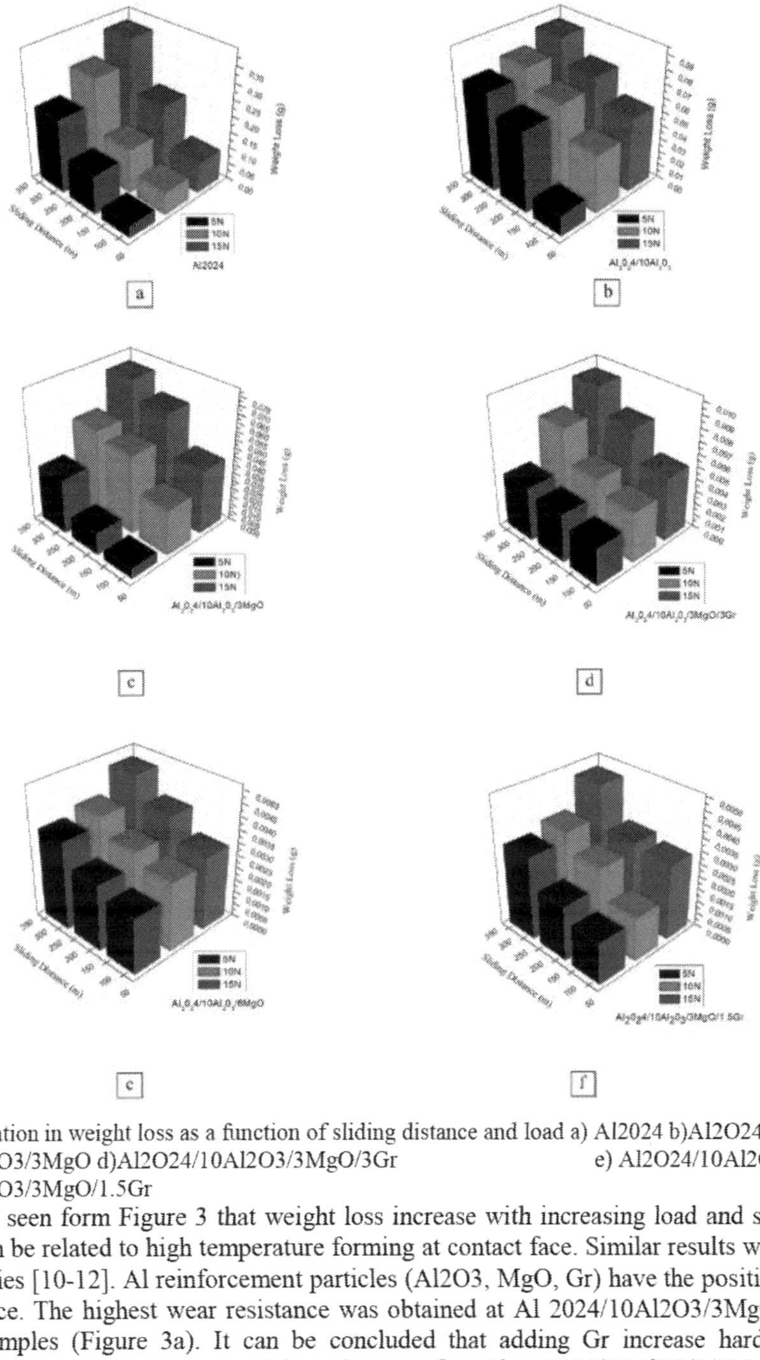

Figure 3. Variation in weight loss as a function of sliding distance and load a) Al2024 b)Al2O24/10Al2O3 c) Al2O24/10Al2O3/3MgO d)Al2O24/10Al2O3/3MgO/3Gr e) Al2O24/10Al2O3/6MgO f) Al2O24/10Al2O3/3MgO/1.5Gr

As can be seen form Figure 3 that weight loss increase with increasing load and sliding distance. The result can be related to high temperature forming at contact face. Similar results were obtained by previous studies [10-12]. Al reinforcement particles (Al2O3, MgO, Gr) have the positive effect on the wear resistance. The highest wear resistance was obtained at Al 2024/10Al2O3/3MgO/1.5Gr hybrid composite samples (Figure 3a). It can be concluded that adding Gr increase hardness and wear resistance when compared weight loss of Al 2024/10Al2O/3/3MgO and Al 2024/10Al2O/3/3MgO/3Gr. It is well known that Gr also have lubricant effects. The effects also increase the wear performance of metal matrix composite. Graphite layer are formed by adding Gr. Riahi and Alpas [13] find that wear resistance of composites reinforced with graphite are higher when compared with unreinforced A356 and nongraphitic A356/20SiCp samples. The similar higher wear resistance results were obtained at previous studies which used ceramics and graphite as reinforcement [14-15].

3.2 Effect of the reinforcement type and %wt. on Hardness of HMMCs

The hardness property is widely used in order to evaluate the effects of reinforcement type and %wt. on the wear resistance. The hardness of the unreinforced and HMMCs samples are given in the Figure 4.

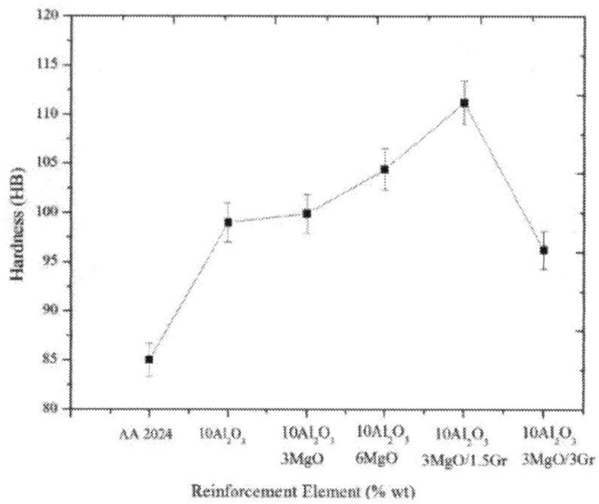

Figure 4. The hardness of the HMMCs samples

The lowest hardness was obtained for Un-reinforcement Al2024 samples at 85 HB while the highest hardness was obtained for Al 2024/10Al2O3/3MgO/1.5Gr at 111.2 HB. It can be explained that harder reinforcement particles increased to compression and tensile strength of the matrix material [16]. Some previous studies show similar results for effects reinforcement on the hardness [13-15].

REFRENCES

1. A.R. Riahi and A.T. Alpas, Wear, **251** p 1396-1407 (2001)
2. S. Mohan, J.P. Pathak, R.C. Gupta, and S. Srivastava, Z. Metallkd, **93(12)** p 1245-1251 (2002)
3. S. Das, S.V. Prasad, and T.R. Ramachandran, Wear, **133** p 173-187 1(989)
4. N. Ch. Kaushik, R. N. Rao, Trib. Inter.l, **103** 298–308 (2016)
5. V. M. Ravindranath, M.Yerriswamy, S. V. Vivek, G. S Shiva Shankar, N. G Siddesh Kumar, Mat. Tod.: Proc., **4** 9898–9901(2017)
6. M. Uthayakumar, S. Aravindan, K. Rajkumar, Mat.and Des., **47** 456–464 (2013)
7. T. Lokesh, U. S. Mallik, Mat. Tod.: Proc.s, **4** 11175–11180 (2017)
8. Y. H. Çelik, K. Seçilmiş, Adv. Pow.Tech., **28** 2218–2224 (2017)
9. S. Thirumalai Kumaran, M. Uthayakumar, Pro. of the Ins. of Mech. Eng., Part J: J. of Eng. Trib. 2014 228: 332 originally published online 18 October 2013 DOI: 10.1177/1350650113508103
10. R.L. Deuis, C. Subramanian, and J.M. Yellup, Wear, **201** p 132-144 (1996)
11. J. Goni, Mater. Sci. Technol., **16** p 743-746 (2000)
12. G.J. Howell and A. Ball, Wear, **181- 183** p 379-390 (1995)
13. A.R. Riahi and A.T. Alpas, Wear, **251** p 1396-1407 (2001)
14. S.K. Rhee, SAE Trans., paper **740415**, (1974)
15. P.K. Rohatgi, S. Ray, and Y. Liu, Int. Mat. Rev., **37** p 129- 149 (1992)
16. B. Venkataraman and G. Sundararajan, Wear, **245** p 22-38 (2000)

Development of High Precision Vehicle Dynamic Model with an Intelligent Torque Transfer System (All-Wheel Drive System)

Beom-Joon Pyun[1], Chul-Woo Moon[1,a], Chang-Hyun Jeong[1] and Do-Hyun Jung[1]

[1]*Korea Automotive Technology Institute, 31214 Chungnam, Republic of Korea*

Abstract. High precision vehicle simulation environment is required for development of control system of any newly suggested intelligent system. Hence, a high precision full-vehicle simulation environment integrated with an intelligent torque transfer system should be developed for an advanced control logic for enhancement of vehicle stability. In the perspective of making enhanced AWD system, there are many kinds of methods to make the system. And a controller part of the AWD module is regarded as a major part of the system development in consideration of enhancement of the vehicle stability with the suggested AWD system. Therefore, in this study, high precision full-vehicle simulation environment is developed for the development of an intelligent control system of the AWD module. In order to make models for the simulation, vehicle test is performed with a commercial vehicle, and the several performance tests of the developed AWD system are also conducted in a laboratory. Then, the simulation environment comprised of several models of important sub-systems is developed based on the previously conducted test results, and the developed simulation environment is verified by comparing the simulation results to the test results.

1 Introduction

All-Wheel Drive(AWD) mode is an automatically adjustable transfer case between Two-Wheel Drive(2WD) mode and Four-Wheel Drive(4WD) mode depending on such driving conditions. An intelligent torque varying controlled system can enhance the driving performance and stability because those two types of driving systems, 2WD and 4WD, have their own strengths for such conditions; 2WD mode has superior perspective in consideration of fuel efficiency for urban driving, while 4WD mode can provide stable enhanced driving performance with stability for a condition such as an off-road. AWD mode can control the torque distribution between those two modes, 2WD and 4WD, depending on such driving conditions, hence, a vehicle equipped with intelligent AWD system can have those two types of advantages, simultaneously [1]. In this study, therefore, a high precision simulation environment integrated with AWD system is developed and verified as a part of the development of the novel control software of the developed system. For the development of the simulation environment, the several performance tests are conducted with a target vehicle in KATECH (Korea Automotive Technology Institute of Technology) proving ground, and both the lateral and longitudinal vehicle dynamics are modelled. Except for dominant parts of vehicle

[a] Corresponding author : cwmoon@katech.re.kr

dynamics, minor parts of the vehicle dynamics are developed by using a commercial tool, CarSim. Furthermore, several performance tests of the developed AWD system are also performed by using four-wheel drive dynamometer test, then, the high precision plant model for the developed system is developed with the test results. In this modelling, commercial tool AMESim is used to develop the plant model because fluid property should be considered specifically. As a result, the vehicle model and AWD plant model are integrated, and verified with those test results. This integrated simulation environment allows users to develop and assess control algorithms.

2 Vehicle Simulation Environment based on Vehicle Test

Vehicle dynamics is basically considered in order to develop the simulation environment based on the vehicle test results [2]. Well-understanding of the vehicle dynamics can make not only the vehicle test effective, but also the simulation environment better [3].

2.1 Powertrain System

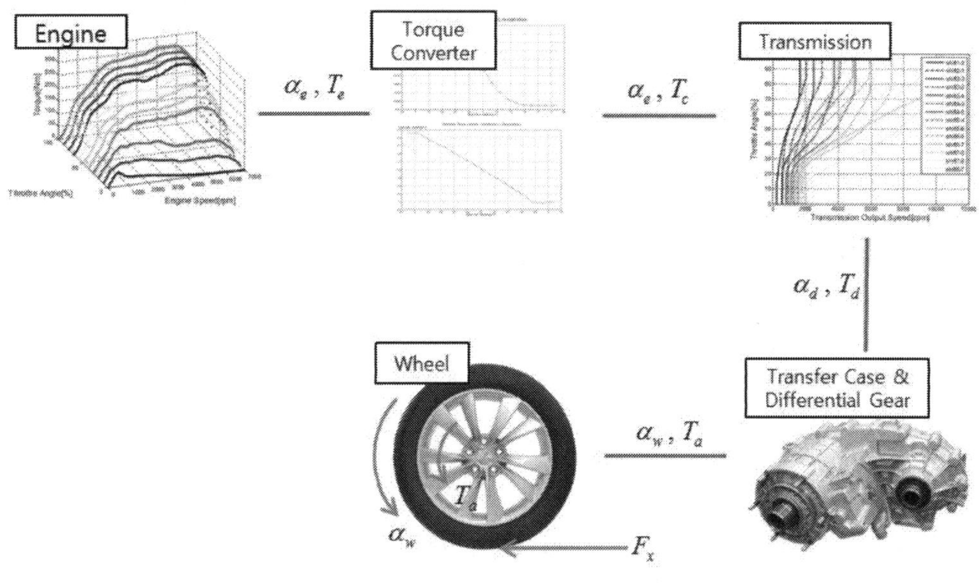

Figure 1. Powertrain Power Flow Diagram

$$I_e\alpha_e = T_e - T_c \tag{1}$$

$$I_t\alpha_e = T_c - \frac{1}{N_t}T_d \tag{2}$$

$$I_d\alpha_d = T_d - \frac{1}{N_f}T_a \tag{3}$$

$$I_w\alpha_w = T_a - rF_x \tag{4}$$

$$F_x = [-I_w\alpha_w - N_{tf}{}^2(I_e + I_t)\alpha_w + N_{tf}T_e - N_f{}^2 I_d\alpha_w]/r \tag{5}$$

Wherein, $\alpha_e = N_t\alpha_d$, $\alpha_d = N_f\alpha_w$, $\alpha_e = N_{tf}\alpha_w$, I_e is engine inertia[kgm^2]; α_e is angular acceleration of engine (rad/s); T_e is engine torque(Nm); T_c is torque of torque converter (Nm); I_t is transmission inertia (kgm^2); N_t is gear ratio of transmission (-); T_d is differential Torque (Nm); I_d is differential gear inertia (kgm^2); α_d is angular acceleration of differential gear (rad/s); N_f is gear ratio of differential gear (-); T_a is wheel torque (Nm); I_w is wheel inertia (kgm^2); α_w is angular acceleration of wheel (rad/s); r is wheel effective radius (m); F_x is traction force (N).

Figure 1 is a process of power transfer from the engine to wheels [4]. As seen in figure 1, many aspects should be considered for powertrain system modelling: engine map, torque converter properties, transmission scheduling map, type of transfer case, gear ratio of differential gear, wheel torque, etc. And acceleration test is needed with varying throttle input to find these powertrain parameters. In this modelling of powertrain system, parameters are adapted, and the developed model is verified.

2.2 Vehicle Lateral Dynamics

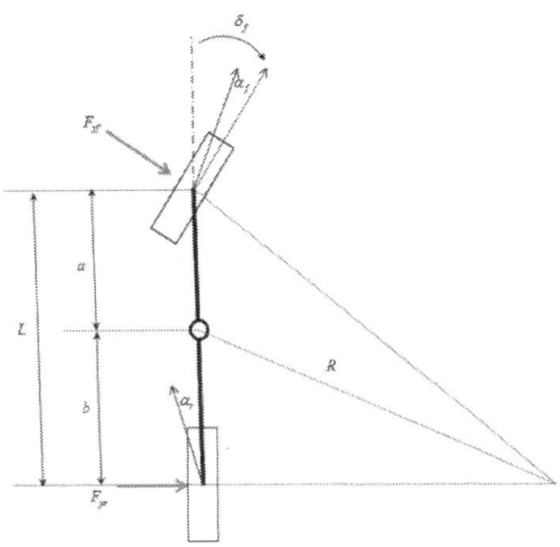

Figure 2. Lateral Vehicle Dynamics (Bicycle Model)

$$F_{yf} = \frac{W_f V^2}{gR}, \ F_{yr} = \frac{W_r V^2}{gR} \tag{6}$$

$$\alpha_f = \frac{F_{yf}}{C_{\alpha f}}, \ \alpha_r = \frac{F_{yr}}{C_{\alpha r}} \tag{7}$$

$$\delta = 57.3\frac{L}{R} + \alpha_f + \alpha_r \tag{8}$$

Wherein, δ is steering angle from driver (deg); L is length of the vehicle wheelbase (m); R is road radius of curvature (m); α_f, α_r are slip angle of front and rear tire (deg); F_{yf}, F_{yr} are lateral force of front and rear tire (N); $C_{\alpha f}$, $C_{\alpha r}$ are cornering stiffness of front and rear tire (N/deg); W_f, W_r are weight of front and rear (N); V is longitudinal velocity (m/s); g is acceleration of gravity (m/s^2).

Figure 2 is a lateral dynamics described by bicycle model. Formulas (6)~(8) are about lateral dynamics. It tells that there exist slip angle which opposes steering angle in over certain longitudinal velocity. Formula (7) shows that slip angle is a function of lateral force and cornering stiffness, and formula (6) shows that lateral force is a function of weight, curvature and longitudinal velocity. Therefore, as seen in lateral dynamics formula (8), because of prevailing constant values, considering fixed parameters is most important work in lateral modelling.

2.3 Vehicle Longitudinal Dynamics

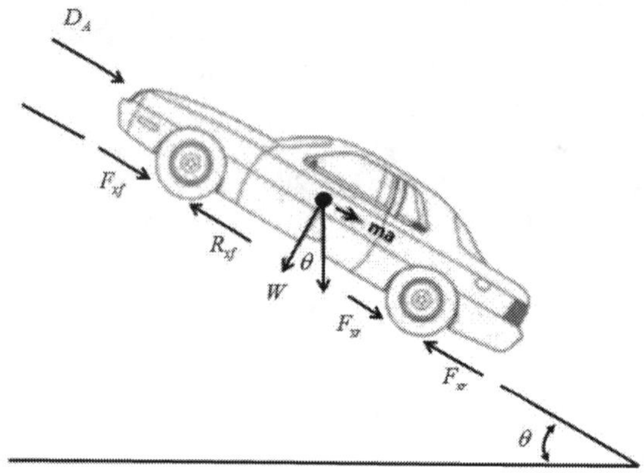

Figure 3. Longitudinal Vehicle Dynamics

$$F_x = F_{xf} + F_{xr} \tag{9}$$

$$R_x = R_{xf} + R_{xr} \tag{10}$$

$$D_A = \frac{1}{2}\rho C_d A V^2 \tag{11}$$

$$\frac{W}{g}a_x = F_{xf} + F_{xr} - D_A - (R_{xf} + R_{xr}) - W\sin\theta \tag{12}$$

Wherein, F_x is traction force (N); R_x is rolling resistance (N); D_A is drag force (N); a_x is vehicle acceleration (m/s^2); θ is road slope (rad); ρ is atmospheric pressure (Pa); C_d is form factor (-); A is projection surface area (m^2).

Figure 3 shows that longitudinal dynamics is composed of traction force and the opposition forces. Formula (9) is the traction force of a vehicle, and formula (10) and (11) are the opposition forces. Formula (12) is the longitudinal dynamics formula considering these forces. In this longitudinal dynamics formula, acceleration terms are dominant variables. Therefore, sensitive variation should be predicted in this dynamics.

3 High Precision Model of an Intelligent Torque Transfer System AWD

This simulation environment is proposed for a newly developed AWD system. AWD system is tested by dynamometer test which gives a transfer torque with varying input of clutch pressure. AWD system is composed of hydraulic piston, clutch pack, gear set and return spring. Therefore, each part is considered with a commercial tool AMESim as shown in Figure 4.

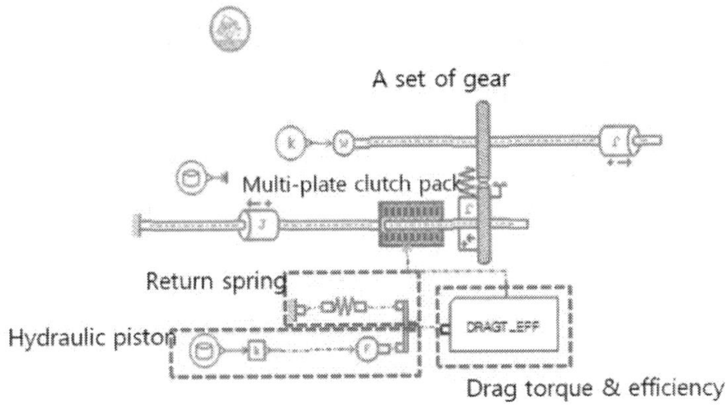

Figure 4. Formula based AWD system model

$$R_c = \frac{2}{3} \times \frac{D_{out}^{\ 3} - D_{in}^{\ 3}}{D_{out}^{\ 2} - D_{in}^{\ 2}} \tag{13}$$

$$F_{piston} = P \times A \tag{14}$$

$$F_{axial} = F_{piston} - F_{spring} \tag{15}$$

$$T_c = \mu_c \times R_c \times F_{axial} \times n_c \tag{16}$$

Wherein, D_{out} is clutch outer diameter (mm) ; D_{in} is clutch inner diameter (mm) ; R_c is effective radius of clutch (mm) ; P is piston pressure (bar) ; A is area of piston (m^2) ; F_{piston} is piston force (N) ; F_{spring} is return spring force (N) ; F_{axial} is axial force of clutch (N) ; T_c is clutch Torque (Nm) ; n_c is clutch contact surfaces ; μ_c is friction coefficient.

The reason why AWD system is considered specifically is that development object is operating by hydraulic power. And AMESim is a superior tool to analyse the hydraulic power. This tool is based on theoretical formulas [5]. Formula (13) is related to clutch capacity, formula (14) is related to piston, formula (15) is related to force equilibrium and formula (16) is related to clutch transfer torque. For verification of this modelling, the input conditions for the dynamometer test is used as an input conditions for the correlation analysis of the developed model.

4 Integrated Simulation Environment

This integrated simulation environment is developed based on the test results of vehicle integrated with AWD module. Figure 5 is a test vehicle for the integrated vehicle test. The vehicle test follows the protocol of acceleration test and slalom test. The acceleration test is for verification of longitudinal vehicle modelling, and the slalom test is for verification of lateral vehicle modelling. Because this simulation environment's goal is for verification of control logic, in this test, AWD module is controlled with fully engaged (4WD) mode and fully disengaged (2WD) mode.

Figure 5. Integrated Vehicle Test

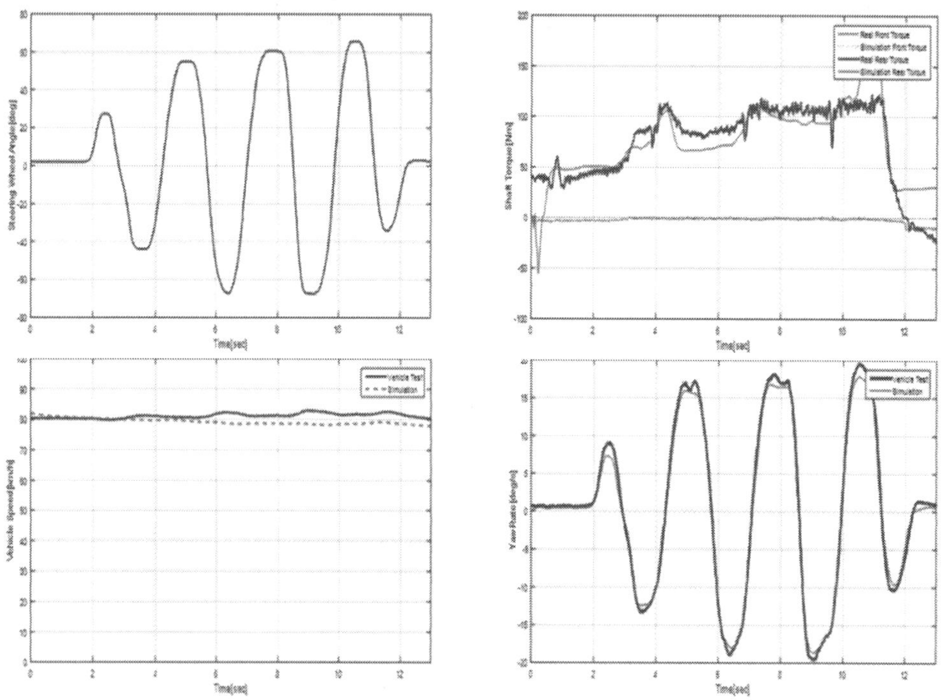

Figure 6. Lateral Simulation Model Verification

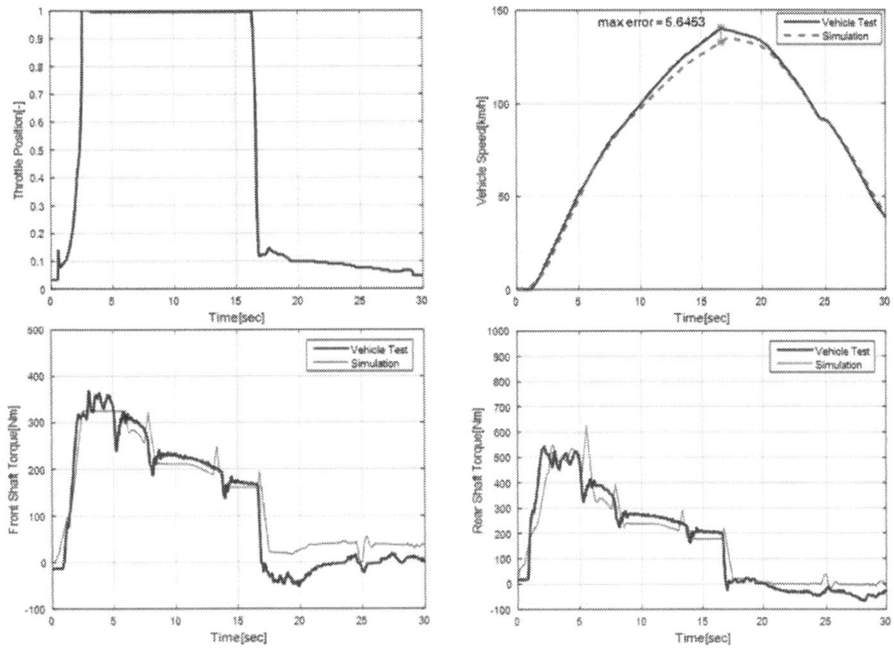

Figure 7. Longitudinal Simulation Model Verification

Figure 6 and 7 are comparison between integrated vehicle test and integrated simulation environment. Generally lateral vehicle modelling verification is conducted by comparison of lateral acceleration and yaw rate [6], and figure 6 tells that correlation is over than 99%. Moreover, longitudinal vehicle modelling verification is conducted by comparison of longitudinal velocity, and figure 7 shows that the correlation is over than 99% [7]. In conclusion, high precision vehicle simulation environment for AWD control logics is developed.

References

1. C. Senatore, C. Sandu, M.M. T*orque distribution influence on tractive efficiency and mobility of off-road wheeled vehicles* (Journal of Terranmechnics, 2011)
2. T. D. Gillespie, *Fundamentals of Vehicle Dynamics,* (SAE, 1992)
3. Thomas E. Scott, Power Transmission (Prentice Hall, 2000)
4. J. Reimpell, H. Stoll, J. W. Betzler, The Automotive Chassics : Engineering Principles 2nd Edition (Butterworth-Heinemann, 2001)
5. Katsuhiko Ogata, System Dynamics (Prentice Hall, 2004)
6. U. Kiencke, L. Nielsen, Automotive Control Systems : For Engine, Driveline and Vehicle 2nd Edition (Springer, 2004)
7. Julius S. Bendat, Allan G. Piersol, Random Data : Analysis and Measurement Procedures (WILEY, 2010)

MATEC Web of Conferences **167**, 02012 (2018) https://doi.org/10.1051/matecconf/201816702012
IC4M & ICDES 2018

Design of Hydraulic Bushing and Vehicle Testing for Reducing the Judder Vibration

Youngman Kim[1], Sangjin Jeong[1], Van-Quyet Nguyen[1], Kwangsuck Boo[2] and Heungseob Kim[2,a]

[1]*Graduate school of Mechanical Engineering, Inje University, 50834 Gimhae, Gyoungnam, Korea*
[2]*HSV-TRC Center, Inje University, 50834, Gimhae, Gyoungnam, Korea*

Abstract. Generally, judder vibration is a low-frequency vibration phenomenon caused by a braking force imbalance that occurs when a vehicle is lightly decelerated within a range of 0.1 to 0.2g at a speed of 120 to 60 km/h. This comes from the change in the brake disk thickness (DTV), which is mainly caused by the side run-out (SRO) and thermal deformation. The adoption of hydro-bushing in the low arm G bushings of the vehicle front suspension has been done in order to provide great damping in a particular frequency range (<20Hz) in order to prevent this judder vibration from being transmitted to the body. The hydro bushing was formulated using a lumped parameter model. The fluid passage between the two chambers was modelled as a nonlinear element such as an orifice, and its important parameters (resistance, compliance) were measured using a simplified experimental setup. The main design parameters are the ratio of the cross-sectional area of the chamber to the fluid passage, the length of the fluid passage, etc., and their optimal design is such that the loss angle is greater than 45 ° in the target frequency range of 10 to 20 Hz. The hydro bushing designed for reducing the judder vibration was prepared for the actual vehicle application test and applied to the actual vehicle test. In this study, the proposed hydro bushing was applied to the G bushing of the low arm of the front suspension system of the vehicle. The loss angle of the manufactured hydro bushing was measured using acceleration signals before and after passing through the bushing. The actual vehicle test was performed on the noise dynamometer for the performance analysis of the judder vibration reduction.

1 Introduction

Generally, in vibration theory, it is known that high stiffness and large damping are easy to cut off low frequency and large displacement vibration, low stiffness and small damping are advantageous for high frequency and small displacement vibration. The conventional rubber bushing meets the contradictory requirements of sufficient attenuation for large displacement vibrations in the rigid body mode below 100 Hz and high decoupling of the structural noise in the high frequency range. The requirements of this contradictory bushing design have required the development of hydrobushing which can enhance the damping force through the fluid and control the stiffness and damping in the frequency range of interest. The principle of fluid damping technology is already known in engine mount technology, but there are constraints in design and material selection for application in chassis

[a] Corresponding author : mechhsk@inje.ac.kr

components. Nevertheless, the suspension application of the hydro-bushing was applied to the lower arm G bushing of the front suspension in Fig.1 to reduce the steering wheel judder vibration due to the braking force imbalance. [1~2]

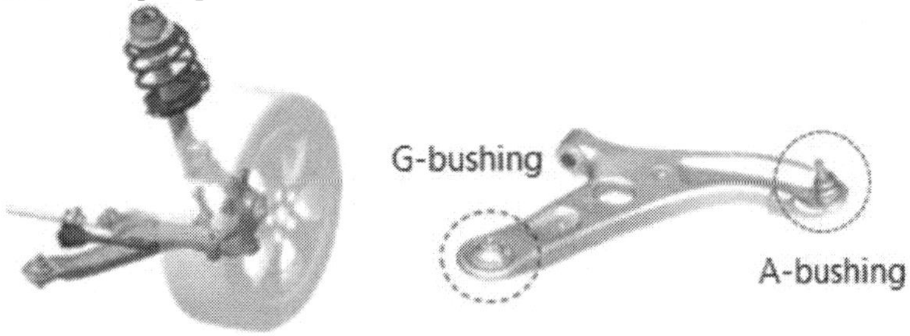

Figure 1. G bushing position of lower arm in the MacPherson suspension

Judder vibration is a low-frequency vibration phenomenon in the range of 10 ~ 20Hz caused by unbalance of braking force when the vehicle is braking at a light deceleration of 0.1 ~ 0.2g at a speed of 120km/h, which is a highway driving speed. Judder is known to be caused by disc thickness variation (DTV), which is caused by side run-out (SRO) during manufacturing and thermal deformation resulting from the braking process. Compared to conventional rubber bushings, hydro bushings can have high viscous damping in certain frequency ranges, but have amplitude and frequency dependencies. The design core of the hydro bushing is to determine the peak frequency and magnitude of the loss angle. The main design parameters are rubber spring constant, fluid characteristics, and the shape of the chamber and inertia track.

In this study, the frequency characteristics of the hydro bushing applied to the G bushing of the front suspension low arm were designed to effectively reduce the judder vibration. Also, the performance of the hydro bushing was verified by performing the actual vehicle test on the noise dynamo using the vehicle with the hydro bushing manufactured.

2 Characteristic design of hydrobushing

The lumped parameter model (LP model) as shown in Fig.2 is widely used to predict the characteristic analysis of the hydrobushing [3~4]. The hydro bushing consists of two chambers connected by fluid passages, rubber elements and a metal sleeve on the inside and on the opposite side. The rubber elements consist of a Kelvin-Voigt model with frequency-invariant stiffness K_R and viscous damping coefficient C_R. The two chambers are defined by compliance C_1, C_2 controlled by the rubber portion and the fluid portion including the air. The fluid passage is modeled by a resistance element R_i and an inertance element I_i.

When dynamic displacement is applied to the inner sleeve of the hydrobushing, the dynamic pressure of both chambers is

$$C_1\dot{p}_1(t) = -A_1\dot{x}(t) - q_i(t) \quad (1)$$

$$C_2\dot{p}_2(t) = A_2\dot{x}(t) + q_i(t) \quad (2)$$

Where A_1, A_2 represents the effective pumping area of each chamber, and q_i represents the volumetric flow rate of the fluid flowing through the fluid passages connecting the chambers. The conservation of momentum between the chamber and the fluid passage can be expressed as Equation (3).

$$p_1(t) - p_2(t) = I_i\dot{q}_i(t) + R_iq_i(t) \quad (3)$$

The forces transmitted by the two paths to the outer sleeve are as follows.

$$f_T(t) = [C_R\dot{x}(t) + K_Rx(t)] + [A_2p_2(t) - A_1p_1(t)] \quad (4)$$

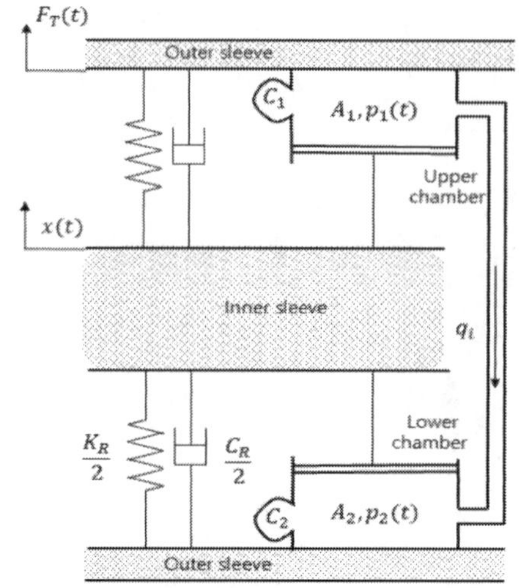

Figure 2. Lumped parameter model of hydrobushing

The resistance $R_i(q_i)$ of the fluid passage can be modeled nonlinearly using the same orifice model as Eq. (5) [5] .

$$R_i(q_i) = \frac{\Delta p}{q_i} = \lambda \frac{\rho}{2A_i C_d^2} |q_i(t)| = c_R |q_i(t)| \tag{5}$$

Where C_d is the expansion coefficient, A_i is the cross-sectional area of the fluid passage, and c_R is the resistance coefficient. λ is a coefficient that adjusts the difference between experimental and theoretical models. The hydro bushing with nonlinear characteristics for sinusoidal input is tried to obtain the transfer function for the dynamic stiffness and loss angle. If the fluid channel is made nonlinear as shown in Eq. (5), Eq. (3) becomes as shown in Eq. (6). If the two chambers have the same cross-sectional area, Eq. (7) can be obtained by combining Eq. (1) and Eq. (2).

$$p_1(t) - p_2(t) = I_i \dot{q}_i(t) + c_R |q_i(t)| q_i(t) \tag{6}$$

$$I_i \ddot{q}_i(t) + 2c_R q_i(t) \dot{q}_i(t) sgn[q_i(t)] + k q_i(t) = \alpha_x \dot{x}(t) \tag{7}$$

Where k = $1/C_1 + C_2$, α_x = - $[A_1/C_1 + A_2/C_2]$. Assuming that the input displacement of the hydrobushing inner sleeve is $x(t) = X \sin(\omega t - \varphi)$, and the flow rate of the resulting hydraulic passage is $q(t) = Q_I \sin(\omega t)$, then Eq. (7) is

$$\begin{aligned} kQ_1 \sin(\omega t) + c_R Q_1^2 \omega \sin(2\omega t)\, sgn[\sin(\omega t)] - I\omega^2 Q_1 \sin(\omega t) \\ = \alpha_x X\omega[\cos(\omega t) cos\varphi + \sin(\omega t) \sin \varphi] \end{aligned} \tag{8}$$

Eq. (8) can be simplified by using the Fourier series for $\sin(2\omega t)\, sgn[\sin(2\omega t)]$, as shown in Eq. (9).

$$\varphi = tan^{-1} \frac{\sqrt{2}(k - I\omega^2)}{\sqrt{-(k - I\omega^2)^2 + \sqrt{(k - I\omega^2)^4 - 4\left(\frac{8\alpha_x X c_R \omega^2}{3\pi}\right)^2}}} \tag{9}$$

$$Q_1^2 = \frac{-(k - I\omega^2)^2 + \sqrt{(k - I\omega^2)^4 - 4\left(\frac{8\alpha_x X c_R \omega^2}{3\pi}\right)^2}}{\frac{2(8c_R\omega)^2}{(3\pi)^2}} \tag{10}$$

The flow rates in Eq. (9) and Eq. (10) are as follows.

$$Q(\omega) = Q_1 cos\varphi + iQ_1 sin\varphi \tag{11}$$

And the dynamic stiffness and loss angle can be obtained by Eq. (4).

$$K_d(\omega) = K_R + i\omega C_R + \gamma + \left(\frac{A_1}{C_1} + \frac{A_2}{C_2}\right)\frac{Q(\omega)}{iX\omega} \tag{12}$$

In Eq. (12), the magnitude and phase of $K_d(\omega)$ correspond to the tangent and loss angles, respectively.

3 Simulation for predicting the performance characteristics

A simulation for predicting the dynamic stiffness and loss angle of hydro bushings with nonlinear elements such as fluid passages was performed. The developed hydro bushing is designed to reduce judder vibration caused by instability caused by friction during braking process. It is designed to have a loss angle of 30 ° or more in the frequency range of 10 ~ 20Hz.

3.1 Design parameters

The developed hydro bushing is shown in Fig. 3. The damping force generated during the flow of the fluid due to the long fluid passage is utilized. Table 1 shows the main design parameters of the hydrobushing. The design is designed to maximize the loss angle in the frequency range of the judder oscillation through the optimal combination of fluid passages and chambers.

Figure 3. Developed hydrobushing with long flow passage

Table 1. Geometric and material properties

Part	Parameter	Value
Rubber	Spring constant, [N/mm]	430
	Damping coefficient, [Ns/mm]	0.1
Inertia track	Length, [mm]	532.1
	Cross sectional area, [mm^2]	8.4
Chamber	Pumping area, [mm^2]	5609
Fluid	Viscosity, [Ns/mm^2]	2.5e-8
	Density, [Ns^2/mm^4]	1.1e-9

3.2 Frequency characteristics

The characteristics of the hydrobushing were obtained by increasing the sinusoidal amplitude of 1 mm to 5~ 50Hz at intervals of 1Hz and measuring the dynamic stiffness and loss angle at each frequency. The results are compared with those predicted by Eq. (12), and the comparison results are shown in Fig. 4. In Fig. 4(a), the predicted results for the dynamic stiffness differ from the experimental values in the region above 15 Hz, but the prediction of the loss angle in Fig. 4(b) accurately predicts the position and the value of the maximum loss angle. Therefore, it was confirmed that the algorithm for predicting the hydro-bushing dynamic characteristics was accurately created.

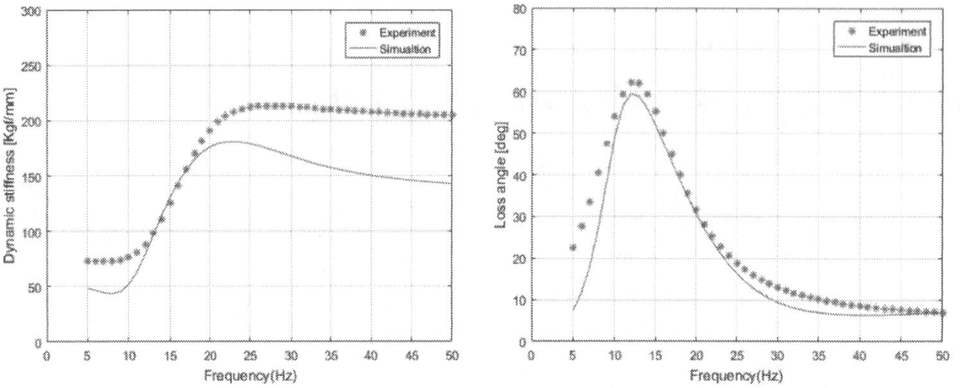

a) Dynamic stiffness b) Loss angle

Figure 4. Comparisons between experimental and predicted results of the dynamic characteristics

4 Conclusions

In order to predict the accurate dynamic performance of the hydrobushing modeled by the lumped parametric model, this study conducted a measurement experiment on the resistance of the fluid passages and chamber compliance with nonlinear characteristics and reflected them in the prediction algorithm. It is found that the predicted results for the dynamic stiffness and loss angles can predict the peak frequency and magnitude accurately with experimental results. Especially, it was confirmed that a large loss angle value was obtained in a desired frequency range so as to reduce the judder vibration.

ACKNOWLEDGEMENT

This research was supported by the Ministry of Trade, Industry & Energy (MOTIE), Korea Institute for Advancement of Technology (KIAT) through the Encouragement Program for The Industries of Economic Cooperation Region. (R0004519). This research was supported by The Leading Human Resource Training Program of Regional Neo industry through the National Research Foundation of Korea (NRF) funded by the Ministry of Science, ICT and future Planning. (2016H1D5A1910490)

References

1. W. Sauer, Y. Guy, SAE 2003-01-1475, (2003)
2. L. Min, T. Yougjun, and W. Enjung, SAE 2010-01-0506, (2010)
3. T. Chai, J. T. Dreyer, and R. Singh, Mechanical Systems and Signal Processing, 56, 92 (2015)
4. W. B. Shangguan, C. Xu, Journal of Vibration and Shock, 26, 7 (2007)
5. T. Chai, R. Singh, and J. Dreyer, SAE 2013-01-1924 (2013)

Geometry Modification of Helical Gear for Reduction of Static Transmission Error

Jeonghyun Park[1] Changjun Seo[2], Kwangsuck Boo[2] and Heungseob Kim[2,1]

[1]*Graduate school of Mechanical Engineering, Inje University, 50834 Gimhae, Gyoungnam, Korea*
[2]*HSV-TRC Center, Inje University, 50834, Gimhae, Gyoungnam, Korea*

Abstract. Gear systems are extensively employed in mechanical systems since they allow the transfer of power with a variety of gear ratios. So gears cause the inherent deflections and deformations due to the high pressure which occurs between the meshing teeth when transmit power and results in the transmission error. It is usually assumed that the transmission error and variation of the gear mesh stiffness are the dominant excitation mechanisms. Predicting the static transmission error is a necessary condition to reduce noise radiated from the gear systems. This paper aims to investigate the effect of tooth profile modifications on the transmission error of helical gear. The contact stress analysis was implemented for different roll positions to find out the most critical roll angle in view point of root flank stress. The PPTE (peak-to-peak of transmission error) is estimated at the roll angles by different loading conditions with two dimensional FEM. The optimal profile modification from the root to the tip is proposed.

1 Introduction

Gear system is being widely used for the rotational speed reduction and acceleration as well as converting the operational direction. Tooth deflection and bending stress in root are generated when a pair of gear is meshing. It causes radiated noise from the gear case by transmission error which excites driving shafts and bearings mounted on them.

Akerblom [1] and Smith [2] found the correlation of noise induced by excitation of transmission error in the entire system. Smith studied about reducing gear transmission error according to contact ratio changes. Park [3] and Perret-Liaudet [4] classified transmission error into static transmission error (STE) and dynamic transmission error (DTE). Many traditional scientists studied method to reduce transmission error in gear system. MarkoviĆ and FranuloviĆ [5] studied on adjust of contact ratio of gear and Rigaud and Barday [6] implemented study on reduction of transmission error by modifying geometry of gear. In addition, Paul and Bhole [7] researched on reduction of transmission error by minimizing the contact stress at the end of tooth flank. Sivakumar [8] researched load transmission in high contact ratio gear system and Sawalhi Nader [9] studied about simulation method in rotation bearing fault phenomenon by rotation shaft vibration. Endo Hiroaki and Sawalhi Nader [9] studied about bearing vibration measurement simulation model at rotating gears. In this paper gear simulator was designed to measure rotation

[1] Corresponding author : mechhsk@inje.ac.kr

angle of pinion and gear and estimate transmission error via difference between both angles.

In this study, method on reduction of general helical gear was researched by finite element analysis. First of all, data of tooth tip deflection was collected in rotational situation of gear pair. Based on it, gear profile modification curve was decided. Finite element analysis model was generated by commercial software like as Hypermesh and Abaqus, to carry out finite element analysis in different loading torque conditions. Then, gear contact characteristics such as transmission error and bending stress was compared and analyzed.

In addition, characteristics analysis of transmission error along the loading torque was studied by experimental method. The test bench was set in order to measure transmission error, and frequency characteristics of transmission error was compared and analyzed in different three loading conditions.

2 FEA for static transmission error

2.1 Helical gear design

Target gear in Fig.1 was designed for diesel engine balance unit operating at maximum 450Nm. gear specification is shown as Table 1, and elastic modulus and poisson's ratio were applied as 208GPa and 0.3. Each gear model consists of finite elements of570 thousands.

Table 1. Helical gear design parameters for gear profile modificatio

Description	Each gear
Number of teeth	43
Module(mm)	1.5
Helix angle(deg)	26
Pressure angle(deg)	15
Pitch diameter(mm)	71.8
Face width(mm)	12
Addendum(mm)	1.5
Dedendum(mm)	2.025

Figure 1. Helical gears for gear profile modification

2.2 Finite Element Analysis conditions

Finite element analysis conditions for model that its profile was modified is that rotational speed of driving gear was applied as shown in Fig.2 and each loading torque, 450Nm, 1000Nm, and 150Nm, was applied to driven gear. Friction coefficient is 0.1, and translational and rotational movement except for rotational axis were constrained to move.

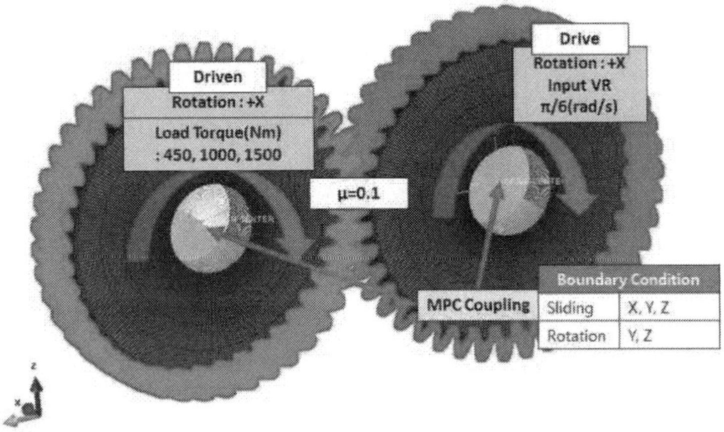

Figure 2. Boundary condition for gear profile modification

3 Simulation for prediction of static transmission error

3.1 Transmission error reduction along the tooth profile modification

Contact patterns varies along the contact ratio (CR) as gear pair is rotating. Contact ratio is calculated by Eq. (1), so this gear pair has approximately 3.7 contact ratio. Therefore, three or four teeth contact patterns are periodically shown Fig.3.

$$\mathrm{CR} = \frac{\sqrt{(r_1 + a)^2 - r_1^2 \cos^2\alpha}}{\pi m \cos\alpha} + \frac{\sqrt{(r_2 + a)^2 - r_2^2 \cos^2\alpha} - (r_1 + r_2)\sin\alpha}{\pi m \cos\alpha} \cong 3.7 \quad (1)$$

Where r_1, r_2 is pitch radius of pinion and gear, a is addendum width, α is pressure angle, and m is module. Paul and Bhole [7] researched on gear profile modification method by minimizing the contact stress at the end of tooth flank and they applied some relief at teeth tip. When gear pair rotates, deflection of ith tooth is maximum. When gear pair rotates, deflection of ith tooth is maximum. Then, deflection characteristics is not straight line but parabolic line until meshing finish, because i+1~3th tooth mesh simultaneously. As a result, profile was modified as curve from tooth tip to root, 39.8μm and 38.4μm respectively. In this paper gear profile was modified as involute curve. Both side of a teeth has same profile curve.

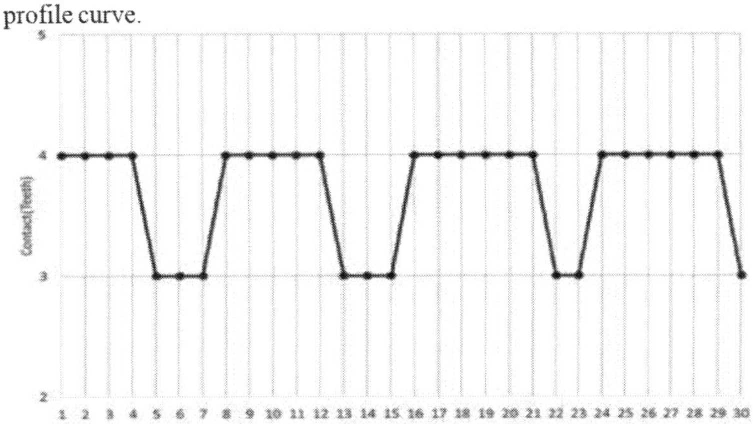

(a) Contact teeth number of rotation angle

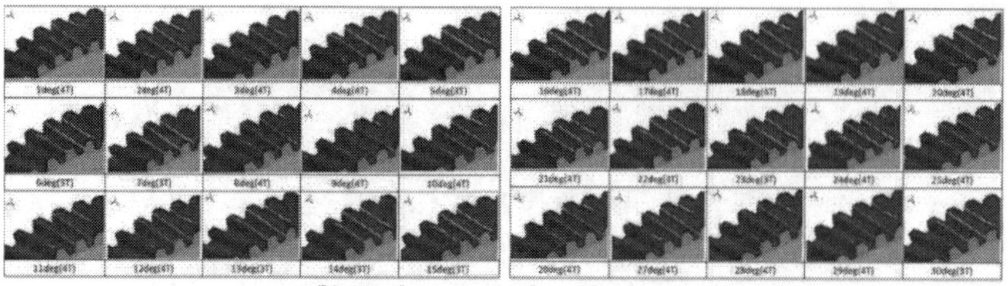

(b) Teeth contact of rotation angle

Figure 3. Contact pattern

Transmission error for teeth modification model was calculated by Eq. (2), and it was reduced to maximum 23.5% and minimum 13.2% via profile modification in Fig.4 and Table 2. The formulation means difference between drive gear and driven gear on slow rotation velocity as under 20RPM. In formula, thetas mean rotation angle(rad) and R means pitch radius. It is possible to reduce gear rotation vibration characteristics as transmission error by applying tip relief under teeth deflection propensity. Gear profile modification as tip relief that is about 40μm is in manufacture error region. So it is very important method to manufacture gear tooth.

$$\text{Trans mission error} = \theta_{gear} - \frac{R_{pinion}}{R_{gear}}\theta_{pinion} \tag{2}$$

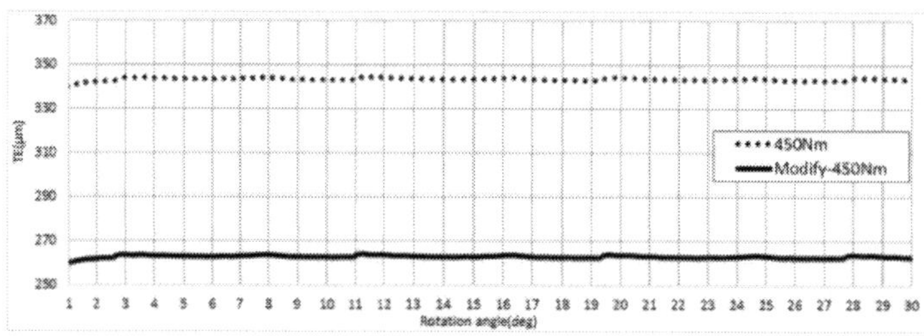

(a) Torque 450Nm

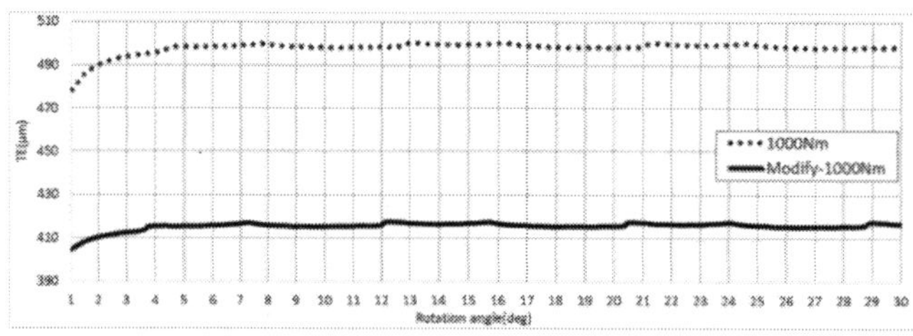

(b) Torque 1000Nm

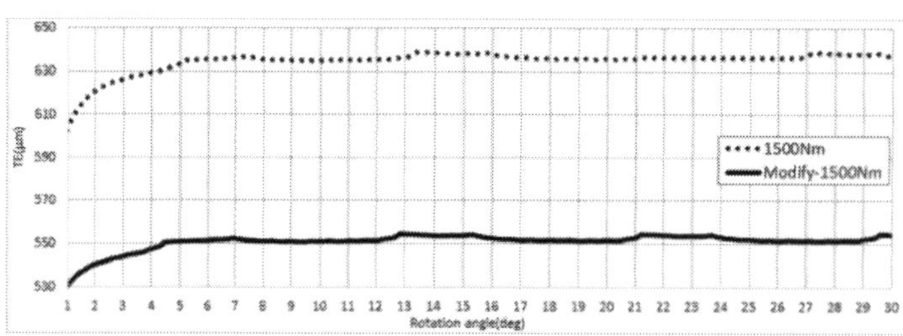

(c) Torque 1500Nm

Figure 4. Transmission error via gear profile modification

Table 2 Root stress of gear profile modification

Transmission error		Root stress(MPa)
450Nm	Original	1276
	Modify	1194
1000Nm	Original	2760
	Modify	2552
1500Nm	Original	4032
	Modify	3262

Load sharing was calculated at each rotational angle as shown Eq. (3). Contact force which is a target tooth flank versus total contact force on contact area was calculated for each rotational angle. It is found load sharing increased after teeth profile modification as shown Fig. 5. It has same definition as the above description for contact force dispersion. Particularly, when single tooth rotates, load sharing increased by 10~12% after 15~16 degree rotation that contact force is maximum. It means that difference of contact force of contact tooth was reduced before maximum contact force occurred, and it can be also considered that many teeth rotated with stronger contact force than before modifying tooth profile. As shown below load sharing, this is because i+1th tooth keeps up its meshing as soon as meshing of ith tooth reached maximum through tooth profile modification.

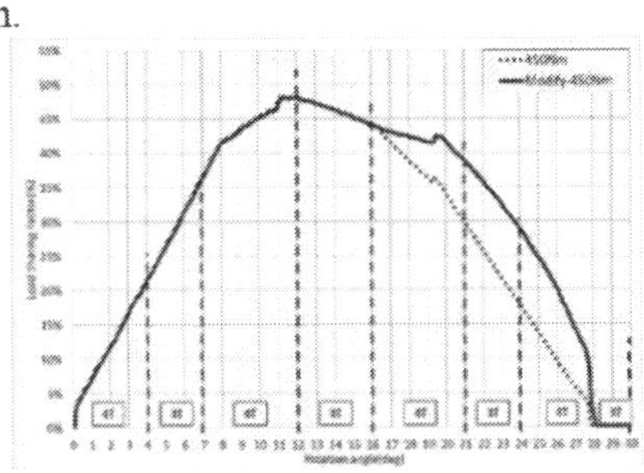

(a) Torque 450Nm

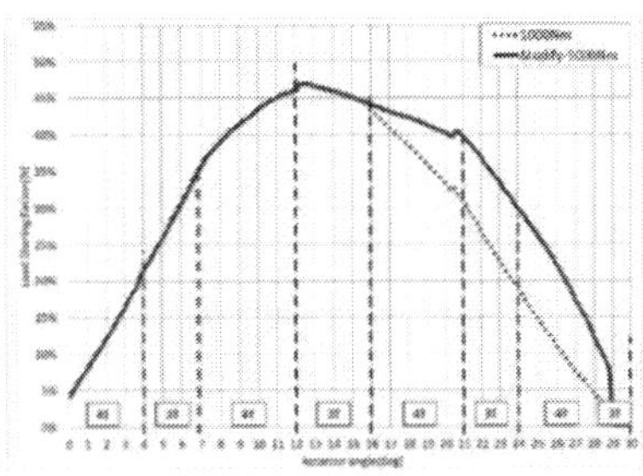

(b) Torque 1000Nm

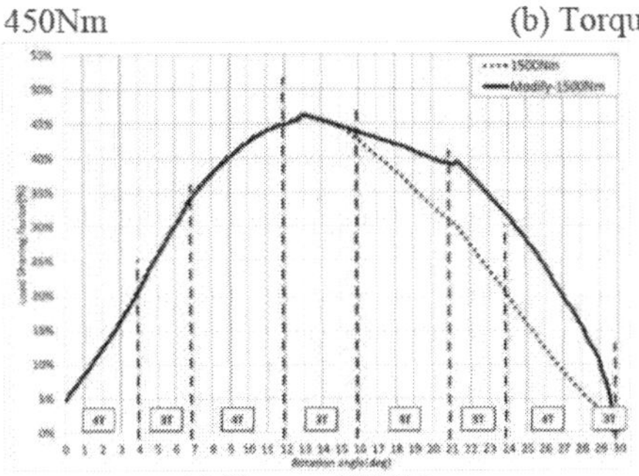

(c) Torque 1500Nm
(dotted line : original, solid line : modified)

Figure 5. Load sharing improvement by profile modification

4 Conclusion

In this study, we found the method to reduce transmission error in gear system. Gear profile modification method was applied based on tooth deflection. It was found that this method has better load sharing effect on reduction of transmission error and bending stress of tooth in standard loading torque (450Nm) and higher loading conditions. Therefore, these results definitely show gear profile modification method is effective to reduce transmission error. In further study, application method on correlation between transmission error with shaft vibration and tip relief by tooth profile modification will be researched.

Acknowledgement

This research was supported by The Leading Human Resource Training Program of Regional Neo industry through the National Research Foundation of Korea (NRF) funded by the Ministry of Science, ICT and future Planning. (2016H1D5A1911140).

This research was supported by The Leading Human Resource Training Program of Regional Neo industry through the National Research Foundation of Korea (NRF) funded by the Ministry of Science, ICT and future Planning. (2016H1D5A1910490).

References

1. M. Akerblom, *Volvo Construction Equipment Components* ABSE-631 85(2001)
2. J. D. Smith, *Gear Noise and Vibration, CRC Press,* (2003)
3. J. I. Park, D. H. Cho, *Trans. Korean Soc. Mech. Eng. A,* **28(9)**, 1359(2004)
4. J. C. Perret-Liaudet, E. A. Rigaud, and B. Nelain, SAE 2014-01-2090, (2014) 5. K. MarkoviĆ, M. FranuloviĆ, *UDC 621.833.15:62-233.5,* 19 (2011)
6. E. Rigaud, D. Barday, HAL-00121847, 1961 (1999)
7. I. D. Paul, G. P. Bhole, ICIEOM Spring conference, 197(2010)
8. M. Rameshkumar, G. Venkatesan, P. Sivakumar, AGMA Technical Resources, (2010)
9. Endo Hiroaki, Sawalhi Nader. *Gearbox Simulation Models with Gear and Bearing Faults* (2012)

Heat Transfer Characteristics of Flooded Type Evaporator for Seawater Cooling System

Jung-In Yoon[1], Chang-Hyo Son[1], Hyung-Min Han[2], Kwang-Seok Lee[2,a], Jung-Mok Lee[3], In-Duck You[3]

[1]*Department of Refrigeration and Air-conditioning Engineering, College of Engineering, Pukyong National University, San 100, Yongdang-dong, Nam-gu, Busan, 608-739, South Korea*
[2]*Graduate School, Pukyong National University, San 100, Yongdang-dong, Nam-gu, Busan, South Korea*
[3] *Korea Seal Tech, 26, Sinhosandan 3-ro 88beon-gil, Gangseo-gu, Busan, South Korea*

Abstract. In these days, many studies on ocean fisheries have been conducted for developing their own industries. Especially, the efforts of using seawater cooling system were actively conducted to keep the caught fisheries fresh, increase the catches, and decrease the manpower. The purpose of this study is to suggest design criteria of seawater cooling system using the flood-type evaporator. In this study, the seawater cooling system using flooded-type evaporator was manufactured as a prototype capacitating 15kW. This study examined performance of the seawater cooling system and then compared the performance with the system capacitating 163kW and 238kW, which is actually being loaded on the fishery boats. In addition, heat transfer characteristics of the flooded-type evaporator used in the equipment are analysed.

1 Introduction

Due to production decrease with purse seine fishery and unstable supply of fish, it is suggested to find ways to improve freshness of fish to increase fishery amounts and decrease manpower. It is important to keep temperature constant in fish storage in order to improve freshness of the caught fish. Hence, many researches on ice making machine and seawater cooling system have been actively conducted. Normally, there are two ways to refrigerate the fish storage. The first one is to use fresh water ice, which is very popular way. However, this may damage the fish due to the weight and sharpness of ice, and also it is difficult to adjust salinity in the storage. The second way is to utilize seawater cooling system on the boats and it doesn't damage the fish. This also can lengthen the fishing hours and decrease manpower to keep the caught fish fresh. Previous research on seawater cooling system used flooded-type evaporator to enhance the performance of the system. These relevant studies are as follows.

Yoon et al. [1] evaluated performance characteristics of flooded type evaporator for fishery vessels using R22 and R134a. They analysed heat transfer coefficient of manufactured tube with respect to heat flux, chilled water temperature and velocity of chilled water etc. As a result, the study suggested guideline of chilled seawater velocity, temperature and heat transfer coefficient of manufactured tube for the flooded type evaporator.

[a] Corresponding author : suwe1862@gmail.com

Jung et al. [2] conducted the heat transfer coefficient experiment with respect to working fluid. Also, the study compared heat transfer characteristics of plain tube, 26fpi low-fin tube, Turbo-B tube, and Thermoexcel-E tube with regard to heat flux by using electric heater.

Kim [3] examined the effect of oil on boiling in flooded-type evaporator by experiment. As a result, when oil occupies less than 1% of the portion, overall heat transfer efficiency decreased dramatically. However, the overall heat transfer coefficient decreased gradually when oil occupies over 1% of the portion.

Based on these studies, applying seawater cooling system on the boat is more efficient to keep temperature constant in fish tank. This research also analysed heat transfer characteristics with regard to the tube materials of flooded type evaporator through the experiments. In addition, the research examined the system performance of the seawater cooling system applied to the flooded type evaporator. Therefore, this research suggests that the heat transfer characteristics of flooded type evaporator need to be examined and designing the seawater cooling system applied flooded type evaporator is critically important.

2 Experiment and Analysis condition

2.1 Experimental apparatus

2.1.1 Flooded type evaporator

Fig.1 shows experiment apparatus of the flooded type evaporator applied to seawater cooling system and piping and instrumentation schematic diagram. Compositions of the flooded type evaporator apparatus are flooded type evaporator, double pipe type condenser, receiver and pump. Operating principle of the apparatus is as follows. The refrigerant at the receiver flows into the evaporator by pump, and evaporates by exchanging heat with the seawater inside the evaporator. The evaporated gaseous refrigerant flows into the double pipe type condenser located on the evaporator, and is condensed by cooling water.

Figure 1. Piping and instrumentation schematic diagram and Experimental apparatus of the flooded type evaporator.

2.1.2 Seawater cooling system

Fig. 2 shows experiment apparatus and piping and instrumentation schematic diagram of the seawater cooling system applied the flooded type evaporator. The apparatus are composed with the flooded type evaporator, open-type reciprocating compressor, condenser, receiver, expansion valve, oil recovery heat exchanger, surge drum, separator and oil return collector. Working principle of the system is as follows. Gaseous refrigerant with high temperature and high-pressure from the compressor outlet flows into the condenser. Liquid refrigerant condensed by the cooling water flows to the oil recovery heat exchanger, and evaporates at the evaporator. After the evaporator, the refrigerant flows to compressor inlet.

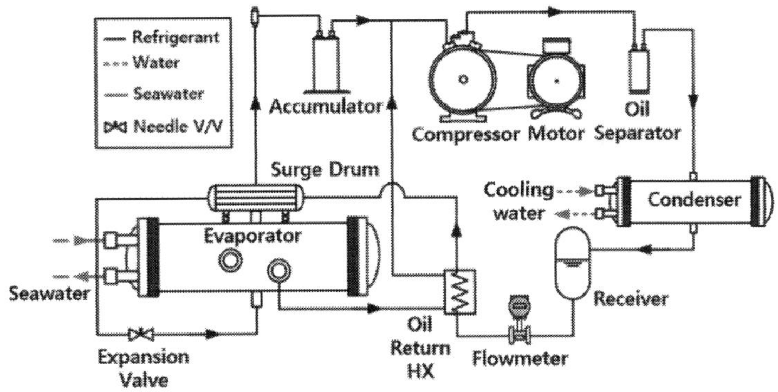

Figure 2. Schematic diagram of seawater cooling system.

2.2 Experimental and analysis condition

Table 1 shows analysis condition and the experimental condition of both evaporator heat transfer performance characteristics and the seawater cooling system. Working fluids are R134 and R22, and the experiment conditions are set up as follows. The seawater inlet and outlet temperature are set up as 5°C and 3°C respectively in the experiment of the seawater cooling system. The outlet temperature depends on kinds of fish in fish tank. This study assumed to refrigerate external blue colored fish in fish tank, so storage temperature is set up as 3°C by default. The cooling water temperature has range of temperature from 19°C to 29°C based on surface seawater mean temperature 25°C. Also, if compressor operates with same rotation speed for both refrigerants, compressor overload due to increase of condensation pressure. So in order to prevent compressor overload and maintain same condensation pressure for both refrigerants, the compressor rpm should be set up differently for refrigerant. Therefore, compressor rpm is set up as 40Hz for R22, 60Hz for R134a.

Table 1. Experimental and analysis condition.

Parameter	Flooded type evaporator	HYSYS simulation(15kW) [4]	Seawater cooling system(15kW)
Refrigerant [-]	R134a, R22	R134a, R22	R134a, R22
Chilled seawater inlet temperature [°C]	5~25	5	5
Chilled seawater outlet temperature [°C]	-	3	3
Evaporating temperature [°C]	-5~15	-	-
Cooling water inlet temperature [°C]	-	19~29	19~29
Velocity of chilled seawater [m/s]	1.8	-	-
Inverter frequency [Hz]	-	-	60, 40

2.3 Data analysis

The properties of the refrigerants used in this study are based on the REFPROP program of NIST [5], and the following equations were applied to determine the heat transfer characteristics of the flood-type evaporator.

First, Eq. (1) was applied to calculate the evaporation capacity. In the equation, the mass flow rate of the seawater flowing into the tube was multiplied by the temperature difference between the inlet and outlet. Eq. (2) was used to calculate COP of the seawater cooling system.

$$Q_e = G_{csw} C_{p,csw} (T_{csw,in} - T_{csw,out}) \tag{1}$$

$$\text{COP} = \frac{Q_e}{W} \tag{2}$$

Eq. (3) was used to calculate the overall heat transfer coefficient. Here, the evaporation capacity (Q_e) can be obtained from Eq.(2) above, and A_o can be obtained from the specification data of the tube.

$$U_o = \frac{Q_e}{A_o \cdot \Delta T_{LMTD}} \tag{3}$$

The overall heat transfer coefficient (U_o) of the flood-type evaporator can be obtained from Eq. (4), and the resistance of heat conduction (R_w) of the tube can be computed using Eq. (5).

$$\frac{1}{U_o A_o} = \frac{1}{h_i A_i} + R_w + \frac{1}{h_o A_o} \tag{4}$$

$$R_w = \frac{\ln(d_o/d_i)}{2\pi L k} \tag{5}$$

The heat transfer coefficient (h_i) of the seawater can be obtained from the Dittus-Boelter equations shown in Eq.(6)~(8), which illustrate the correlation applicable to the convection heat transfer of a single phase liquid with turbulent flow. The heat transfer coefficient h_o can be calculated using Eq.(4).

$$Nu_i = 0.023 Re^{4/5} Pr^{0.3} \tag{6}$$

$$Re_i = \frac{\rho v L}{\mu} \tag{7}$$

$$Nu_i = \frac{h_i L}{k} \tag{8}$$

3 Experiment result

3.1 Heat transfer coefficient experiment with respect to evaporating temperature

The flooded type evaporator experiment was conducted with the condition that inlet temperature and velocity of the chilled seawater were each 10°C, 1.8m/s. Also, R134a and R22 were used as working fluids. Aluminum-brass tube and copper-nickel tube were applied to this experiment, and the heat transfer coefficients of those were obtained. As a result, when evaporating temperature decreases, heat transfer coefficient tends to increase as shown in Fig. 3. Also, the heat transfer coefficient increased 76.6% on average as the evaporating temperature decreased from 5°C to -5°C. It was analyzed as the cause that increment of the evaporation capacity was larger than that of the mean temperature difference. Besides, difference of heat transfer coefficient for refrigerants was 0.35kW/m^{2}K when same material tube used. Difference of the coefficient for tube materials was 0.25kW/m^{2}K when same refrigerant used. These values were not presented largely, because the heat capacity was calculated by short tubes of 1m.

3.2 Evaporating and seawater outlet temperature with respect to cooling water temperature for flooded type evaporator

Fig. 4 shows the seawater outlet temperature and evaporating temperature with respect to cooling water temperature in flooded type evaporator experiment. The chilled seawater outlet temperature showed range from 2.2°C to 3.4°C for R134a, and the average value was 2.85°C. For R22, range of the temperature had from 2.8°C to 3.2°C and the mean temperature was 3.06°C. These range of the chilled

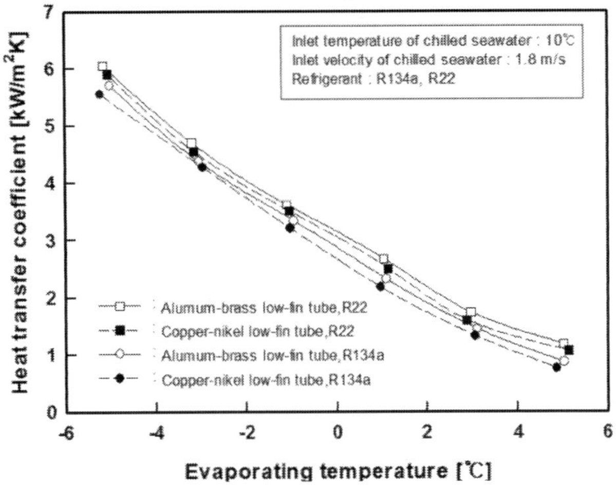

Figure 3. Heat transfer coefficient with regard to evaporating temperature for flooded type evaporator experiment.

seawater outlet temperature satisfied quantitative desired range from 2°C to 4°C. Furthermore, the evaporating temperature was set as saturation temperature from REFPROP program by evaporation pressure measured. Consequently, all values of the evaporating temperature satisfied the quantitative desired value of that under 0°C. In the case using R134a, minimum of evaporating temperature was -2.6°C and maximum of that was -1.8°C, and the mean value was -2.2°C. On the other hand, the case using R22 showed -5.01°C as value of minimum and -2.8°C as value of maximum. The average evaporating temperature for R22 was -3.9°C.

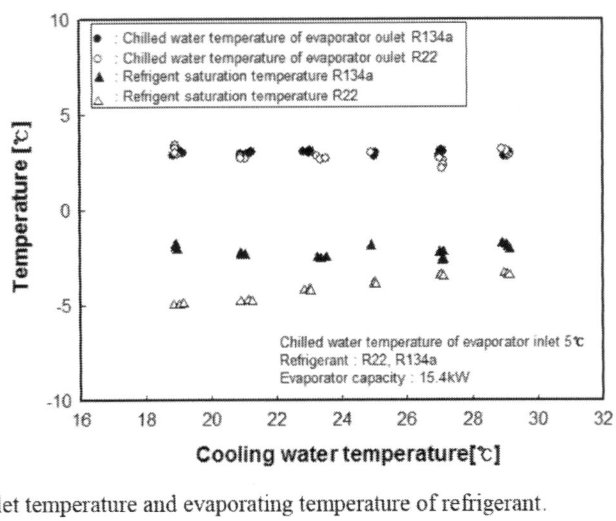

Figure 4. Seawater outlet temperature and evaporating temperature of refrigerant.

The COP of the 15kW seawater cooling system is shown at Fig. 5. When the cooling water temperature increases, evaporation capacity for R134a tended to increase slightly. Moreover, compressor consumption power rose up about 14.4%. As a result, COP tended to decrease, and the decrement presented 12% and 11% from the experiment and HYSYS analysis each. For R22, while the evaporation capacity increased approximately 10%, the increase of the compressor consumption power was larger 16.2%. As a result, COP tended to decrease for R22 and the decrement showed 19% and 10% from the experiment and HYSYS analysis each.

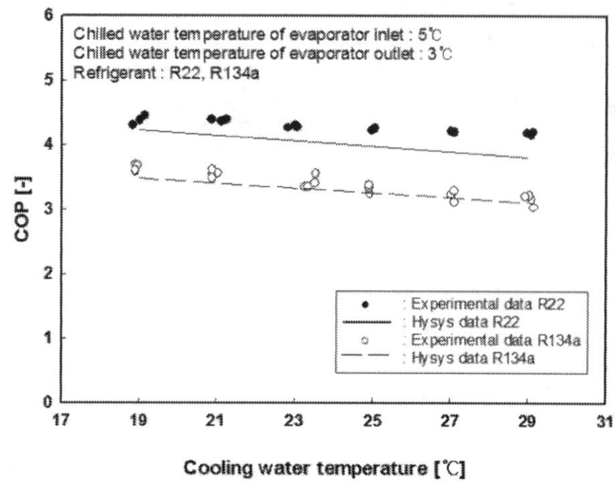

Figure 5. COP of 15kW seawater cooling system with respect to cooling water temperature.

4 Conclusions

The summary of the experimental results is as follows.

From the heat transfer coefficient experiment with respect to evaporating temperature for flooded type evaporator, aluminium-brass low-fin tube had higher heat transfer performance than copper-nickel low-fin tube. Also, R22 as working fluid showed higher heat transfer performance than R134a.

From the heat transfer coefficient experiment with regard to evaporating temperature for 15kW Seawater cooling system, heat transfer performance of aluminium-brass low-fin tube changed as the temperature level of both evaporating temperature and chilled seawater inlet temperature changed. The heat transfer was occurred actively at high temperature level, and the molecular motion and the surface tension were factors to effect on the heat transfer critically.

From the experiment on evaporating and seawater outlet temperature with respect to cooling water temperature for flooded type evaporator, the result of the chilled seawater outlet temperature satisfied the quantitative desired chilled seawater outlet temperature of 3°C at all range of the cooling water temperature. Also, the result of the evaporating temperature presented under 0°C as the quantitative desired range as well.

The COP with respect to cooling water temperature for seawater cooling system was examined in this experiment, and showed decrease of 19% for R22 from 19°C to 29°C of the cooling water temperature.

Acknowledgement

This research was a part of the project titled 'Introduction study of seawater cooling system for fishing-ships', funded by the Ministry of Oceans and Fisheries, Korea.

References

1. J. I. Yoon, C. H. Son, K. H. Choi, S. H. Jung, H. M. Han, S. H. Seol, Heat transfer engineering(to be published)
2. D. S. Jung, K. Y. Park, J. S. Park, Intl. J. of Refrigeration, 27, 8, 710-720 (2001)
3. S. G. Kim, MS thesis, Pusan University, Busan, Korea (2011)
4. Aspentech HYSYS 8.6
5. REFPROP, Version 8.0, NIST Inc, 2007.

Extended Application Of Time-Varying Observer For Rigid Rotors Unbalance Estimation During Acceleration

Xunxing Yu[1,a], Kuanmin Mao[1] and Yaming Zhu[1]

[1]*School of Mechanical Science and Engineering, Huazhong University of Science and Technology, Wuhan, Hubei, 430074, China*

Abstract. Unbalance is one of essential problems for modern rotating machines. In this work, an improved time-varying observer is proposed to estimate the unbalance of rigid rotor during acceleration. In order to fitting different speed acceleration laws, the unbalance forces have been included in an new designed augmented states, meanwhile the state space model of rigid rotor has been also developed. The developed state space model is transformed to a canonical transformation and a new designed time-varying observer can be obtained. The estimated unbalances can be directly obtained by using this time-varying observer. This method would be very helpful for active balancing control strategy during acceleration.

1 Introduction

Modern rotating machines are designed highly-intelligent and highly-integration with speeds-up to achieve excellent quality and efficiency. Excessive vibration would deduced by unbalance forces which are proportional to the square of the spin speed, thus unbalance is an essential problem in rotordynamics.

Traditional off-line balancing was always taken after the system power off, which is time-consuming and costly. In order to perform balancing scheme without interrupting the operation of the rotor system, researchers have done valuable works for speed-varying situation. Active magnetic bearings have been applied in active balancing system for the attenuation of unbalance induced responses during speed-varying stage by interpolate the influence coefficients between different speeds [1]. Mass redistribution balancers using gain scheduling strategy during acceleration have been conducted based on instantaneous influence coefficients, which are calculated from the fileted speed-synchronous response [2]. However, the non-synchronous vibration excited by the acceleration might cause difficulties if the rotor system is in high acceleration or small damping when applying the above method.

An alternative way to estimate the system unbalance by using time-varying observer for rigid rotor system during acceleration. The estimated result can be used in active balancing for rigid rotors [3]. However, this method is based on constant acceleration assumption. In order to applying this method for different speed laws such as exponential law, a new method is proposed by developed the dynamic model.

The main purpose of this paper is to improving the adaption the time-varying observer for different spin acceleration laws. An improved rigid model for different spin acceleration laws will be presented

[a]Corresponding author: yuxx@hust.edu.cn

in section 2, the procedures of the time-varying observer design is given in section 3. A simulation is used to validate the performance of the proposed method. The results shown that the proposed method could fitting different speed acceleration laws.

2 Dynamic Model of Rigid Rotors

For many practical rotors such as spindles are usually operating below the first critical speed for accuracy and safety, moreover, flexible rotors also need to balancing in low speeds, in such conditions, four freedom rigid rotor model can be used. The rigid rotors model is shown in figure 1.

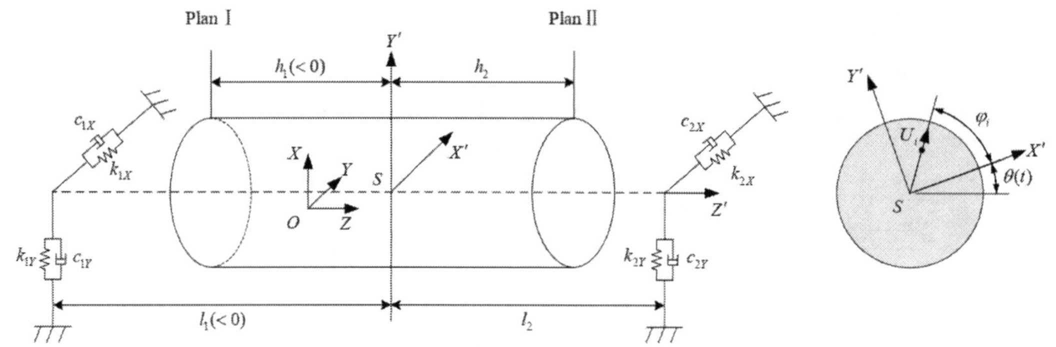

Figure 1. Four freedom rigid rotor model.

As shown in figure 1, S is the mass center of the rigid rotor, $SX'Y'Z'$ is the rotor-fixed frame and $OXYZ$ is the inertial frame, the $SX'Y'Z'$ coordinate system is coincides with the $OXYZ$ coordinate system when the rotor is at rest. The bearings are modelled as a set of linear springs and dampers in two orthogonal directions. Two unbalances in plan I and plan II are used to represent the inherent dynamic unbalance of the rotor. The motions are described by displacements x_G, y_G of the mass center with respect to the inertial reference, as well as the small angles α and β rotation about the X and Y axis.

The dynamic equation of motion of a rigid body can be obtained [4]

$$\begin{cases} m\ddot{x}_G + k_{1X}x_1 + c_{1X}\dot{x}_1 + k_{2X}x_2 + c_{2X}\dot{x}_2 = \mathrm{Re}\left(-\dfrac{d^2}{dt^2}e^{j\theta}\sum_{i=1}^{2}U_i\right) \\ m\ddot{y}_G + k_{1Y}y_1 + c_{1Y}\dot{y}_1 + k_{2Y}y_2 + c_{2Y}\dot{y}_2 = \mathrm{Im}\left(-\dfrac{d^2}{dt^2}e^{j\theta}\sum_{i=1}^{2}U_i\right) \\ J_t\ddot{\alpha} + J_p\dot{\theta}\dot{\beta} + J_p\ddot{\theta}\beta - (k_{1Y}y_1 + c_{1Y}\dot{y}_1)l_1 - (k_{2Y}y_2 + c_{2Y}\dot{y}_2)l_2 = \mathrm{Im}\left(\dfrac{d^2}{dt^2}e^{j\theta}\sum_{i=1}^{2}U_ih_i\right) \\ J_t\ddot{\beta} - J_p\dot{\theta}\dot{\alpha} - J_p\ddot{\theta}\alpha + (k_{1X}x_1 + c_{1X}\dot{x}_1)l_1 + (k_{2X}x_2 + c_{2X}\dot{x}_2)l_2 = \mathrm{Re}\left(-\dfrac{d^2}{dt^2}e^{j\theta}\sum_{i=1}^{2}U_ih_i\right) \end{cases} \tag{1}$$

where $\mathrm{Re}(\bullet)$ and $\mathrm{Im}(\bullet)$ means the real and imaginary part of a complex number, $j=\sqrt{-1}$, $U_i = |U_i|e^{j\varphi_i}$ $(i=1,\ 2)$ are complex numbers as unbalances in the coordinate system, J_p and J_t and the diametric moments of inertia of the rotor, respectively. $\theta(t)$ is the time-varying spin speed. x_1, y_1, x_2, y_2 represent the displacements at the location of the bearings in the inertial frame.

Here exist such relation that

$$\begin{bmatrix} x_1 \\ x_2 \\ y_1 \\ y_2 \end{bmatrix} = \begin{bmatrix} 1 & 0 & 0 & l_1 \\ 1 & 0 & 0 & l_2 \\ 0 & 1 & -l_1 & 0 \\ 0 & 1 & -l_2 & 0 \end{bmatrix} \begin{bmatrix} x_G \\ y_G \\ \alpha \\ \beta \end{bmatrix} \tag{2}$$

Substitute equation (2) into (1)

$$\mathbf{M}\ddot{\mathbf{q}} + \left(\mathbf{C} + \dot{\theta}\mathbf{G}\right)\dot{\mathbf{q}} + \left(\mathbf{K} + \ddot{\theta}\mathbf{G}\right)\mathbf{q} = \mathbf{F} \tag{3}$$

where $\mathbf{q} = \begin{bmatrix} x_G & y_G & \alpha & \beta \end{bmatrix}^T$, matrices $\mathbf{M}$, $\mathbf{C}$, $\mathbf{G}$, $\mathbf{K}$, $\mathbf{F}$ are the mass, damping, gyroscopic, stiffness matrices and unbalance forces vector, and are given as

$$\mathbf{M} = \begin{bmatrix} m & 0 & 0 & 0 \\ 0 & m & 0 & 0 \\ 0 & 0 & J_t & 0 \\ 0 & 0 & 0 & J_t \end{bmatrix}, \mathbf{G} = \begin{bmatrix} 0 & 0 & 0 & 0 \\ 0 & 0 & 0 & 0 \\ 0 & 0 & 0 & J_p \\ 0 & 0 & -J_p & 0 \end{bmatrix}, \mathbf{F} = \begin{bmatrix} \mathrm{Re}(f_1) \\ \mathrm{Im}(f_1) \\ \mathrm{Re}(f_2) \\ \mathrm{Im}(f_2) \end{bmatrix} \tag{4}$$

$$\mathbf{C} = \begin{bmatrix} c_{1x} + c_{2x} & 0 & 0 & c_{1x}l_1 + c_{2x}l_2 \\ 0 & c_{1y} + c_{2y} & -c_{1y}l_1 - c_{2y}l_2 & 0 \\ 0 & -c_{1y}l_1 - c_{2y}l_2 & c_{1y}l_1^2 + c_{2y}l_2^2 & 0 \\ c_{1x}l_1 + c_{2x}l_2 & 0 & 0 & c_{1x}l_1^2 + c_{2x}l_2^2 \end{bmatrix} \tag{5}$$

$$\mathbf{K} = \begin{bmatrix} k_{1x} + k_{2x} & 0 & 0 & k_{1x}l_1 + k_{2x}l_2 \\ 0 & k_{1y} + k_{2y} & -k_{1y}l_1 - k_{2y}l_2 & 0 \\ 0 & -k_{1y}l_1 - c_{2y}l_2 & k_{1y}l_1^2 + k_{2y}l_2^2 & 0 \\ k_{1x}l_1 + k_{2x}l_2 & 0 & 0 & k_{1x}l_1^2 + k_{2x}l_2^2 \end{bmatrix} \tag{6}$$

where $f_1 = C_1 \dfrac{d^2}{dt^2} e^{j\theta}$, $f_2 = C_2 \dfrac{d^2}{dt^2} e^{j\theta}$, C_1 and C_2 are complex numbers and defined as $C_1 = -\sum_{i=1}^{2} U_i$, $C_2 = j\sum_{i=1}^{2} U_i h_i$.

The angular acceleration law of the rotational motion may be not constant, such as exponential law, in order to corporates with different kinds of speed-varying law, an complex number is introduced as $g = e^{j\theta}$ and

$$\frac{d}{dt}\mathbf{g} = \underbrace{\begin{bmatrix} 0 & 0 & 1 & 0 \\ 0 & 0 & 0 & 1 \\ 0 & -\ddot{\theta} & 0 & -\dot{\theta} \\ \ddot{\theta} & 0 & \dot{\theta} & 0 \end{bmatrix}}_{\mathbf{A}_1(t)} \mathbf{g} \tag{7}$$

where $\mathbf{g} = \begin{bmatrix} \mathrm{Re}(g) & \mathrm{Im}(g) & \mathrm{Re}(\dot{g}) & \mathrm{Im}(\dot{g}) \end{bmatrix}^T$, obviously, f_1 and f_2 are the results of $\ddot{g}$ multiplied by complex constants C_1 and C_2 relatively, complex variables $g_1 = C_1 g$ and $g_2 = C_2 g$ also satisfy equation (7). For the sake of convenience, two vectors are defined as

Then representing the rigid rotor system into the state space model by including the unbalance force $\mathbf{F}$ to the state vector

$$\frac{d\mathbf{z}}{dt}=\underbrace{\begin{bmatrix}\mathbf{0} & \mathbf{I}_{4\times4} & \mathbf{0} & \mathbf{0}\\ -\mathbf{M}^{-1}\left(\mathbf{K}+\ddot{\theta}\mathbf{G}\right) & -\mathbf{M}^{-1}\left(\mathbf{C}+\dot{\theta}\mathbf{G}\right) & \mathbf{M}^{-1}\mathbf{Y}_1(t) & \mathbf{M}^{-1}\mathbf{Y}_2(t)\\ \mathbf{0} & \mathbf{0} & \mathbf{A}_1(t) & \mathbf{0}\\ \mathbf{0} & \mathbf{0} & \mathbf{0} & \mathbf{A}_1(t)\end{bmatrix}}_{\mathbf{A}(t)}\mathbf{z} \tag{8}$$

$$\mathbf{y}=\underbrace{\begin{bmatrix}\mathbf{I}_{4\times4} & \mathbf{0}_{4\times12}\end{bmatrix}}_{\mathbf{B}}\mathbf{z}$$

where $\mathbf{z}=\begin{bmatrix}\mathbf{q}^T & \dot{\mathbf{q}}^T & \mathbf{g}_1^T & \mathbf{g}_2^T\end{bmatrix}^T$ is a augmented state, $\mathbf{y}=\mathbf{q}$, $\mathbf{B}$ is a constant matrix, $\mathbf{Y}_1(t)$ and $\mathbf{Y}_2(t)$ are defined as

$$\mathbf{Y}_1(t)=\begin{bmatrix}0 & -\ddot{\theta} & 0 & -\dot{\theta}\\ \ddot{\theta} & 0 & \dot{\theta} & 0\\ 0 & 0 & 0 & 0\\ 0 & 0 & 0 & 0\end{bmatrix};\ \mathbf{Y}_2(t)=\begin{bmatrix}0 & 0 & 0 & 0\\ 0 & 0 & 0 & 0\\ 0 & -\ddot{\theta} & 0 & -\dot{\theta}\\ \ddot{\theta} & 0 & \dot{\theta} & 0\end{bmatrix} \tag{9}$$

Using this modified model, the state estimation problem can be used for different speed laws. As shown in the state space model, displacement can be measured, however, the implicitly included unbalances U_i cannot be measured directly. An observer will used to estimate the states from incomplete states measurements.

3 Time Varying Observer

The rigid dynamic system presented in the state space model as

$$\begin{aligned}\dot{\mathbf{z}}(t)&=\mathbf{A}(t)\mathbf{z}(t)\\ \mathbf{y}(t)&=\mathbf{B}\mathbf{z}(t)\end{aligned} \tag{10}$$

Introduce a state estimator in the following form

$$\dot{\hat{\mathbf{z}}}(t)=\mathbf{F}(t)\hat{\mathbf{z}}(t)+\mathbf{G}(t)\mathbf{y}(t) \tag{11}$$

where $\hat{\mathbf{z}}(t)$ is the estimated state vector, it has been proven by the Identity State Estimator Lemma, if $\mathbf{F}=\mathbf{A}(t)-\mathbf{G}(t)\mathbf{B}$ is a constant matrix with stable eigenvalues, then the system described by (11) is an asymptotic identity state estimator [5]. Then the relation of the true state $\mathbf{z}$ and its estimated state $\hat{\mathbf{z}}$ can be obtained

$$\dot{\hat{\mathbf{z}}}-\dot{\mathbf{z}}=\mathbf{F}(\hat{\mathbf{z}}-\mathbf{z}) \tag{12}$$

The design of asymptotic state estimators procedures are described as follows:

- Obtain a Lyapunov transformation matrix $\mathbf{P}(t)$ by using the method stated in [5].
- Compute $\bar{\mathbf{A}}(t)=\mathbf{P}^{-1}(t)\left[\mathbf{A}(t)\mathbf{P}(t)-\dot{\mathbf{P}}(t)\right]$ and $\bar{\mathbf{B}}(t)=\mathbf{B}\mathbf{P}(t)$, it has been shown in [5] that the matrix $\bar{\mathbf{A}}(t)$ and $\bar{\mathbf{B}}(t)$ are canonical forms. Now the original state space model (10) has been transformed into canonical structures

$$\begin{aligned}\dot{\bar{\mathbf{z}}}(t)&=\bar{\mathbf{A}}(t)\bar{\mathbf{z}}(t)\\ \mathbf{y}(t)&=\bar{\mathbf{B}}(t)\bar{\mathbf{z}}(t)\end{aligned} \tag{13}$$

- Choose distinct desired eigenvalues to design matrix $\bar{\mathbf{F}}$ as stated in [3], $\bar{\mathbf{F}}$ is a constant matrix with stable eigenvalues.
- Using the relation $\bar{\mathbf{F}} = \bar{\mathbf{A}}(t) - \bar{\mathbf{G}}(t)\bar{\mathbf{B}}(t)$ to obtain $\bar{\mathbf{G}}(t)$.
- Hence, an asymptotic identity state estimator of the equivalent system (13) is given by

$$\dot{\bar{\mathbf{z}}}(t) = \bar{\mathbf{F}}\bar{\mathbf{z}}(t) + \bar{\mathbf{G}}(t)\mathbf{y}(t) \tag{14}$$

where $\bar{\mathbf{z}}(t)$ is the equivalent estimated state, $\mathbf{y}(t)$ is the output of original system (8).

- The original estimated state is obtained by $\hat{\mathbf{z}} = \mathbf{P}(t)\bar{\mathbf{z}}(t)$.

Following the above procedures, a desired time-varying observer for the rigid rotor can be design. The estimation error will go to zero as $t \to \infty$. So the estimation of implicitly included unbalances can be obtained.

$$\begin{bmatrix} \mathrm{Re}(U_1) \\ \mathrm{Im}(U_1) \\ \mathrm{Re}(U_2) \\ \mathrm{Im}(U_2) \end{bmatrix} = \underbrace{\begin{bmatrix} -\cos\theta & \sin\theta & -\cos\theta & \sin\theta \\ -\sin\theta & -\cos\theta & -\sin\theta & -\cos\theta \\ -h_1\sin\theta & -h_1\cos\theta & -h_2\sin\theta & -h_2\cos\theta \\ h_1\cos\theta & -h_1\sin\theta & h_2\cos\theta & -h_2\sin\theta \end{bmatrix}}_{\mathbf{H}(t)}^{-1} \begin{bmatrix} \mathrm{Re}(\hat{g}_1) \\ \mathrm{Im}(\hat{g}_1) \\ \mathrm{Re}(\hat{g}_2) \\ \mathrm{Im}(\hat{g}_2) \end{bmatrix} \tag{15}$$

Since $\det(\mathbf{H}) = (h_1 - h_2)^2$ is always positive unless the two unbalance correction plans coincide, which can be avoid by proper choosing the location of the correction plans.

4 Simulation

A simulation is presented to illustrate the above procedures and methods. The rigid rotor shown as figure 1 has a length of 0.5 m and a diameter of 0.2 m and is made from steel with a density of $7810\ \mathrm{kg/m^3}$, the locations in the rotation axis are $l_1 = -0.4$ m, $l_2 = 0.3$ m, $h_1 = -0.2$ m and $h_2 = 0.2$ m. The unbalances are set to $(-0.03 + j*0.01)$ kgm at plan I and $(0.05 + j*0.04)$ kgm at plan II. The stiffness in both directions are 1 MN/m at bearing 1 and 1.3 MN/m at bearing 2. The damping values in in both directions are 100 Ns/m at bearing 1 and 130 Ns/m at bearing 2. The critical speeds of the rigid rotor are calculated as 21.79 Hz, 21.79 Hz, 44.93 Hz and 55.85 Hz.

The unbalance estimation under linear acceleration is implement at first, the angular acceleration is set as a constant value of 100 rad/s^2, as shown in figure 2, the true values of the unbalances in the two correction plan are shown by dashed lines, the solid lines are the estimated results which converges to the inherent unbalances quickly.

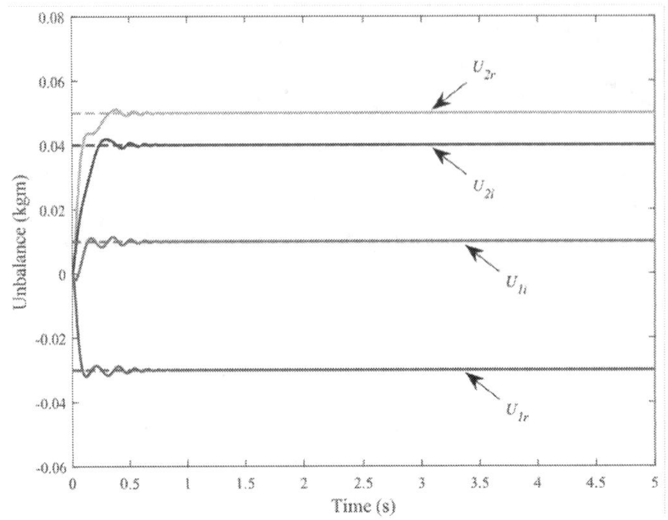

Figure 2. Estimated result under linear acceleration.

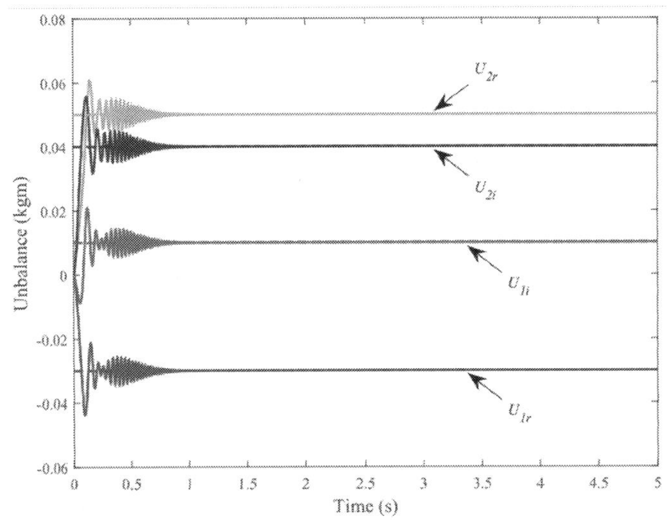

Figure 3. Estimated result under exponential acceleration.

Then another common speed law is used for test, the angular speed is exponential accelerated as $1000*(1-e^{-0.5t})$ rad/s, the great angular acceleration in the beginning may excite undesired transient response as shown in figure 3,, which results that the estimation oscillate around its true values, however, after passing this region the estimated results also converges to the inherent unbalances quickly. Those results prove that the extended time-varying observer works well for different acceleration law, and the performance is desire for active balancing.

5 Conclusions

This work aims to extend the method of applying time-varying observer for unbalance estimation in order to improving the adaption for different spin acceleration law. By assuming the dynamic parameters are known and the displacements and spin speed can be measured, the unbalances are viewing as the outputs of states of the dynamic rigid rotor system, then a new time-varying observer obtained without constant angular speed acceleration assumption. The simulation results show that the estimation converges to the inherent unbalances quickly under different law of acceleration.

Funding

This work is supported by the Hi-Tech Research and Development Program (863) of China (Grant NO. 2015AA043302).

References

1. Knospe C R and Tamer S M 1997 Experiments in robust control of rotor unbalance response using magnetic bearings *Mechatronics* **7** 217-29
2. Zhou S, Dyer S W, Shin K-k, Shi J and Ni J 2004 Extended Influence Coefficient Method for Rotor Active Balancing During Acceleration *Journal of Dynamic Systems, Measurement, and Control* **126** 219-23
3. Zhou S Y and Shi J J 2001 Imbalance estimation for speed-varying rigid rotors using time-varying observer *J Dyn Syst-T Asme* **123** 637-44
4. Genta G 2005 *Dynamics of rotating systems* (New York: Springer)
5. Cuong N and Ting L 1985 Design of a state estimator for a class of time-varying multivariable systems *IEEE Transactions on Automatic Control* **30** 179-82

MATEC Web of Conferences **167**, 02016 (2018) https://doi.org/10.1051/matecconf/201816702016
IC4M & ICDES 2018

Effects of Injection Molding Process Conditions on Physical Properties of EPDM Using Design of Experiment Method

Young Shin Kim[1], Euy Sik Jeon[2,a] and Eui Seob Hwang[3]

[1]*Industrial Technology Research Institute, Kongju National University, Korea*
[2]*Department of Mechanical Engineering, Graduate School (Industrial Technology Research Institute), Kongju National University, Korea*
[3]*New Business department Daesung Hi Tech, Korea*

Abstract. The process variables such as time and temperature during the EPDM-injection molding not only change the physical properties of the raw material but also affect the insertion and separations forces when a grommet product is molded and mounted on a car body. Using the design of experiments method, the major factors in the injection molding process were considered to analyze their effects on the physical properties of the obtained EPDM materials. Test pieces were prepared under different process conditions, tensile strength and elongation were measured, and their correlation was analyzed.

1 Introduction

Grommets are a major component of automobiles, where they provide a proper setting of the external and internal cables. They are mainly made of maleated polymer-grafted ethylene propylene diene monomer (EPDM) rubber. EPDM, a terpolymer composed of irregularly bonded ethylene, propylene, and diene, is employed in various products, as its mechanical properties, water resistance, and heat resistance are superior compared with those of conventional rubber [1–3].

EPDM has been extensively investigated with a focus on the reliability of products containing EPDM and analysis of their physical properties, which can be integrated in composite materials [4–6].

In contrast to plastic, EPDM exhibits high flexibility, elasticity, and elongation. Using injection molding, EPDM-based grommets can be fabricated. The process variables such as time and temperature during the EPDM-injection molding not only change the physical properties of the raw material but also affect the insertion and separations forces when a grommet product is molded and mounted on a car body. Using the design of experiments method [7–9], the major factors in the injection molding process were considered to analyze their effects on the physical properties of the obtained EPDM materials. Test pieces were prepared under different process conditions, tensile strength and elongation were measured, and their correlation was analyzed.

2 Experimental setup

[a] Corresponding author : osjun@kongju.ac.kr

https://doi.org/10.1051/matecconf/201816702016

In this study, test pieces were produced using a rubber injection molding machine with EPDM to analyze the physical properties of EPDM as a function of the injection molding process conditions. The physical properties of the produced test pieces were analyzed using a universal testing machine.

2.1 Process variables

In injection molding of a rubber material, the major process variables include temperature, time, and degassing conditions. After conducting a basic test on various process conditions, three process variables that can be easily regulated and have a significant effect on the test results were selected. An experiment plan was established using the design of experiment method. In particular, the full factorial design method was employed to identify the effects and relationships between the factors; a 3-factor 2-level 2-time repetitive test was performed. Table 1 shows the established experiment plan based on a total of 16 experiments using three process variables. Based on the obtained experimental condition table, rubber injection molding was performed 16 times. The total number of produced rubber test pieces was 32, as there are two test pieces for each experimental condition (Fig. 1 (a)). The prepared rubber test pieces were cut for further studies of their physical properties, as shown in Fig. 1 (b).

Table 1. Experimental conditions for different process variable values according to DOE

Run	Temp. (°C)	Time (sec)	Degassing	Run	Temp. (°C)	Time (sec)	Degassing
1	160	200	O	9	160	200	X
2	160	200	O	10	160	200	X
3	160	600	O	11	160	600	X
4	160	600	O	12	160	600	X
5	180	200	O	13	180	200	X
6	180	200	O	14	180	200	X
7	180	600	O	15	180	600	X
8	180	600	O	16	180	600	X

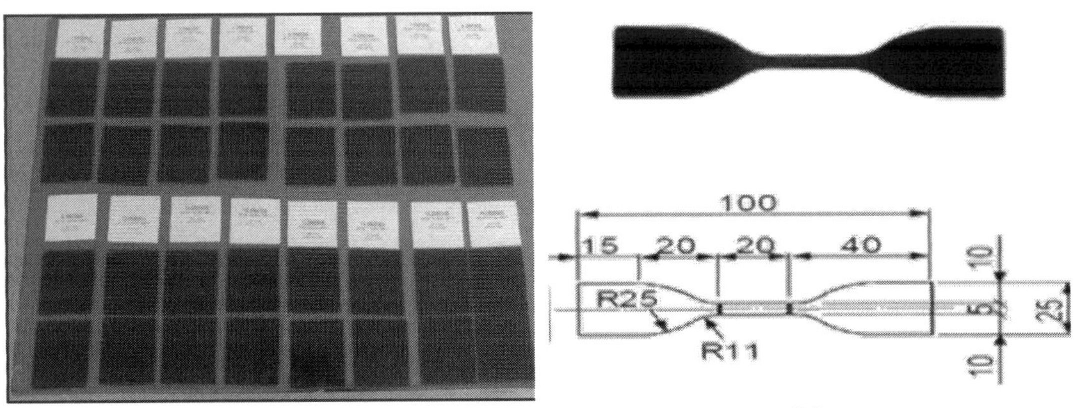

(a) (b)

Figure 1. Specimens produced by various process conditions

2.2 Physical property test

In order to evaluate the physical properties of the materials, tensile strength and elongation were measured. In the tensile strength test, the tensile strength was calculated by measuring the maximum load at the moment when the test piece broke apart in the universal testing machine, and then dividing this value by the area of the cross section. The elongation was calculated by measuring the scale distance at the test piece. The test was performed at a speed of 500 mm/min. Figure 2 shows a photograph of the experimental setup where the tensile strength and elongation of the test pieces were measured, along with its conceptual diagram. Table 2 shows the results of the tensile strength and elongation measured for each process condition.

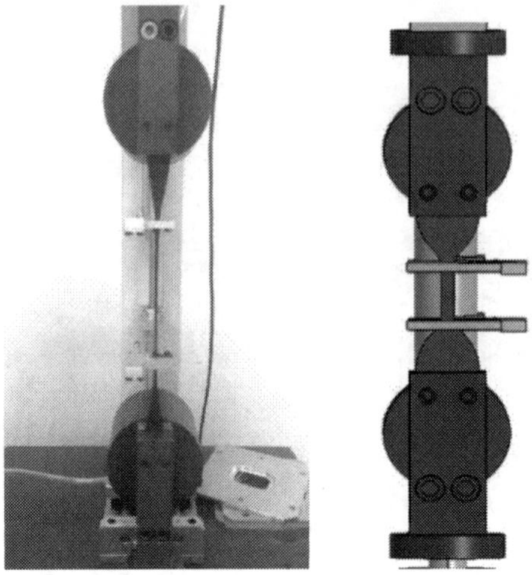

Figure 2. photograph of the experimental setup

Table 2. Formatting sections, subsections and subsubsections.

Run	Temp. (℃)	Time (sec)	Degassing	Tensile strength (kgf/㎠)	Elongation (%)	Run	Temp. (℃)	Time (sec)	Degassing	Tensile strength (kgf/㎠)	Elongation (%)
1	160	200	O	114.4	969.0	9	160	200	X	107.0	968.3
2	160	200	O	113.3	960.4	10	160	200	X	110.3	944.8
3	160	600	O	136.6	853.9	11	160	600	X	137.7	851.5
4	160	600	O	131.4	815.2	12	160	600	X	132.4	821.2
5	180	200	O	134.2	840.5	13	180	200	X	130.5	838.7
6	180	200	O	133.9	836.0	14	180	200	X	134.7	844.8
7	180	600	O	137.0	775.0	15	180	600	X	132.9	749.7
8	180	600	O	139.3	792.3	16	180	600	X	135.2	767.1

3 Analysis of the results

An analysis of the main effects and their relationships was performed according to the experiment plan obtained using the design of experiment method. The factors that mainly affect the tensile strength and elongation were identified and their relationships were analyzed.

3.1 Tensile strength as a function of the temperature, time and degassing conditions

The effects of the temperature, time, and degassing conditions on the tensile strength were evaluated along with their relationships, as shown in Fig. 3(a). The results showed that the factor that has the biggest effect on the tensile strength was the time, followed by the temperature; degassing did not significantly affect the tensile strength. Figure 3(b) shows the relationships between the factors. It reveals that the temperature and time correlate with each other. Figure 3(c) shows the residual plot, which reveals that the residues are normally distributed, as the plot can be represented by almost a straight line. The model obtained by the analysis can be regarded to be valid as the plot showed a consistent distribution based on the average line. The distribution was normal, as no linear and constant terms were observed from the plot of residues and fitted values. The adjusted coefficient of determination of the experimental results was 93.2%. Table 3 shows the results of the analysis of variance (ANOVA), where the correlations of the tensile strength with the process variables were evaluated.

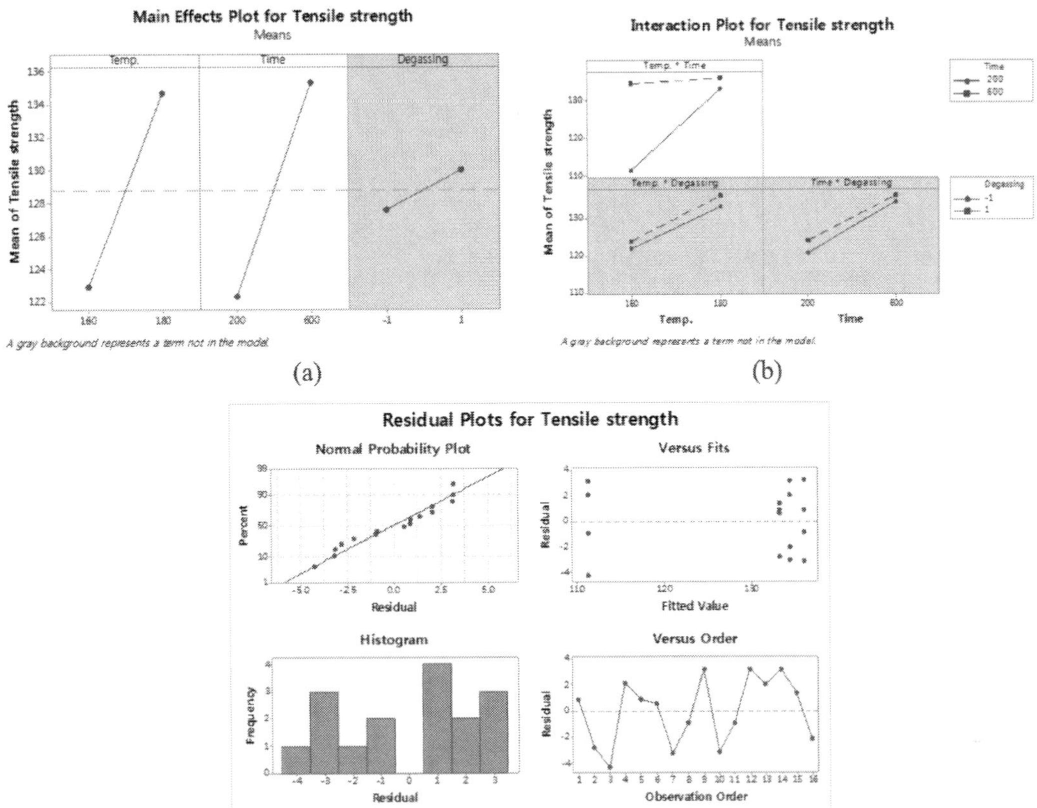

Figure 3. (a)Main effect plot, (b) Interaction plot, (c) Residual plot for tensile strength

Table 3. ANOVA of tensile stress

Source	DF	Adj SS	Adj MS	F-Value	p-Value
Model	3	1658.17	552.725	69.95	0.000
Linear	2	1237.93	618.963	78.33	0.000
Temp.(°C)	1	559.32	559.322	70.78	0.000
time(s)	1	678.6	678.602	85.88	0.000
2-Way interactions	1	420.25	420.25	53.18	0.000
T(°C)*Time(sec)	1	420.25	420.25	53.18	0.000
Error	12	94.83	7.902		
Total	15	1753			

3.2 Elongation as a function of the temperature, time and degassing conditions

Figure 4(a) shows the plot of the main factors that determine the elongation. The results reveal that both time and temperature affect the elongation, whereas degassing had only a small effect on the elongation. Figure 4(b) shows their correlations; it reveals that the temperature and time correlated with each other. Figure 4(c) shows the residual plot, which demonstrates the validity of the model. The adjusted coefficient of determination of the experimental results was 94.12%. Table 4 shows the ANOVA results, obtained by analyzing the correlations of the elongation with the process variables. The results corresponding to a P-value larger than 0.05 were pooled into the error term, and only the significant terms were presented.

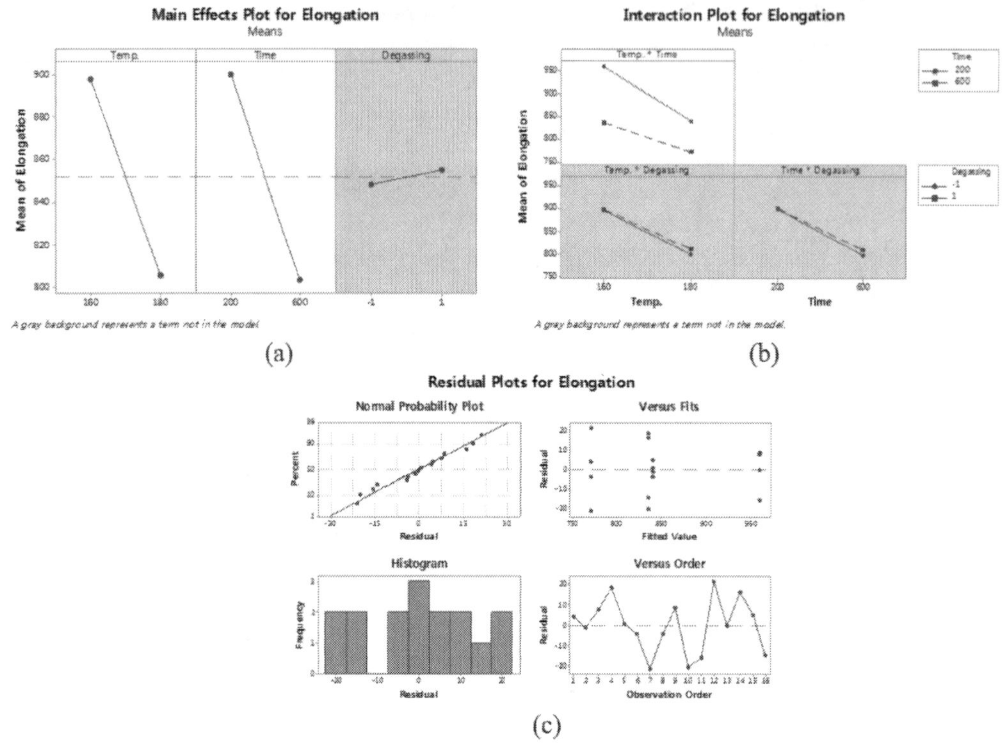

Figure 4. (a)Main effect plot, (b) Interaction plot, (c) Residual plot for elongation

Table 4. ANOVA of elongation

Source	DF	Adj SS	Adj MS	F-Value	p-Value
Model	3	75096.2	25032.1	116.87	0.000
Linear	2	71937.7	35968.9	167.93	0.000
T(°C)	1	34243.5	34243.5	159.88	0.000
time(s)	1	37694.2	37694.2	175.99	0.000
2-Way interactions	1	3158.4	3158.4	14.75	0.034
T(°C)*Time(sec)	1	3158.4	3158.4	14.75	0.002
Error	12	2570.2	214.2		
Total	15	77666.4			

4 Conclusions

Using the design of experiment method, the main factors in the injection molding process were considered to analyze their effects on the physical properties of EPDM. Test pieces were produced at different process conditions, the tensile strength and elongation were measured, and their correlation was analyzed. Both temperature and time affected the tensile strength and elongation; however, the effect of degassing was insignificant. We aim to perform additional studies to reveal the optimal process conditions for achieving the desired values of tensile strength and elongation.

ACKNOWLEDGEMENTS

This research was financially supported by the Ministry of Trade, Industry and Energy (MOTIE), Korea, under the "Regional Specialized Industry Development Program" supervised by the Korea Institute for Advancement of Technology (KIAT). (P0000164)

References

1. Oh, J. H., Kim, S. D., and Kim, J. S., J. of Korean Inst. of Resources Recycling, 12(5) (2003)
2. Song, H. S., and Hyun, J. C., Clean Tech., 7(2) (2001)
3. Folkes, M., and Hope, P. S., Blackie Academic Press, New York, (1992)
4. S.J Part, J.Y Kang, S.K. Hong, Polymer, 29,2 (2005)
5. K.Singh, T. Nanda, R Mehta, Composites Part A: Applied Science and Manufacturing, 103,(2017)
6. L.Cao,A.Zheng, X.Cao, D.Yuan,C.Xu,Y.Chen, Polymer Testing,62,(2017)
7. M. Brahmi, M. Ba, Y. Hidri, A. Hassen, J. of Molecular Structure, (2018)
8. M.Kala, M. V.Shaikh, M. Nivsarkr, J. of Pharmacological and Toxicological Methods, 82 (2016)
9. C.Chen, M.Chuang, Y. Hsiao, Y. Yang, C. T, Expert Systems with Applications, 36 (2009)

MATEC Web of Conferences **167**, 02017 (2018) https://doi.org/10.1051/matecconf/201816702017
IC4M & ICDES 2018

Optimization of Design Parameters of Spiral Spring for Active Headrest Deployment

Yunsik Yang[1], Euy Sik Jeon[2,a] and Dae Ho Park[1]

[1]*Department of Mechanical Engineering, Graduate School, Kongju National University, Korea*
[2]*Department of Mechanical Engineering, (Industrial Technology Research Institute) Kongju National University, Korea.*

Abstract. Several studies have been conducted to prevent neck injury in rear-end collision. The headrest of the seat which suppresses the relative motion of the head and the torso can suppress the extension of the head, thereby alleviating the injury. The active headrest has a mechanism that supports the head by deploying the headrest at the rear-end collision. The spring remains compressed or twisted until a collision signal is generated and the headrest is deployed after the collision signal. Depending on the shape and deployment structure of the spring, a spring design with a high resilience that is acceptable to the headrest is required. In this paper, design parameter of spiral spring suitable for the structure of the developed headrest is selected, prototypes are fabricated, and development parameters such as development time and development distance are checked and optimal design parameters of the spiral spring are derived. The feasibility of the headrest with the designed spiral spring was verified by the finite element analysis.

1 Introduction

In addition to the comfort of passengers, it is essential to ensure safety in order to minimize injury during accidents. It is reported that rear-end collision accident accounts for about 30% of all traffic accidents and more than 50% of drivers are suffered neck injury.[1] Several studies have been conducted to prevent neck injuries, and structural studies are under way to control the relative motion of head and body.[2-3] Especially, the headrest of seats is an effective preventive tool to prevent the movement of the head.[4] The deployed headrest has a mechanism that supports the head when the headrest is deployed at the real end collision.[5] Although the driving power source is varied, the structure utilizing the elastic force of the spring is popularly used.[6] The spring remains compressed or twisted until the collision signal is generated, and the headrest is deployed at the same time the deployment signal is generated. The development speed is proportional to the magnitude of the spring force, but it is necessary to design the spring force according to the shape and development structure of the spring to the allowable range of the headrest structure.

In this paper, the design parameters of the spring suitable for the development structure of the developed headrest were selected, and prototypes were fabricated, and the development characteristics such as the development time and the development distance were checked, and the optimum design

[a] Corresponding author : osjun@kongju.ac.kr

factors of the spring were derived. The finite element analysis of the headrest using the designed spring was performed and the structural validity of the headrest was verified by checking the stress distribution.

2 Mechanism of rear-end collision and neck injury

A person's neck is an important organ that connects the torso and head. It is a passage through the nerves, and relative movement is possible by the joints and muscles. Sudden shock or vibration can cause pain and damage to muscles and nerves.

The head is anatomically capable of movement such as Flection, Extension, Rotation, and most of the neck injury caused by the Extension of the head in the rear-end collision. As the extension accelerates suddenly in the rearward collision of the vehicle, the upper body of the occupant moves with the seat at the same speed as the vehicle, but the head is not supported firmly with the upper body and moves at a speed 0.5 to 0.6 times slower than the upper body. Therefore, as shown in Figure 1, the upper body and the head are subjected to different movements, resulting in injury.[7-8]

Whiplash was first introduced by Crow in 1982 and generally presents with multiple, complex symptoms such as spinal pain, headache, and speech disturbance due to muscle and ligament strain as a result of deceleration or acceleration in a short time within 50ms or noncontact.[9-10]

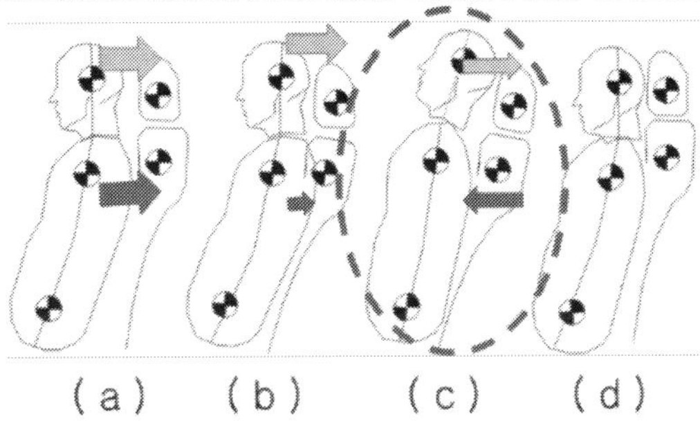

Figure 1. The four phases of head motion as seen from the vehicle headrest

3 Design of the Headrest structure

3.1 Mechanism of headrest

Figure 2 shows a schematic view of the deployment headrest. Upon receiving the collision signal, the actuator unlocks the spiral spring. The link then rotates and the headrest is deployed. After releasing by the trigger signal, the rotational force of the spring becomes the angular velocity of the end point through the link, which is the deployment speed of the headrest.

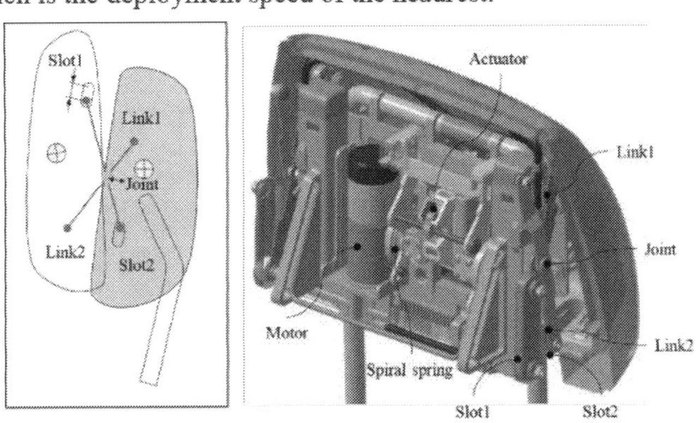

Figure 2. Schematic diagram of the deployment headrest

Figure 3 Is the motion trajectory of the link to select the deployment distance of the deployment headrest.

It expands forward and upward, and the deployment distance is determined by three links. The stroke is moved upward by the rotational force of the spiral spring in a straight line.

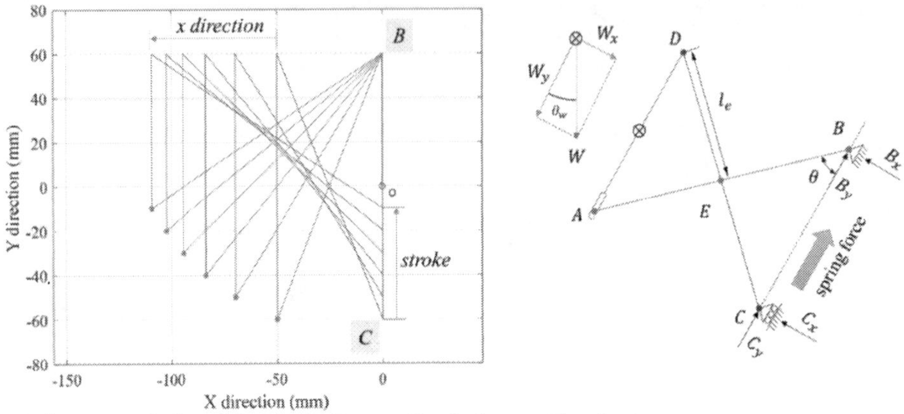

Figure 3. Trajectory analysis and free body diagram for deployment headrest

3.2 Headrest design parameter

The designed unfolded headrest is developed by the movement of the crossed links and is influenced by the weight acting on the front headrest according to the tilt angle of the vehicle under free load conditions.

Figure 4 shows the free body diagram of the unfolded headrest. W is the weight acting on the front headrest, θ is the expansion angle, and l_e is the distance of the link from the cross point.

The point *B* is fixed and the headrest moves forward when the point *C* approaches the point *B* by slid motion due to the elastic force of the spring. The minimum elastic force required for the headrest to deploy is expressed as:

$$W = \frac{W_x}{sin\theta_w} = \frac{W_y}{cos\,\theta_w} \tag{1}$$

$$C_y = \frac{-W_x}{tan\,\theta} = spring\ force \tag{2}$$

$$C_x = W_y\,tan\,\theta - \frac{W_x}{2} \tag{3}$$

$$B_y = \frac{W_x}{tan\,\theta} + W_y \tag{4}$$

At this time, friction occurring in each joints and slots was ignored.

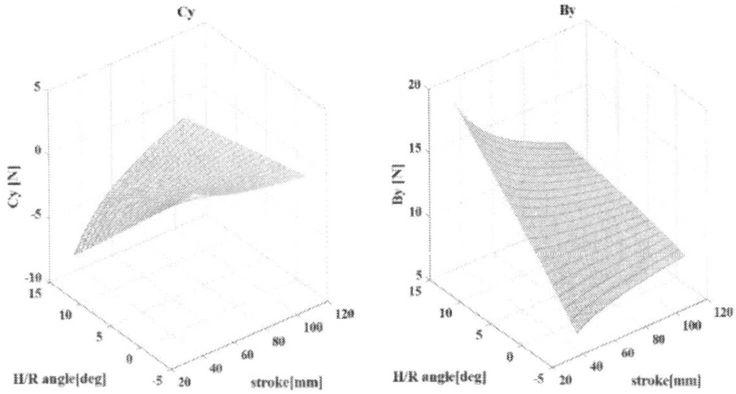

Figure 4. Change of elasticity according to headrest deployment angle and deployment distance

As shown in Figure 4, the change of the minimum required elasticity according to the change of headrest deployment angle and deployment distance was confirmed.

The range of the elastic force was selected by applying the safety factor to the derived elastic force in consideration of the dynamic motion and the friction coefficient.

4 Theoretical analysis of expansion spring

4.1 Theoretical analysis of spiral spring

Spiral springs are generally used because they can efficiently utilize the characteristics and space using twist. One end is fixed and the other end is twisted to use the spring action by the twist deformation at that time. The elastic energy to be stored per unit volume is large and lightweight, does not take up much space, has a simple shape, and has an advantage that the spring characteristic is in good agreement with the calculated value. Since the linear relationship can be maintained in the relationship between the torsion angle and the torque, the formula considering various conditions should be used.

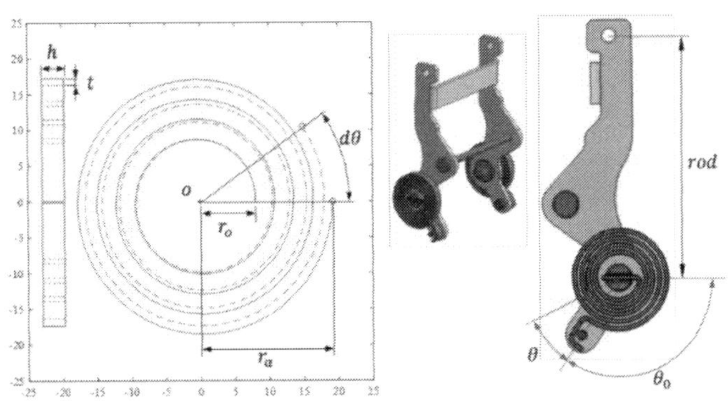

Figure 5. Shape of spiral spring and active mechanism

As shown in Figure 5, $r = m(\theta + \dot{\theta}_0) + r_0$ where r_0 is the inner diameter of the spiral portion of the spiral spring, and r_a is the distance from the center O to the outer end. Where, $m = \frac{P}{2\pi}$, P is the pitch, θ_0 is the angle(rad) from the position fixed to the inner shaft of the spiral to the *x-axis* measured in the counter clockwise direction, and the length l is expressed as follows.

$$L = \int_0^L ds = \int_0^L \sqrt{(dr)^2 + (rd\theta)^2} = \int_0^{\theta_a} \sqrt{m^2 + r^2}d\theta \qquad (5)$$

The strain angle θ can be expressed as follows by the Castilian's Theorem.[11]

$$U = \int \frac{M^2 dx}{2EI} \tag{6}$$

Where, $M = Fr$, and the deformation angle of the end is expressed as follows.

$$r\theta = \frac{\partial U}{\partial F} = \int_0^l \frac{\partial}{\partial F}\left(\frac{F^2 r^2 dx}{2EI}\right) = \int_0^l \frac{Fr^2}{EI} dx \tag{7}$$

Where, $I = \frac{ht^3}{12}$,

$$d\theta = \frac{12ML}{Eht^3}\frac{180}{\pi} \tag{8}$$

$$k = \frac{M}{d\theta} = \frac{Eht^3}{12L}\frac{\pi}{180} \tag{9}$$

The elastic force of the torsion spring is proportional to the third square of the thickness and is inversely proportional to the length, so it is selected as the main factor in designing and manufacturing the spring.

4.2 Design parameters of spiral spring affecting moment

The deployed headrests are typically powered by a spring-loaded structure, because it is simple to structure and allows for high-acceleration movements of large loads with very high loads in a very short time. The elasticity of the spiral spring and the length of the rod determine the deployment speed and deployment distance. It is very important to analyse and optimize the dynamic characteristics of the spring, because the dynamic performance of the spring determines the expansion performance of the headrest.

5 Analysis of headrest deployment test

5.1 headrest deployment test

Since the spring rotates momentarily in a twisted state, it is necessary to analyze the dynamic behavior of the spring as a test.

In this study, a simple product was made to confirm that the headrest was developed by a spiral spring. As shown in Figure 6, the model of the spiral spring was fabricated using pw-2 material. The inner diameter was fixed, and the test piece was fabricated by changing the moment of inertia, effective winding number and outer diameter

Fix the notch rod of the headrest and give a signal to the actuator to analyse the dynamic behaviour using a high-speed camera when the spring rotates and the link rotates at high speed. As shown in Figure 7, the development of the headrest was taken at 1000 fps. The measurements were made under the same conditions and showed errors within 2%.

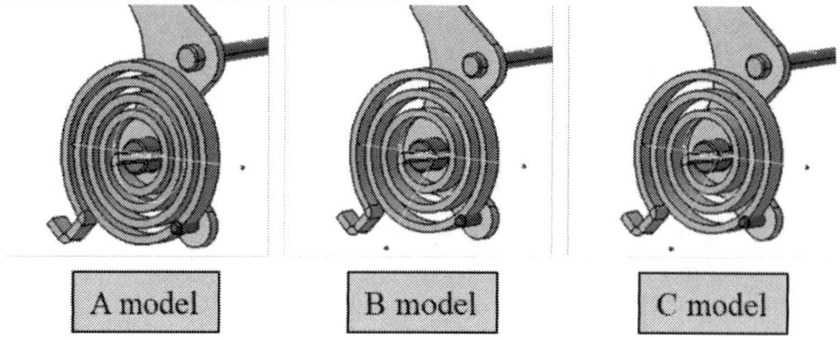

Figure 6. Model for spiral springs

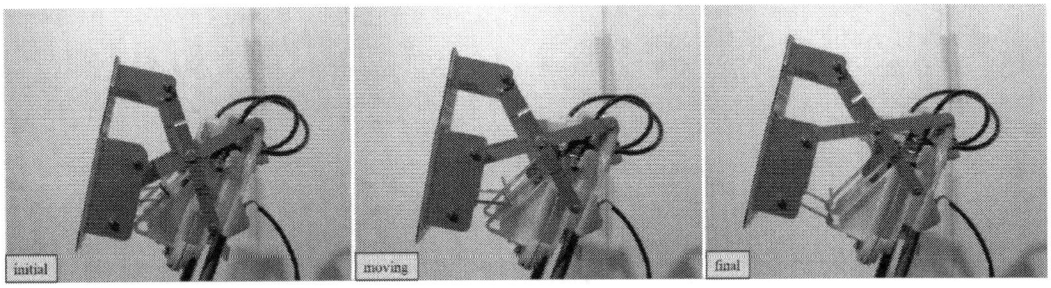

Figure 7. Active headrest motion

5.2 Analysis of Relation between Deployment Speed and Moment

When the headrest was deployed, the displacement and the deployment time of the headrest end of the test were confirmed. It is assumed that the moment acting on each part is constant and no external force other than moment acts.

As shown in Table 1, the development time is inversely proportional to the moment as the moment of each specimen increases. Overall, it was confirmed that the shape of the prototype was inversely proportional to the design parameters.

In the theoretical analysis, the stiffness coefficient is found to be proportional to the moment of inertia and inversely proportional to the effective length, number of active coils. The design parameters of the spiral spring were selected as shown in Figure 8, and the range of the development time was estimated according to the change of the moment.

Table 1. Experimental conditions for spiral spring

Element	Unit	A model	B model	C model
Hight (h)	mm	5	5	5
Thickness (t)	mm	2.0	2.0	2.2
Number of active coil	-	4	3	3
Moment	Nm	43	101	134
Active time	μs	100	40	30

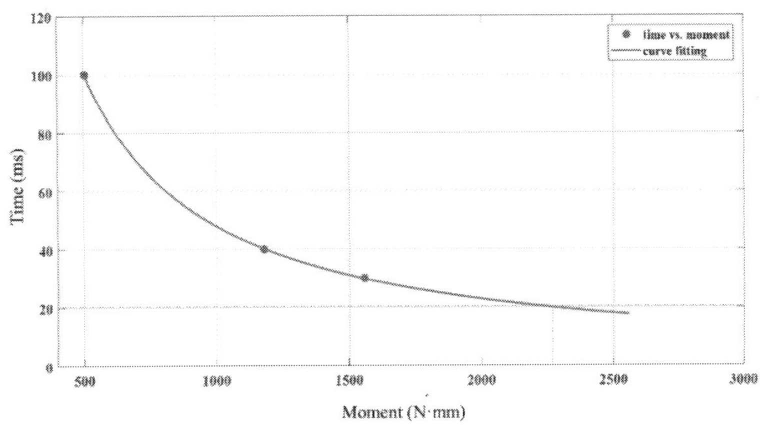

Figure 8. Estimation of deployment time using moments

In order to obtain the value of the experimental value close to the theoretical value, different tests such as the moment of inertia, length, and number of active coils were performed.

From the experimental results, we can confirm the power approximation function such as Equation 10 using the target development time and spiral spring moment and set the design goal of the spring that satisfies the development time of 20ms or less.

$$y = 77654.69x^{-1.07} \tag{10}$$

Based on the test results, we will use torsion and torque as data to estimate the design factors affecting each deployment time.

6 Finite element analysis

Structural analysis was performed to determine the effect on the headrest body when spiral spring was selected according to the range of design parameters. As Shown in Figure 9, the elastic force and the stresses distribution of the supporting structure are confirmed and the fatigue characteristics at the repeated operation are also confirmed. Finite element analysis showed no effect on the headrest frame.

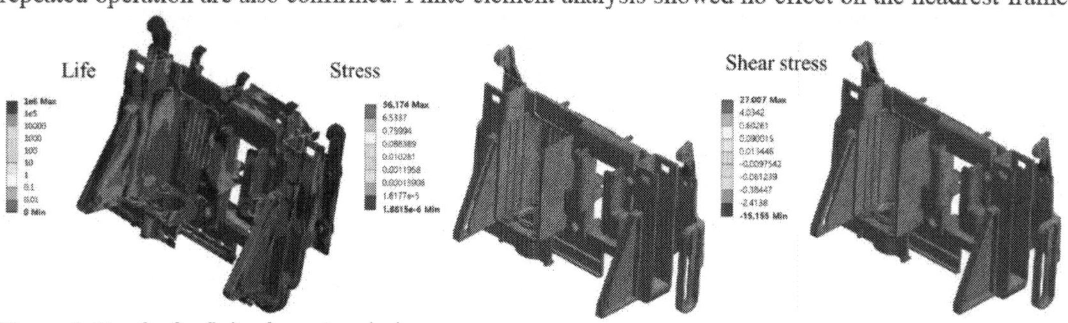

Figure 9. Results for finite element analysis

7 Conclusions

We confirmed the headrest deployment characteristics to mitigate neck injuries and confirmed the design parameters. Through the kinematic analysis of the active headrest, the proposed design parameters were confirmed and the development structure was designed based on this.

The design parameters of the spiral spring for the headrest to accurately and quickly support the head of the passenger are proposed. Spiral spring and headrest prototypes were fabricated and a simple test was carried out to confirm the design range of the spring. The designed headrest was subjected to a structural analysis of the corresponding spring to verify the design validity of the headrest.

ACKNOWLEDGEMENTS

This research was supported by the Ministry of Trade, Industry & Energy(MOTIE), Korea Institute for Advancement of Technology(KIAT) through the Encouragement Program for The Industries of Economic Cooperation Region(R0004149)

References

1. P. Ardoino, Whiplash injuries, (1996)
2. W. H. Castro, M. Schilgen, S. Meyer, M. Weber, C. Peuker, K. Wortler, whiplash injuries, **6**, 6 (1997)

3. R. M. Greendyke, Jama, **195**, 7 (1966)
4. A. J. Watts, D. Atkinson, C. J. Hennessy, Lawyers & Judges Publishing Co., Tuscon, USA (1996)
5. A. Nygren, H. Gustafsson, C. Tingvall, (1985)
6. M. Panjabi, J. Wang, N. Delson, *Proceedings of the International Research Council on the Biomechanics of Injury Conference*, **27** (1999)
7. H. Fischer, H. Malsch, Active Headrest for a Vehicle Seat, (2003)
8. M. Farquhar, M. Humer, K. Grewal, K. Low, K. McQueen, N. V. Yetukuri, (2004)
9. J. Schilling, M. Böhmer, D. Hippel, D. Wrobel, H. Klein, G. Ruess, et al., (2006)
10. K. Boes, M. Schmid, T. Frank, A. Tscherbner, (2011)
11. J. Chen, I. Chen, *Int. J. of Solids and Structures*, **64** (2015)

Electromagnetic Field Simulation in a Microwave Chamber with Multiple Waveguides

Jaekyung Kim[1] and Euysik Jeon[2,a]

[1]*Department of Mechanical Engineering, Kongju National University College of Engineering, Korea*
[2]*Department of Mechanical Engineering, Industrial Technology Research Institute, Kongju National University College of Engineering, Korea*

Abstract. While microwaves have many features and advantages, problems may occur, including non-heating, partial overheating, and fire due to damaged magnetrons caused by reflected waves, when they are used without a proper understanding of the permittivity of the object to be heated, the electromagnetic field distribution, the matching between the chamber and the waveguide, and the reflected electromagnetic waves. Simulation was performed using the Ansys HFSS tool. Conditions for the uniform electromagnetic field were derived using the distance from the waveguides to the ceramic material as well as the microwave energy intensity as major parameters.

1 Introduction

Microwaves are powerful energy sources with excellent reproducibility that can be applied to many applications. Microwave energy has been applied in a variety of industries. It has reduced work hours and improved product quality by using uniform heating. [1]

In addition to the four frequency bands used in industrial, scientific, and medical applications, 2.45 GHz is widely used because it has an appropriate penetration thickness to penetrate laboratory reaction capacity and there are energy sources that can produce microwaves in this frequency band. [2]

In the case of ceramic processes including water, a significantly fast drying speed can be achieved and the non-uniformity of humidity, depending on the position, can be improved using microwaves because the temperature distribution is opposite to that of the existing method.

In addition, when microwaves are used for the burning of ceramic, the temperature rises from the inside of the material. Therefore, if microwaves and surface heating technology are combined, a very uniform temperature distribution can be obtained.

As the reaction between microwaves and materials can heat only specific parts, such as the interface, the application of microwaves to bonding processes can yield faster and better performance than the existing method (joining) and can control the characteristics of composite materials at the interface.

While microwaves have many features and advantages, problems may occur, including non-heating, partial overheating, and fire due to damaged magnetrons caused by reflected waves, when

[a] Corresponding author: osjun@kongju.ac.kr

they are used without a proper understanding of the permittivity of the object to be heated, the electromagnetic field distribution, the matching between the chamber and the waveguide, and the reflected electromagnetic waves.[3] As such, many studies have been conducted on the uniformity of an electromagnetic field to avoid hot spots that could damage magnetrons or destroy the material to be heated. It has been reported that various materials, including zirconia, alumina, and nylon, can be damaged by heat concentration. [4-5]

The penetration depth of microwaves into a material is dependent on their electrical characteristics, and heat is generated accordingly. Electromagnetic wave absorption causes macroscopic, microscopic, and structural changes in the material shape, and the electric field distribution is a major element that affects the process of heating (processing) the material in a microwave chamber.

This study analyzed and determined the installation locations of the waveguides through which microwaves are introduced into a chamber by analyzing the characteristics of microwaves and the electromagnetic field. Simulation was performed using the Ansys HFSS tool. Conditions for the uniform electromagnetic field were derived using the distance from the waveguides to the ceramic material as well as the microwave energy intensity as major parameters.

2 A microwave chamber with six waveguides

The microwave simulation chamber was 400mm×400mm×192mm in size and made of stainless steel. WR-340 was used for the waveguides in accordance with the international standard. The microwave input power was 1 kW and the frequency was 2.45 GHz, while the heating time was set as a parameter. As the initial simulation settings, the microwave input and physical properties of the ceramic in the chamber were selected.

The typical heating temperature of ceramic materials is between 500 ℃ and 1200 ℃. The heating temperature required in this study was 700 ℃. As heating time can be shortened by the microwave input power, the maximum number of the waveguides was set to six (6 kW). Furthermore, the waveguides were placed symmetrically at constant intervals, considering the interference and influence among them. Figure 1 shows schematic diagram modelled microwave chamber and table 1 shows main properties of materials.

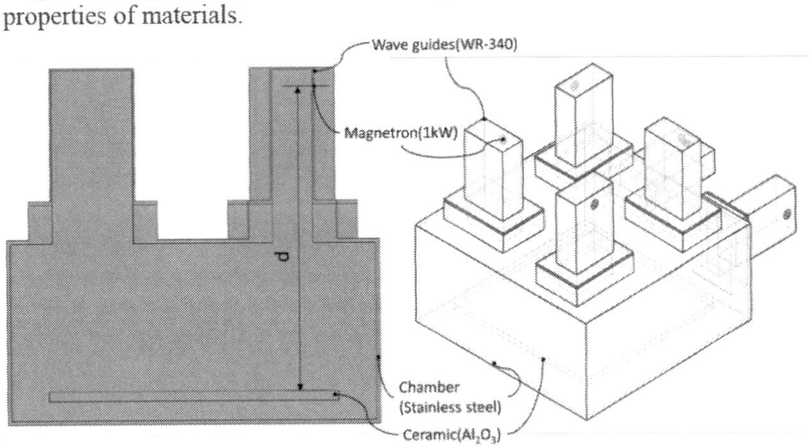

Figure 1. Schematic diagram modelled microwave chamber

Table 1. The material's characteristics

Materials	**Stainless steel**	**Ceramic**
Relative permittivity	1	9.8

Relative permeability	1	1
Thermal conductivity(w/mk)	13.8	35

3 Simulation results

While the frequency change of magnetrons is very important for the electric field distribution inside the chamber, a small change of 10 % occurs and the frequency of the electric field increases mainly due to the wavelength decrease as the frequency increases [6]. Therefore, the simulation frequency was fixed at 2.45 GHz.

3.1 Electric field distribution by Input power

The electric field distribution was simulated using the six installed waveguides according to the microwave intensity (1 - 6 kW). Simulation results showed that the microwave intensity affected the electric field distribution and that higher microwave intensity led to better electric field uniformity. This appears to be influenced by the symmetrical locations of the waveguides. Figure 2 shows electric field distribution by input power.

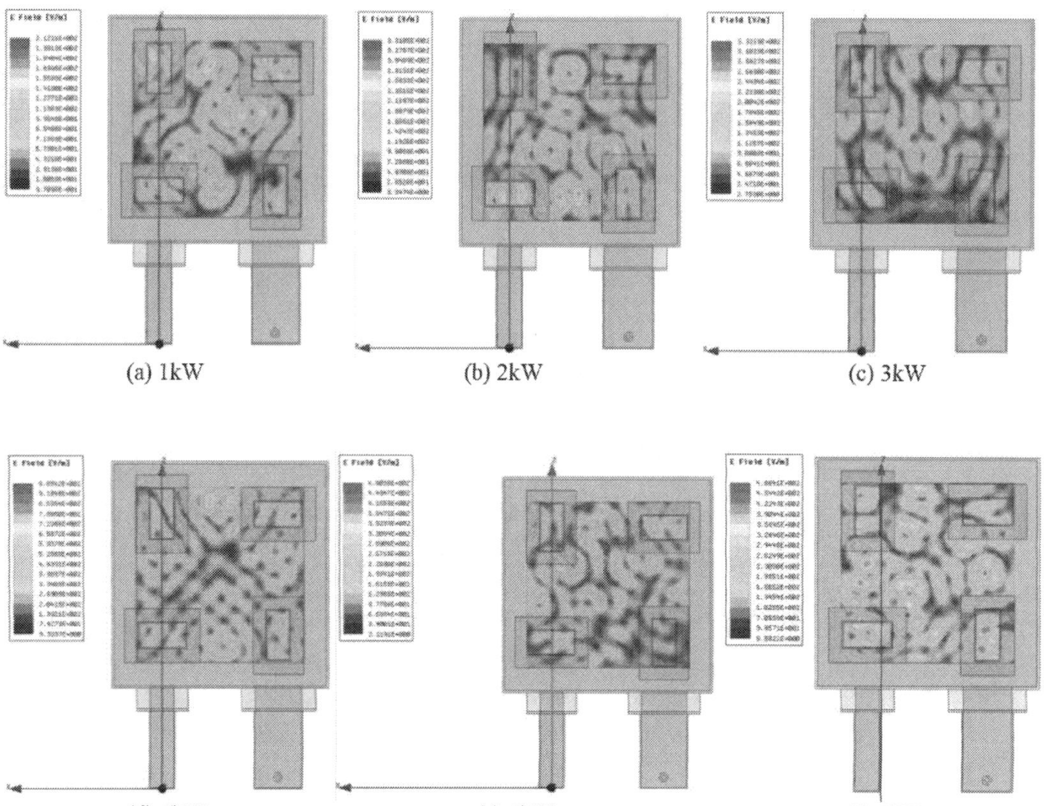

(a) 1kW (b) 2kW (c) 3kW

(d) 4kW (e) 5kW (f) 6kW

Figure 2. Electric field distribution by input power

3.2 Electric field distribution by distance

Simulation was conducted to investigate the electric field distribution according to the distance from the multiple waveguides to the ceramic material. As a result, it was confirmed that changes in the distance from the magnetrons to the ceramic material affected the electric field distribution.

The distance between the magnetron and the ceramic material was set as a ratio. When the distance is 60% and 80%, the deviation value of the E-field is small. Such deviations can cause temperature imbalance or temperature concentration. The current ceramic thickness was set to 10 mm, but it is expected that the electric field distribution will be changed according to the size and shape of the material. Therefore, it is considered that the optimal locations must be derived for each experiment according to the chamber size and material shape. Figure 3 show electric field distribution and simulations results by distance between ceramic and magnetron.

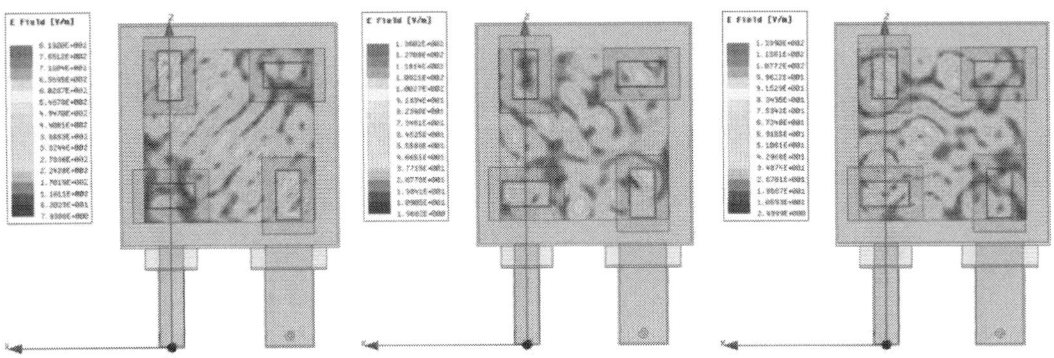

(a) 30% (b) 60% (c) 80%

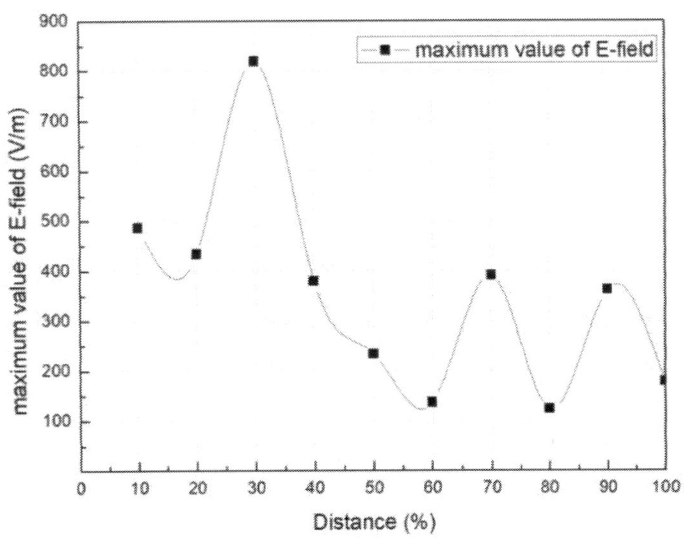

(d) Results of simulation

Figure 3. Simulation results by distance between ceramic and magnetron

4 Conclusions

This study conducted an electric field analysis on a shielded chamber with multiple waveguides using ANSYS HFSS software. The effect of the microwave input power on the electric field distribution was investigated and it was confirmed that the electric field distribution became more uniform as microwave intensity increased when the magnetrons were symmetrically installed. It was also confirmed that changes in the distance from the magnetrons to the ceramic material affected the electric field distribution in the material. Additional simulation will be conducted in the future to investigate the effect of the electric field distribution on the temperature change of the material. Simulation will also be performed on the optimal distance from the ceramic material to the magnetrons and on the electric field distribution according to the patterns of multiple waveguides.

Acknowledgments

This research was supported by Basic Science Research Program through the National Research Foundation of Korea(NRF) funded by the Ministry of Science and ICT(NRF-2017R1A2B4009929)

References

1. Che Woo Seong, Jou. of Korean Soc. of Mechanical Technology, **16**, 1(2014)
2. Adu, B. And Otten, L., Journal of microwave Power and Electro-magnetic Energy, **31**, 1(1993)
3. Han-Ji Ju and Qing Zhao, JOURNAL OF ELECTRONIC SCIENCE AND TECHNOLOGY OF CHINA, **7**, 2(2009)
4. M. Arai, J. G. Binner, G. E. Carr, T. E. Cross, Microwave Processing of Materials III, 269,(1992)
5. D. L. Johnson, D. J. Skamser, M. S. Spotz, Ceramic Transactions: Microwave Theory and Applications in Materials Processing II, **36** (1993)
6. T. Santos, L. C. Costa, M. Valente, J. Monteiro, J. Sousa, Excerpt from the Proceedings of the COMSOL Conference (2010)

A Comparison of FDM Structural Layouts and Implementation of an Open-Source Arm-Based System

David Pollard[1,a], Guido Herrmann[2], Carwyn Ward[3] and Julie Etches[3]

[1]*Bristol Robotics Laboratory, University of the West of England, Coldharbour Lane, Bristol, BS16 1QY, UK*
[2]*Department of Mechanical Engineering, University of Bristol, University Walk, Bristol, BS8 1TR, UK*
[3]*Bristol Composites Institute (ACCIS), University of Bristol, University Walk, Bristol, BS8 1TR, UK*

Abstract. Fused Deposition Modelling (FDM) is a manufacturing process to build components in a layer-by-layer approach through extrusion of polymers from a movable nozzle, allowing for significantly higher degrees of complexity over machined parts. Current FDM systems typically use actuation provided through a gantry or delta structural layout, operating through depositing successive planar layers in a 2.5D process; it has been shown in numerous studies the bonding between layers has significantly lower strength than the homogeneous material or in-plane properties - an issue which can be mitigated through the deposition of curved layers. This paper compares four differing structural layouts of FDM systems (gantry, delta, Stewart Platform, and arm-based) to identify the key advantages of an arm-based method as the increased workspace and manipulability enabling "Additive Finalisation" of components, and suitability for curved layer FDM. Details are then presented of the open-source implementation and evaluation of a 6 degree-of-freedom arm-based FDM printer at the University of Bristol.

1 Introduction

Additive Manufacture (AM) was defined in ISO/ASTM 52900 as a "process of joining materials to make parts from 3D model data, usually layer upon layer" [1]. Components produced through AM are constructed through fusion of raw materials, as opposed to subtractive manufacture where they are machined from a larger block. This provides numerous advantages over the traditionally machined parts, such as reduced material waste, increased design freedom, and suitability for economical low-volume manufacture [2]. Fused Deposition Modelling (FDM) is a method of AM comprising of extrusion of a molten polymer through a movable nozzle, and was pioneered by Stratasys in 1992 [3]. Recent open-source developments and the cheap, non-toxic materials have enabled its use for desktop production of plastic components [2].

A majority of AM processes, including FDM, are based on a slicing procedure to generate a component; a CAD format file is provided, "sliced" into a stack of 2D layers, and a toolpath for the required material deposition is generated for each layer. The effect of layered manufacturing requires support structures for overhanging sections, and the "staircasing", where the layer thickness causes an uneven surface on sloping sections; this effect is visible in FDM where layer heights are typically greater than 0.1 mm [4]. Additionally, the part orientation greatly affects the final properties, as

[a] Corresponding author: d.pollard@bristol.ac.uk

numerous studies have found the bonding between surfaces is significantly weaker than the material properties in the plane of the layer [2, 4].

Previous research efforts have investigated 3D layered manufacturing techniques for FDM. Singamneni et al. [5] evaluated curved layer deposition, with an investigation into toolpath generation for curved layers, and demonstrated improved surface quality and mechanical properties of a curved beam; the compressive load of the curved part under 3-point bending increased by 40% over the traditional 2.5D layered design. Multi-Directional Layered Deposition (MDLD) was explored by Singh et al. [6], where overhangs, which typically require support structures, were eliminated through changing the angle of the slicing plane relative to the build surface. This process allows for the division of a CAD model into supported and unsupported sections and, through varying the slicing orientation between sections, eliminated the requirement for printing of supports for overhanging sections. Such works demonstrate the advantages of printing in multiple dimensions, but limitations imposed by traditional gantry-style mechanisms limit the use of these systems.

This paper details an approach to adaptation of an industrial robotic arm for use in FDM research, to explore aspects of in-situ manufacture. A brief overview of academic multi-dimensional AM is first presented, followed by a comparison of structural layouts of existing FDM machines to identify the use cases for an arm-based FDM system. Section 4 provides an overview of the current implementation at the University of Bristol, using an ABB IRB 140 robotic arm, followed by examples of its performance. Finally, future developments are presented, along with details on access to the open source code developed during this project.

2 Background

To date, there have been a number of systems developed for multi-dimensional FDM based on parallel or serial manipulators. Song et al. implemented a parallel kinematic robot, with the extruder affixed to a 6 degree of freedom (DOF) Stewart Platform [7]; a platform comprising of 6 linear actuators affixed to the extruder. They demonstrated the capability to laser scan a surface, generate a toolpath to maintain an angle normal to the surface, and deposit material on a curved surface. A similar concept was trialled by Dröder et al. [8] demonstrating "Additive Finalisation", a concept where a mass-produced base component has final geometric features added using FDM - with use of a Stewart Platform, they were able to add features onto existing surfaces. A similar project equipped an open-source 3D printer with two additional degrees of freedom, and a laser scanner [9], allowing the identification of a surface, and superimposition of a CAD object to generate a conforming toolpath for deposition onto the scanned surface, and final build – it was noted there is a significantly higher complexity for slicing curved layers over the traditional planar layers due to the possibility of collision with the previously deposited structure.

MIT implanted an early instance of robotic arm-based AM, through use of a Kuka arm with changeable end effectors allowing for foam extrusion, a milling tool to machine the foam to a fine finish, and the FDM of polymers [10]. This robot was operated through generation of the toolpath through a traditional slicing program, and conversion into the Kuka Robot Language file for operation. A second implementation of FDM on a robotic arm is discussed in [11], where a Motoman robot arm demonstrated multi-plane deposition. The commands were produced in the format of GCode, with additional rotational parameters for nozzle orientation, and communicated from the PC to the robot via the MotoCom SDK libraries.

3 Comparison of 3D printing system structures

To identify the merits of robotic arm-based printing, an accuracy comparison was performed between three structural layouts, as presented in Table 1. The Ultimaker 3 [12] and Rostock Max V3 [13] were used as representative machines of the gantry and delta classes respectively, and the 6 DOF ABB IRB 140 [14], as used in Section 4, to compare the arm-based printer. As the Stewart Platform-based printers were only discussed in the literature, data was not available and so discussed below to aid comparison.

Table 1: Comparison of structural layouts suitable for FDM

	Gantry [12]	Delta [13]	Arm [14]
Positioning Accuracy (μm)	125	(variable)	30
Max speed (XY) (mm s^{-1})	300	300	1000 +
Max speed (Z) (mm s^{-1})	5	300	1000 +
Build area / printer footprint	0.356	0.336	22.2

It was noted there are no standardised methods for assessing the accuracy of FDM machines. A common method of stating accuracy is based upon the stepper motor resolution and associated gearing, with the gantry layout achieving 12.5 μm accuracy in the *XY*, and 2.5 μm in *Z*, as described by the manufacturers [12]. However, upon examination of a similar printer, Weiss et al. [15] identified a mean error during a trajectory to be 400 μm at a speed of 90 mm s^{-1}. No data could be found for the accuracy of the delta printer, but uses gearing similar to the XY stage of the gantry printer. The Stewart Platform system used by Song et al. [7] had a stated accuracy of <500 μm, and Dröder et al. [8] note the accuracy of their implemented Stewart Platform printer remained slightly lower than a comparable 3D printer; this may be partially due to the difficulties in implementing inverse kinematics for the high DOF parallel system. In contrast, the robot arm has significantly better positioning capabilities, due to its development for industrial applications. The stated accuracy when following a path is significantly lower, at 670 μm, as discussed in Section5.

Table 1 shows the key difference between the gantry and delta printers is the capability of fast movement in the *Z* axis – as the former has a single stepper motor for each axis, the *XY* positioning is controlled by a belt drive, and the *Z* through a lead screw, increasing the accuracy but decreasing the maximum speed. At typical print speeds, the slower *Z* axis speed of the gantry inhibits the printing of curved layers; it was for this reason Allen et al. to use a delta printer during their investigations into curved layer deposition [16]. Similar to a delta printer, the robot arm is capable of rapid movements throughout its workspace, as an FDM nozzle would typically be within the payload range of the arm.

Finally, the ratio of build area to the printer footprint is relatively low for both the Gantry and Delta robot. The calculation in Table 1 was for the robot arm was performed with the *XY* plane intersecting the base of the robot, as the reachable workspace varies with *Z*, as shown in Figure 1. While the Delta robot is capable of rapid movement in the *Z* direction required for printing curved layers, the inability to position the nozzle normal to the substrate constrains the maximum printed gradient to be dependent on the nozzle geometry. The Stewart Platforms proposed in [7, 8] allow for repositioning, but exhibit a strong positional and orientation coupling.

Through the comparison of the four structural layouts for FDM, use cases for arm-based FDM can be identified as increased manufacturing speed of larger components through utilisation of higher achievable speeds over the workspace – this would require larger diameter nozzles, reducing achievable quality but increasing deposition rates. "Additive Finalisation" would be a second use case, presented by Dröder et al. [8], where the base component could be placed on a conveyor belt adjacent to the robot arm, allowing the arm to locate and print on the surface. A major consideration for its industrial use would be the value added, as a robotic arm would represent a significantly higher investment than a commercial printer, and as such its corresponding capabilities must be higher.

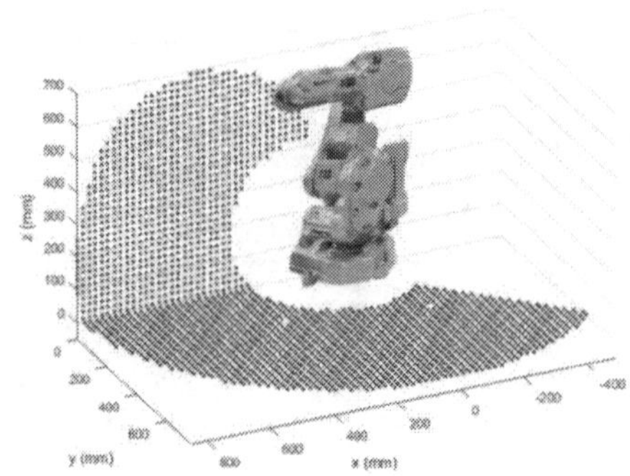

Figure 1: Workspace of IRB 140 robot arm constrained to keeping the end effector normal to the *XY* plane, as used for conventional FDM. Removing the constraint for angle orientation would allow an increased workspace.

4 System design

The robotic system at the University of Bristol was implemented on a 6 DOF ABB IRB 140 robotic arm, which has a 6 kg payload capability, 0.8 m reach, and a pose repeatability of 0.03 mm. The robot cell also contained a 2 DOF IRBP A250 workpiece positioner, and was controlled through an ABB IRC5 controller running RobotWare 5.15. The relevant options enabled on the robot are the 616-1 PC Interface, and a DSQC 652 digital I/O module. An Arduino, connected to a RAMPS 1.4 breakout board, was used to control the extruder temperature and a stepper motor for filament feed. This was connected to the DSQC I/O module via a voltage divider, reducing the voltage from the 24V signal to 4.8V suitable for the Arduino. The extruder is comprised of a BondTech dual drive extruder, directly mounted to an E3D V6 hotend with a 0.6 mm nozzle – selected to enable higher speed printing. Figure 2 shows the system architecture, detailing the key operational functions.

OpenABB [17] was a software package developed for controlling ABB robots through Python, C++, or ROS (Robotic Operating System) released with the MIT License. The implemented code was composed of two parts; a server module written in RAPID, the native language implemented within the IRC5 controller, and a Python class implemented on the controlling PC. To counter the lag of the Ethernet socket communication between the PC and IRC5, movements within a toolpath were sent in a buffer, providing the IRC5 with the coordinates for the execution of a full path to avoid pauses whilst awaiting further position data. As similar paths are often repeated for successive layers in FDM, functions were implemented to save, load, and offset the buffer to reduce time spent communicating coordinates.

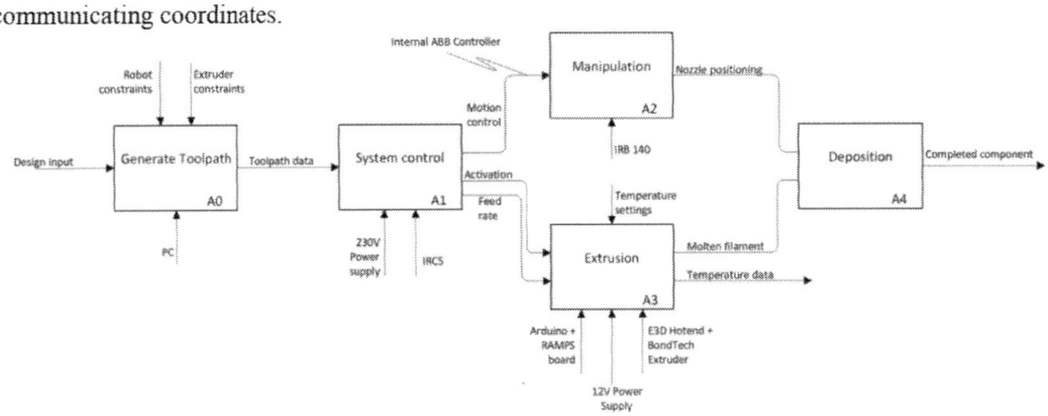

Figure 2: Diagram based on IDEF0 detailing key operations of the robotic system.

The extruder activation signal was supplied by the IRC5 robot controller to ensure synchronisation with the robot motion. To adjust the feed rate for the current path, a signal was passed to the extruder controller through a Group Output; a collection of eight digital signals to representing an 8-bit unsigned integer in the range [0, 255]. The Arduino controlling the extruder was connected to the operator's PC to allow monitoring of extrusion temperature, and adjustment of extrusion settings; this allows the testing of difference control systems, such as those examined by these authors in [18].

5 Results and Discussion

As previously discussed, the stated accuracy of the ABB IRB 140 arm for linear path following is 670 μm, with a pose accuracy of 30 μm [14]. This reduces the theoretical accuracy, but it should be noted that the linear path following accuracy was tested under ISO 9283, where the error was measured and averaged over a range of tool speeds - previous research has shown a significant increase in trajectory following error with speed [19]. Upon implementation, there was no noticeable error along layer lines due to the low theoretical accuracy in the *Z* direction – potentially due to the relatively low speeds used during the printing process, as opposed to the maximum speeds used during the manufacturers tests, allowing for improved trajectory following. The limiting factor on maximum print speed was the inter-layer adhesion, with deposited material peeling from the substrate.

Figure 3 shows two objects produced through the robotic arm FDM process; a rectangular section (left), and a repair patch to a composite panel demonstrating the ability to print within a confined space on an angled surface. During printing, the average time to send a within a buffer was 0.08 s – this could be reduced through consistent orientation throughout a movement to reduce the data sent. A key issue found was the accuracy during the start and stop of the extrusion process, due to the over-extrusion during the IRC5 processing time. These errors were minimised with the addition of a vertical movement at the end of an extrusion process.

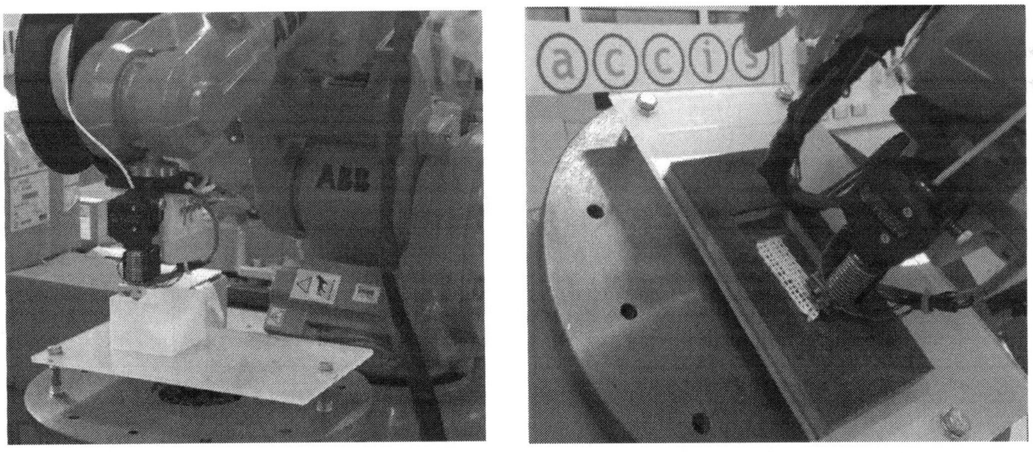

Figure 3: Examples of components produced through FDM with the robotic arm. *Left* shows the printing of a large component, and *right* demonstrates the printing of a repair patch with a composite sandwich panel as a 45°angle.

6 Conclusions and Further Work

This work has examined the different structural layouts used for FDM (gantry, delta, Stewart Platform, and arm), finding the gantry platform has more consistent printing accuracy over the build area, and the delta platform has improved *Z* axis speed, at the cost of an increased volume. The Stewart Platform allows for multi-DOF printing, but at the cost of a reduced workspace, and the arm is capable of rapid movements, higher manipulability, and a workspace significantly larger than its footprint, a factor not exhibited by other structural layouts for FDM. Based on this criteria, it was concluded arm-based FDM printers are suited for higher speed manufacturing of larger components, and "Additive Finalisation".

An implementation of FDM on a 6 DOF robotic arm, in this case an ABB IRB 140, was then presented, describing the hardware setup and a summary of the software communication protocol. Examples of the robot operation were presented, and its flexibility to print on slanted surfaces shown. The control software, based on the OpenABB architecture, used to operate this robot was released under the MIT license, and is available at:

https://github.com/bristolroboticslab/OpenABB-FDM

A number of studies are currently underway utilising the developed platform, such as the adaptation of toolpaths to orient the nozzle for printing on curved surfaces whilst avoiding edge collisions, and work on additive finalisation. Implementation improvements will investigate improvements in extruder control through current tool speed feedback. Additionally, the trade-off between system cost and capability will be explored to identify minimum performance requirements.

Acknowledgements

This work was supported by the EPSRC Centre for Doctoral Training in Future Autonomous Robotic Systems (FARSCOPE) at the Bristol Robotics Laboratory (grant EP/L015293/1).

The date necessary to support the conclusions are included within this paper, and the control software developed over the course of this project is available at https://github.com/bristolroboticslab/OpenABB-FDM.

References

1. ISO/ASTM 52900:2015, (2015)
2. W. Gao et al., Comput Aided Des, **69**, 65-89, (2015)
3. S. S. Crump, US5121329A, (1992)
4. W. Oropallo, L. A. Piegl, Eng Comput, **32**, 1-14, (2016)
5. S. Singamneni, A. Roychoudhury, O. Diegel, B. Huang, J Mater Process Tech, **212**, 27-35, (2012)
6. P. Singh, Y. Moon, D. Dutta, S. Kota, Solid Freeform Fabr Symp Proc (2003)
7. X. Song, Y. Pan, Y. Chen, J. Manuf. Sci. Eng., **137**, 021005-021005-13, 2015.
8. K. Dröder, J. K. Heyn, R. Gerbers, B. Wonnenberg, F. Dietrich, Procedia CIRP, **55**, 122-127, (2016)
9. N. Bausch, D. P. Dawkins, R. Frei, S. Klein, *IFAC-PapersOnLine,* **49**, 583-590, (2016)
10. S. Keating, N. Oxman, Robot Comput Integr Manuf, **29**, 439-448, (2013)
11. I. B. Ishak, J. Fisher, P. Larochelle, IDETC/CIE 2016, (2016)
12. Ultimaker, "Ultimaker 3 specifications," [Online]. Available: https://ultimaker.com/en/products/ultimaker-3/specifications. [Accessed 16 12 2017].
13. SeeMeCNC, "Rostock Max V3," [Online]. Available: https://www.seemecnc.com/products/rostock-max-v3-desktop-3d-printer-rtp. [Accessed 16 12 2017].
14. ABB AB, Robotics, "Product specification IRB 140," (2017)
15. B. Weiss, D. W. Storti, M. A. Ganter, Rapid Prototyping J, **21**, 482-490, (2015)
16. R. J. A. Allen, R. S. Trask, Additive Manuf., **8**, 78-87, (2015)
17. M. Dawson-Haggerty, "OpenABB," [Online]. Available: https://github.com/robotics/open_abb. [Accessed 28 11 2017].
18. D. Pollard, C. Ward, G. Herrmann, J. Etches, Procedia Manuf., **11**, 536-544, (2017)
19. M. Slamani, A. Nubiola, I. Bonev, Ind Rob, **39**, 57-68, (2012)

Touch Screen Calibration Based on Laser Sensors

Seung Byum Seo[1] and Sang-Young Cho[1,a]

[1] *Division of Computer & Electronic Systems Engineering, Hankuk University of Foreign Studies, Republic of Korea*

Abstract. Interactive learning is becoming important in teaching and learning. In particular, the electronic board has educational significance in terms of being able to share and discuss learner's contents with instructors. However, the existing electronic board is not easily used in an educational environment due to the high cost and complicated installation process. This paper propose a mathematical model to calibrate the position of user coordinates by installing two laser sensors and a reflector on a wall. We virtually locate 9 points with software. To validate the algorithm, Raspberry Pi2 is used as a virtual sensor and simulation was performed by PC using RS-232 communication. This algorithm is powerful because up to four pre-saved coordinates are used to provide the higher accuracy.

1 Introduction

Electronic board, as known as SMART Board, is one of the devices for inputting information to a computer. The electronic board recognizes the coordinates or the path of the point pointed by the user and transmits the coordinates to the computer. The importance of electronic blackboard is increasing as a tool to help communication between teachers and learners [1].

However, the existing electronic boards have disadvantages due to high cost and complicated installation environment [2]. These electronic boards inlcude electromagnetic induction system, pressure sensitive system, and ultrasonic system. In this paper, we propose a calibration method that uses laser scanned method. Unlike previous systems, our system is free from unintended obstacles inlcuding acoustic noise and electromagnetic signals. Since our system can be installed anywhere on a wall, it is also affordable and precise [3].

This paper is organized as follows. Section 2 discusses the overview of our proposed system. Section 3 introduces the proposed algorithm. Finally, section 4 discusses the results of this paper and the future works needed.

2 System Overview

[a] Corresponding author: sycho@hufs.ac.kr

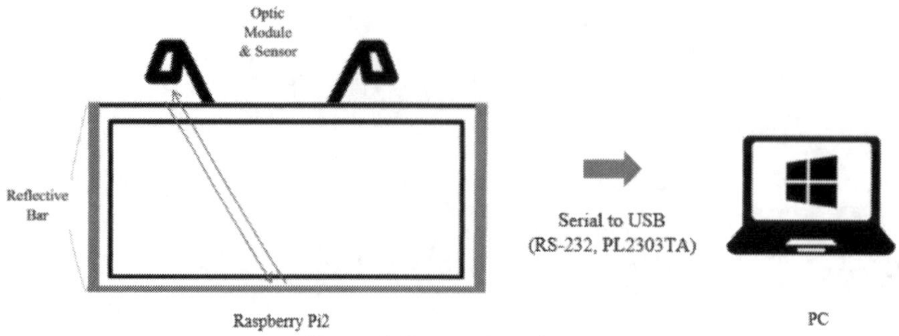

Figure 1. System Overview

Various touchscreen methods are used to implement electronic blackboards [4]. The proposed system consists of optic module, laser sensor, and reflective bar. For simulation, Raspberry Pi2 takes the role of laser sensor and electronic board. PC receives data from Raspberry Pi2 and uses it to process the location entered by the user (Figure 1).

Python was used as the development language for Raspberry Pi2 to simulate the sensor. On the PC, GUI is configured using C # and WPF [5]. RS-232 serial communication was used for communication between PC and Raspberry Pi2.

In this paper, we assumed that the laser sensor always rotates counterclockwise. In addition, the sensor on the left side from user perspective is called right sensor because it is installed on the left side from the sensor standard. Finally, the screen output projected by the projector is called *display*, while the wall where the reflection plate is installed is called *screen*.

3 Mathematical Model

Figure 2. Four stages of algorithm

Figure 2 shows the operation steps. The ultimate goal of the algorithm is to find the coordinates of the input values from the user. In stage1, the algorithm initializes coordinates by arranging nine points in software. In stage2, sector division step, a total of 7 sectors are created based on nine points. In stage3, coordinate search, trigonometric function is used to find the position of the sensor coordinates in each sector. In stage4, two angles (left and right) outputted from the sensor and the sensor coordinates obtained in stage3 are used to find the position of the value inputted by the user

3.1 Coordinate Initialization

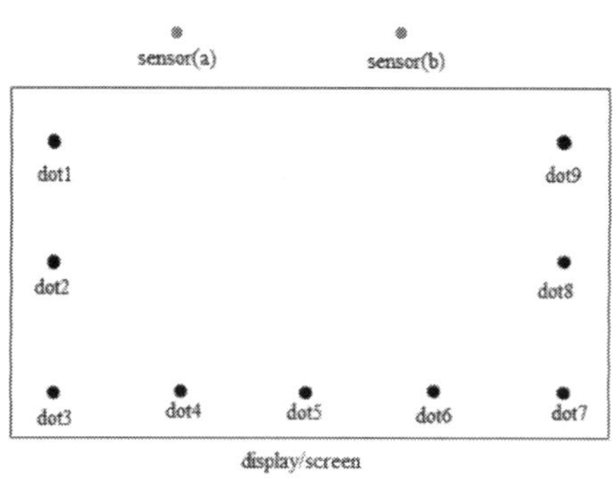

Figure 3. Setting Default Coordinates

Nine points were displayed on the display to determine the position of the sensor at various locations in Figure 3. The position of the coordinates can be set arbitrarily, but it is suggested to be close to the border of the screen. This is because, at coordinates close to the rim, the need for correction is higher because the sensor is most likely to detect errors.

3.2 Sector Division

Based on the above nine coordinates, we divide the area into seven sectors. Then, the coordinates of the *sensor(a)* and the *sensor(b)* are stored in a separate space by using three consecutive coordinates for each sector. For example, in the area corresponding to *dot1*, *dot2*, and *dot3*, the coordinates of the *sensor(a)* and the coordinates of the *sensor(b)* can be known by using coordinate search alogrithm given in stage 3. The coordinates of sensors are saved separately in a list. Table 1 shows the examples of how sectors are divided from sector one to sector seven.

Table 1. Examples of seven sectors and their type

Sector One (Type C)	**Sector Two** (Type B)	**Sector Three** (Type A)	**Sector Seven** (Type C)
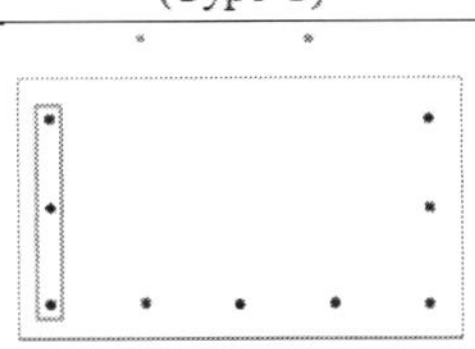	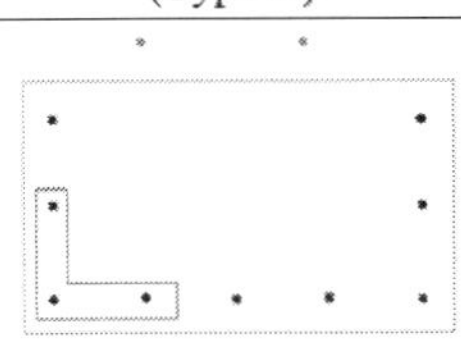	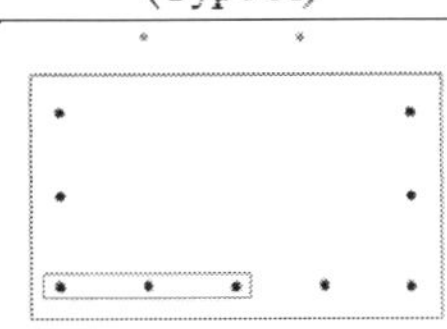	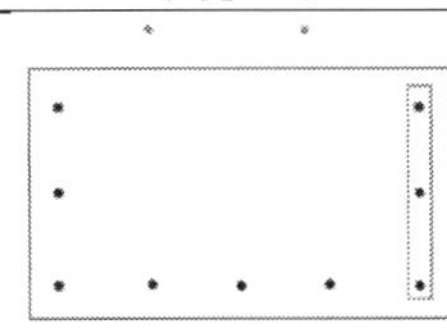

3.3 Sensor Coordinate Search

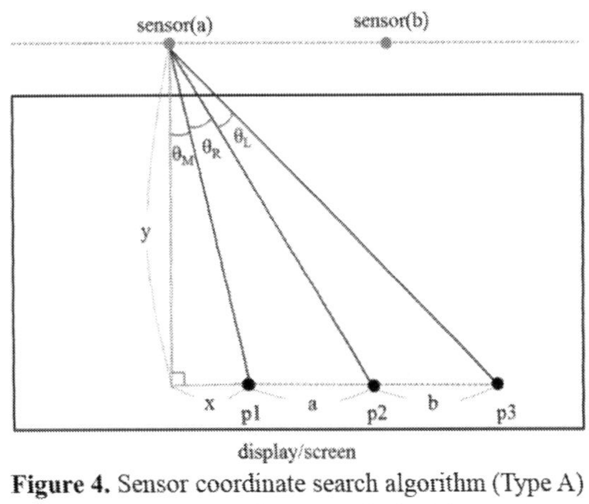

Figure 4. Sensor coordinate search algorithm (Type A)

In each sector, the coordinates of *sensor(a)* and *sensor(b)* in Figure 4 can be obtained through the following algorithms. The seven sectors can be divided into three types. While *sector three, four, five* are arranged horizontally(Type A), *sector one and seven* is arranged vertically (Type C). *Sector two and sector six* are arragned in a perpendicual diretion (Type B).

The method of obtaining the sensor coordinates in type A is as follows. In type A, there are three arbitrary coordinates on a straight line. In this case, the values of θ_R and θ_L can be measured by the angle formed by *sensor(a)* and *p1, p2, p3*. Also, the values of *a* and *b* can be measured as the number of pixels between *p1* and *p2*, and the number of pixels between *p2* and *p3*, respectively. Therefore,

Type A is an algorithm that obtains x and y when θ_R, θ_L, a, and b are given. To solve this problem, the following formula can be obtained.

$$tan\,\theta_M = x/y \tag{1}$$

$$tan(\theta_R + \theta_M) = (a + x)/y \tag{2}$$

$$tan(\theta_L - \theta_M) = (b - x)/y \tag{3}$$

Based on tangent addition theorem, equation (2) and (3) can be simplified as below

$$\frac{tan\,\theta_R + tan\,\theta_M}{1 - tan\,\theta_R\,tan\,\theta_M} = \frac{a}{y} + \frac{x}{y} \tag{4}$$

$$\frac{tan\,\theta_L - tan\,\theta_M}{1 + tan\,\theta_L\,tan\,\theta_M} = \frac{b}{y} - \frac{x}{y} \tag{5}$$

The value of $tan\,\theta_M$ can be obtained by replacing x/y with $tan\,\theta_M$ and simplify the equation in terms of y.

$$tan\,\theta_M = \frac{a\,tan\,\theta_L - b\,tan\,\theta_R}{(a+b)\,tan\,\theta_R\,tan\,\theta_L} \tag{6}$$

By plugging in the value of $tan\,\theta_M$ into equation (2) and (3) respectively, the value of θ_M, x, y are obtained.

$$\theta_M = tan^{-1}\left(\frac{a\,tan\,\theta_L - b\,tan\,\theta_R}{(a+b)\,tan\,\theta_R\,tan\,\theta_L}\right) \tag{7}$$

$$y = \frac{a}{tan(\theta_R + \theta_M) - tan\,\theta_M} \tag{8}$$

$$x = y\,tan\,\theta_M \tag{9}$$

All of the coordinates of *sensor(a)* and *sensor(b)* in type A, B, and C can be obtained in a similar way.

3.4 User Point Search

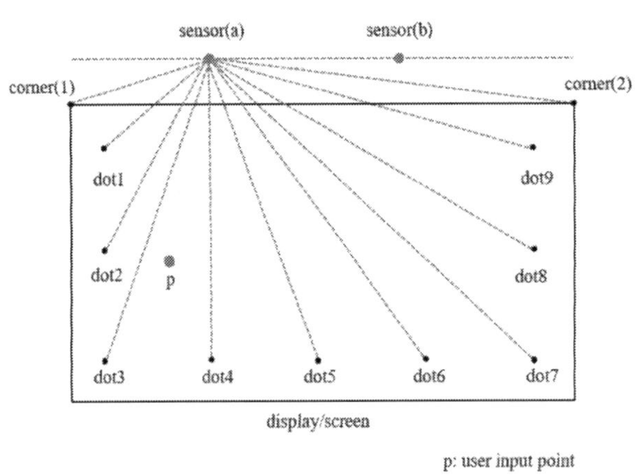

Figure 5. Default angle and region setting

After storing the sensor position coordinates by sectors, the coordinates input by the user can be obtained. To obtain the coordinates input by the user, the following angle should be obtained as shown in Figure 5. *D(n)* where n is an integer from 0 to 9 denotes the default sensor angle.

$$D(0) = \angle corner(1), sensor, dot1$$
$$D(1) = \angle dot1, sensor, dot2$$
$$D\,(2) = \angle dot2, sensor, dot3$$
$$\ldots$$
$$D(9) = \angle dot9, sensor, corner(2)$$

The values from *D(n)* are calculated as above are stored and managed separately. Then sector angle determines which sector does the user point belongs to. Region scope, *R(n)*, can be obtained by subtracting the angle formed by each angle from the angle at which the sensor outputs.

$$R(n) = \theta_{input} - \sum_{k=0}^{n-1} D(k) \quad (1 \leq n \leq 10, \text{n is integer}) \tag{10}$$

Since the region scope is achieved from R*(1)* to R*(10)*, it is possible for system to locate which regions the user point p belongs to. For example, suppose that the user point *p* is located on the touch screen in Figure 5. Since *p* belongs to the region consists of *dot2*, *dot3*, *dot4* and *dot5*, it can be inferred that *p* belongs to *R(4)* and *R(5)*.

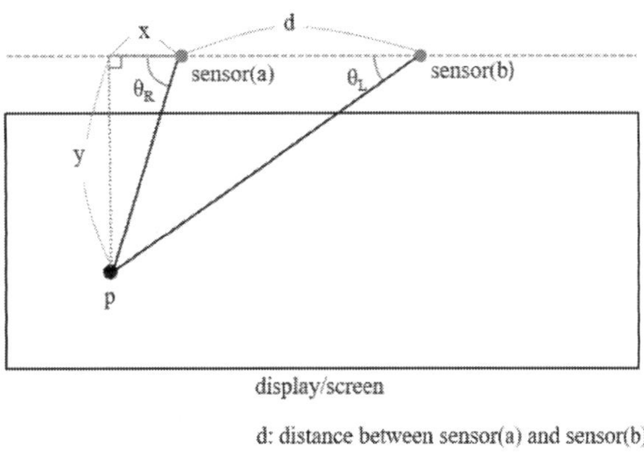

Figure 6. User coordinate search

User point p of Figure 6 can be obtained by solving the equations below.

$$\tan\,\theta_R = y/x \tag{11}$$
$$\tan\,\theta_L = y/(d + x) \tag{12}$$

Therefore, the user point p can be obtained as equation (13) and (14).

$$x = (d \tan\,\theta_L)/(\tan\,\theta_R - \tan\theta_L\,) \tag{13}$$
$$y = x \tan\theta_R \tag{14}$$

This algorithm is powerful because the data up to four coordinates could be used. In the worst situation, the value of p is calibrated by averaging *sensor (a)* and *sensor(b)* from one region. In an ideal circumstance, p is calibrated by averaging *sensor(a)* and *sensor(b)* from two regions which leads to a higher accuracy.

4 Conclusion

Figure 7. Simulation Environment

Figure 8. Simulation result

In this paper, two laser sensors, reflection bar, and nine virtual points were used to calibrate a virtual touch screen. As a result, we developed a total of four stages algorithm and built a test-bed to validate it. For simulation, Raspberry Pi2 was assumed as a laser sensor and was connected to a PC through RS-232 communication method (Figure7 and Figure 8). As a result, we have confirmed that the proposed algorithm is precise and efficient for touch screen calibration based on using a beam projector. For future work, we plan to test the system in a real world setting. At present, the simulation assumed that angles recognized from sensors are precise. However, the external factors from the real world is expected to be more complex. Therefore, future work will include examining unknown dynamics of calibration so that the higher accuracy is guaranteed.

This research was supported by a grant(16CTAP-C117026-01) from Technology Advancement Research Program funded by Ministry of Land, Infrastructure and Transport of Korean government

References

1. M. Long, Input in second language acquisition, ERIC, p 473, (1985)
2. Sungchul Chun, In-soo Koo, "Beam projector calibration System based on Zigbee," *The Journal of the Institute of Internet, Broadcasting and Communication,* vol. 11, pp. 13-19, (2011)
3. F. Richards-Gustafson, "Smartboard vs. Projector," [Online]. Available: http://smallbusiness.chron.com/smartboard-vs-projector-14141.html (2017)
4. Touchscreen, [Online], Available: https://en.wikipedia.org/wiki/Touchscreen (2017)
5. T. Robinson, [Online]. Available: http://libusbdotnet.sourceforge.net/V2/Index.html (2010)

A Study on Multi-body Dynamic Modelling and Optimization for Mechanical Quick Closing Valve

Wang Bo[a], Li Junyan, Wu Lifang and Chen Xuekai

Research Institute of Physical and Chemical Engineering of Tianjin, Tianjin 300180, China

Abstract. The multi-body dynamic modelling and optimization methods of mechanical quick closing valve are studied in this paper. The Lagrange equation is specified characterizing the closing process. Normal contact force and tangent friction force between the components are obtained by using equivalent spring-damper model and no-linear coulomb friction model. The theoretical result of the valve's kinematic response is verified by the experiment. In order to analyse the relationship between structural parameters and the closing behaviour intuitively, a single degree of freedom model is provided. It is shown that the mass of the control rod is the key parameter exerting important influence on the cut-off time of the valve. The cut-off time will be shortened by about 1.2ms while the mass of control rod reduces 0.01kg.

1 Introduction

The mechanical quick closing valve in this paper, as a kind of stop valve with a wide application, is installed between the pressure vessel and the downstream pipe. In this way, should the pressure vessels fail, the valve can protect the equipment connected with downstream pipe from the impact flow. Shortening the cut-off time will improve the protection capability of the valve significantly. And establishing dynamic model to describe the closing behaviour is precondition for the structural optimization.

A multi-body dynamic model of the mechanical quick closing valve is established based on Lagrange equation and contact phenomena modelling method [1-4] in this paper. The effect of the structural parameters on the closing behaviour of the valve in working situation is studied. Then a optimization method of the valve is provided based on the theoretical and experimental result.

2 Multi-body dynamic modelling of the quick closing valve

As shown in Figure 1, the quick closing valve is actuated by a compressed spring. The control rod is linked to pressure vessel by means of a sliding transmission structure. Under normal circumstances, the connecting shaft and control rod provide braced force to keep the valve opening. When the pressure vessel fails, the control rod will be not supported anymore. Then the guiding shaft will slip down immediately because of the force of the compressed spring. So the upstream and downstream pipes will be cut off by the deformed rubber sealing element.

[a] Corresponding author : wang13o@163.com

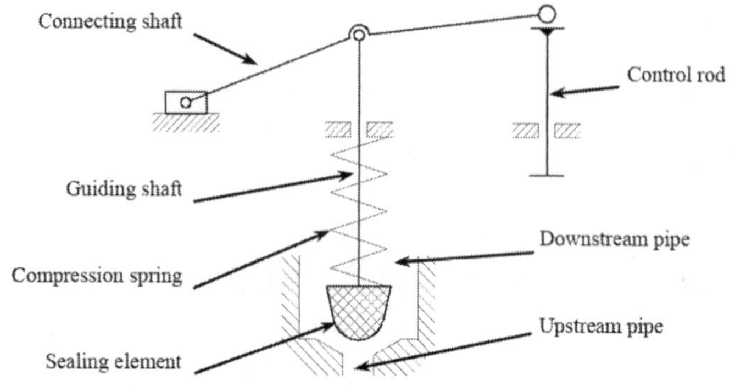

Figure 1. Structure diagram of the quick closing valve

A multi-degree of freedom dynamic model is established to quantitative describe the closing behaviour in this section. The following assumptions are applied:

a. The buffering action caused by the rubber sealing element is not considered. Because its effect on the cut-off time of the valve is slightly according to the experimental result.

b. Every component of the system is regarded as rigid.

Based on the assumptions, the kinematic response of the valve during closing process can be described by using Lagrange equation:

$$\begin{cases} \frac{d}{dt}\left(\frac{\partial E}{\partial \dot{q}_j}\right) - \frac{\partial E}{\partial q_j} + \sum_{i=1}^{n} \frac{\partial \Phi_i}{\partial q_j} \lambda_i = F_i & (i, j = 1 \sim 18) \\ \Phi_i = 0 \end{cases} \tag{1}$$

q_j -- generalized coordinate of the component barycentre (including the translation and rotation of the connecting shaft, guiding shaft and control rod)

E -- kinetic energy of the system

Φ_j -- constraint equation of generalized coordinate, obtained from the geometric structure

F_j -- generalized force striking the direction of generalized coordinate

λ_j -- Lagrange multiplier

Once the valve switch activates, every component is in state of dry friction without lubricant. So it is significant to calculate the contact force correctly. The impact force equation applied in this paper is as follows [5,6]:

$$F_n = K\sigma^e + Cv_n \tag{2}$$

F_n -- normal contact force

K -- Hertz contact stiffness ($K = 10^8 N/m$)

σ -- penetration depth of the contact point

e -- exponent of the contact force ($e = 2.2$)

C -- damping factor ($C = 10^4 N \cdot s/m$)

v_n -- normal velocity of the contact pair

The nonlinear coulomb friction equation:

$$\mu(v_\tau) = \begin{cases} -sign(v_\tau) \cdot \mu_d, |v_\tau| \geq v_d \\ -step(|v_\tau|, v_d, \mu_d, v_s, \mu_s) \cdot sign(v_\tau), v_s < |v_\tau| < v_d \\ step(v_\tau, -v_s, \mu_s, v_s, -\mu_s), -v_s \leq v_\tau \leq v_s \end{cases} \tag{3}$$

v_τ -- slip velocity of the contact pair

v_s -- stiction transition slip velocity ($v_s = 0.01 m/s$)

v_d -- dynamic transition slip velocity ($v_d = 0.1 m/s$)

μ_s -- static friction coefficient ($\mu_s = 0.7$)

μ_d -- dynamic friction coefficient ($\mu_d = 0.45 \sim 0.55$)

The equation step is a cubic polynomial interpolating function that provides the relationship between friction coefficient and slip velocity. It is as follows with given points (x_0, y_0) and (x_1, y_1).

$$step(x, x_0, y_0, x_1, y_1) = y_0 - (y_0 - y_1) \cdot \left(\frac{x - x_0}{x_1 - x_0} \right)^2 \cdot \left(3 - 2\frac{x - x_0}{x_1 - x_0} \right), x_0 < x < x_1 \tag{4}$$

The solution of the dynamic model composed by Equation (1) ~ (4) can be obtained by software conveniently. The main parameters for the model are given in Table 1.

Table 1. Table of the main parameters for the model

No	Name	Value
1	Mass of the guiding shaft / kg	0.024
2	Mass of the connecting shaft / kg	0.077
3	Mass of the control rod / kg	0.093
4	Stiffness of the spring / N/m	6.2×10^3
5	Preload of the spring / N	148.8
6	Hertz contact stiffness / N/m	10^8
7	Exponent of the contact force	2.2
8	Damping factor / N·s/m	104
9	Stiction transition slip velocity / m/s	0.01
10	Dynamic transition slip velocity / m/s	0.1
11	Static friction coefficient	0.7
12	Dynamic friction coefficient	0.45~0.55

In order to verify the theoretical result, the experimental data of guiding shaft falling displacement and velocity during the closing process are obtained by applying high-speed photography. The comparative schemes of them are given by Figure 2.

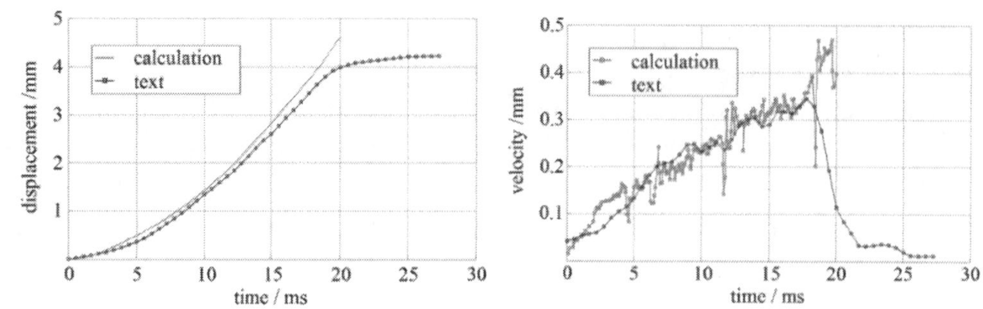

(a) (b)

Figure 2. Theoretical and experimental result of the falling displacement (a) and velocity (b)

According to the experimental data, the valve will be closed when the falling displacement of guiding shaft reaches 4mm. The theoretical result of the cut-off time is 18.16ms, differs by 9.7% from the experimental data (20.11ms). It is because the buffering action caused by the rubber sealing element is not considered. Therefore the model cannot describe the retarded motion of the guiding shaft by the end of the closing operation. However the theoretical model is accurate enough, especially in the accelerated motion phase.

3 Structural optimization and experiment

3.1 Identification of the key structural parameters

According to dynamic model in the previous section, the kinematic response of the valve during closing process is available. However it is not intuitive to analyse the effect of structural parameters on cut-off time. So a single degree of freedom model is established to describe the relationship between structural parameters and the closing behaviour.

The model is simplified further based on two additional assumptions compared to section 2.2:

c. All joints are considered to be ideal constraints. So the system is simplified as a rod mechanism with only one downward degree of freedom of the guiding shaft.

d. The contact forces, includes normal contact force and tangent friction force, are ignored.

Based on the above additional assumptions and equation (1), a concise equation of generalized coordinate and generalized force can be obtained by introducing equivalent moment of inertia J_e:

$$J_e\ddot{q}' + \frac{1}{2}\frac{dJ_e}{dq'}\dot{q}'^2 = F_k - F_r\frac{\dot{y}_r}{\dot{q}'} \tag{5}$$

q' -- generalized coordinate of the guiding shaft

q' -- equivalent moment of inertia of the system

F_k -- force of the compressed spring

F_r, y_r -- inertial force and displacement of the control rod

The equation shows that F_k and F_r are the key factors of the closing operation. On the premise of maintaining the geometric parameters, the cut-off time can be shortened by adjusting the spring stiffness and the mass of the control rod. However increasing stiffness will make the spring larger and also increase the load of other structure. So the discussion will be focused on the effect of control rod mass.

3.2 The effect of control rod mass on the cut-off time

In practice, the mass of control rod is about 0.093kg. Three kinds of light control rod are designed by applying light material and hollow structure. The numerical solution of the dynamic model composed by Equation (1) ~ (4) and experimental data of cut-off time are obtained simultaneous (As shown in Table 2).

Table 2. Table of the cut-off time obtained by different control rod

No	Mass / kg	Experimental result		Numerical solution	
		cut-off time / ms	rate of descent	cut-off time / ms	rate of descent
1	0.093	20.11	——	18.16	——
2	0.060	15.44	23%	13.62	25%
3	0.033	12.58	37%	10.80	41%
4	0.021	11.19	44%	8.88	51%

The numerical solutions are consistent with the experimental result in every scheme. And the conclusion provided in the previous section is verified as well. The cut-off time is negatively correlated with the control rod's mass. The experimental result shows that the cut-off time will be shortened by about 1.2ms while the mass of control rod reduces 0.01kg.

4 Conclusions

The multi-body dynamic model which considers the influence of normal contact and tangent friction force is provided to study the kinematic response of mechanical quick closing valve. Its numerical solution of the valve's cut-off time is verified by the high-speed photography experiment. Based on the given valve in this paper, the dynamic model can describe the closing behaviour accurately, especially in the accelerated motion phase.

A simple model is provided to analyse the effect of structural parameters qualitatively. It is shown that reducing the mass of the control rod is an effective method to shorten the cut-off time and improve the protection capability of the valve thereby. Based on the numerical solution and experimental result, there is a good linear relationship between them. The cut-off time will be shortened by about 1.2ms while the mass of control rod reduces 0.01kg.

References

1. H. Rahnejat, P I MECH ENG C-J MEC, **214**, 149 (2000)
2. G. Gilardi, I. Sharf, MECH MACH THEORY, **37**, 1213 (2002)
3. S.W. Kim, *Contact dynamics and force control of flexible multi-body systems* (Ph.D. Dissertation, Canada, McGill University, 1999)
4. Paul, B. MECH MACH THEOR, **10**, 481(1975).
5. K.E. Harald, S. Frantisek, M. Ante, PARTICUOLOGY, **8**, (2010)
6. B. Shi, *A study on dynamic modelling and virtual prototype of 600MW supercritical steam turbine reheat-stop-valve* (Ph.D.Dissertation, China, Shanghai Jiao Tong University, 2008)

Research on Systematic Design and Installation Based on Explosion Protection Technology Combined with Traction Elevator

Changming Li[1] ,Qibing Wang[1,a], Yonggang Leng[1] and Yu Li [2]

[1] *Tianjin University, Tianjin, China, 300072;*
[2] *Zhejiang Machinery Industry Information Research Institute, Zhejiang, China, 310009*

Abstract: According to GB, NEC, ASME17.1, ANSI A10.4 and EN 81.1 standard, there is a danger of explosive gas or dust. And then for elevator explosion protection, we design a device type up to requirements of NEC Class 1, Division 1, Group C&D as well as NEC Class 1, Division 2. It is also a controllable device type working well in highly explosive and hazardous work areas. We keep the GEDA explosion-proof lift fully fixed to protect workers and materials to the maximum extent. In addition, explosion-proof elevators in China must meet the standards of GB3836.1/2, GB50257, GB7588, etc. Combining the engineering practice over the years, the authors put forward in this paper a method of design and development as well as components based on explosion protection combined with traction elevator, which opens a new window for the systematic design and installation optimization of explosion-proof elevator.

1 Preface

Flammable and explosive dangerous media is common in the production environment of industries such as coal, oil, metallurgy, chemical, textile and pharmaceutical. With rapid development of these industries and people's increasing safety awareness of explosion protection in China in recent years, electrical apparatus for explosive atmospheres is widely used in relevant hazardous areas. For example, explosion-proof traction passenger-freight elevator and explosion-proof hydraulic passenger-freight elevator have become standard equipment in all explosion-proof places of industrial categories such as cereal, food, pesticide, fertilizer, wood processing, mine, textile, papermaking rubber, plastics, chemical, etc. [1] However, there are not many enterprises focusing on the research, development and production of explosion-proof elevator in China, which are also uneven. What's worse, explosion-proof elevators are different from ordinary passenger-freight elevators in performance and requirements, so the operation place and some standardized technical requirements for ordinary passenger-freight elevators can not be fully applied to explosion-proof elevators. Combining the engineering practice over the years, the authors briefly analyze and discuss the particularity of explosion-proof elevator in this paper in the case of explosion-proof traction elevator design, which applies to explosive gas environment area 1 and IIBT4.

2 Design and Installation of Electrical Components

[a]Corresponding author : 43410697@qq.com

2.1. Design of Electrical Components

2.1.1. Design of Electric Control Cabinet

As the core of explosion-proof elevator design, electric control cabinet, located in the elevator room, decides whether the manufacturer has the ability to design and produce explosion-proof elevators. First of all, it is required to design the shell of the electric control cabinet to meet requirements of dIIBT4. Its strength and explosion-proof surface design must meet the requirements of GB3836.1/2. The larger the net volume of the shell, the longer the flameproof joint surface L, which means the strength requirements of the shell are higher. As there are a lot of parts in the electric control cabinet for elevators, if we design it into a single-cavity structure, then taking into account the requirements of the strength of the shell and the length of the flameproof surface, the control cabinet shell is bound to be very thick and then its weight as well as cost will be quite large. [2] Therefore, we can design it into a two-cavity structure, one for wiring and the other one for placing the intrinsically safe circuit, control circuit, inverter and so on. As the inverter is installed in the aluminum alloy control cabinet, the temperature rise will be relatively high. As for this, we can install an aluminum type radiator in the upper part of the control cabinet so that we can ensure that the inverter works normally in the explosion-proof control cabinet and that the elevator will not be affected by the poor heat dissipation. It is notable that the explosion-proof elevator control circuit has the requirements of the intrinsically safe circuit, which not only needs to meet requirements of the intrinsically safe circuit in the explosion-proof terminology, but also must meet the requirements of the elevator function essential safety.

2.1.2. Design and Selection of Other Electrical Components

For most parts of the explosion-proof elevator, we can use the existing mature explosion-proof products, for example, traction machine with explosion-proof three-phase asynchronous motor, explosion-proof telephone, explosion-proof brakes, door machines, fans and so on. However, we have to take the three points into consideration during our design:
First of all, the type of explosion protection shall apply to hazardous areas in Zone 1, so we can select the type according to the regional category based on Article 5.2 in GB3836.15-2000 standard.
Secondly, select the type according to temperature group [3].

Table 1. Relationship among Temperature Group, Surface Temperature and Ignition Temperature

Temperature group of the electrical equipment	Maximum surface temperature of the electrical equipment	Ignition temperature of gas or steam
T1	450℃	>450℃
T2	300℃	>300℃
T3	200℃	>200℃
T4	135℃	>135℃
T5	100℃	>100℃
T6	85℃	>85℃

As shown in the table 1, equipment T4-T6 applies to dangerous places T4 requires.
Thirdly, select the type according to equipment category.

Table 2. Relationship between Gas/Steam Classification and Equipment Category

Gas/steam classification	Equipment category
IIA	IIA, IIB or IIC
IIB	IIB or IIC
IIC	IIC

As shown in the table 2, equipment graded IIB and IIC applies to dangerous places IIB requires. There are also some simple electrical parts or components, such as trip switches, limit switches and so on. For these components, it's not necessary to use mature explosion-proof products because they are just a switch contact, with only resistance circuit internally. According to requirements of GB3836.4-2010 "explosive environment Part 4: equipment protected by the intrinsically safe 'I'", power supply directly though explosion-proof safe shed is enough to meet the requirements of explosion protection. [4] As this type of components is in large consumption in explosion-proof elevators, this design will thus save a lot of cost.

2.2. Installation of Electrical Components

The installation of explosion-proof electrical equipment shall comply with the requirements of GB50257-1996 "Codes for construction and acceptance of electrical devices in electrical device installation project explosion and fire hazardous environment" and GB3836.15-2000 "Electrical equipment in explosive gas atmospheres (Part 15 - electrical installation in dangerous places (except coal mines) ". According to the authors' years of experience, the following questions are common in the installation of explosion-proof electrical equipment:

- The intrinsically safe wiring doesn't use shielded cable and is not marked blue.
- Flameproof, increased safety products tightening nuts are mixed up and redundant port is not effective blocked.
- Use raw tape on the explosion-proof surface to meet the sealing requirements.
- The safe shed is placed in the safety-type junction box.
- The metal housing reserved for the ground is not earthed.

The basic electrical equipment made by Ex i explosion-proof technology has the advantages of light weight, small size, low cost, electrical maintenance and application in hazardous area 0, so this technology is the best protection technology in fieldbus. It is the trend of the fieldbus practice theory and overcomes the restriction of the Ex i explosion-proof fieldbus. [5]

3 Explosion-Proof Design of Non-Electrical Parts

3.1. Requirements for Materials

The selection of materials should be taken to prevent the occurrence of mechanical sparks and static electricity, which requires the supplier to provide the necessary material reports or supporting documents, such as the selection of materials for the floor of the room, the pit, the car and the car wall. [6]

3.2. Design of Speed Governor

Explosion-proof elevators are more and more widely used in the flammable and explosive places. Due to the particularity of the environment, in the design of explosion-proof elevators, various methods need to be taken to achieve the purpose of explosion protection, such as flameproof and intrinsically safe circuit. In addition, it is important to pay special attention to the collision and friction between the various parts of the elevator. Apart from the elevator traction pulleys and lifting ropes, the speed governor pulleys and ropes are also the main components that are prone to generate heat when friction occurs. The governor-tensioning device shall contain a device for preventing the wire rope from being disengaged and the device shall be of a sparkless material. The tensioning device shall be set to ensure that the tensioning device will not strike the ground when the rope is broken. [7] And the actual temperature can be achieved through the fast inverse algorithm as a boundary condition. The proposed method applies to a composite temperature sensor with inner and outer two-layer sheath and composite heat transfer process. [8]

3.3. Design of Buffer

A spring bumper or polyurethane bumper can be used and the rubber cushion can be added to the surface of the spring bumper against the counterweight or car. [9]

3.4. Design of Other Parts

Measures to prevent sparks shall be taken for parts with a potential risk of spark in normal use, such as traction wheel, guide wheel and reverse sheave rope, brake block and clamp, safety tongs, shaft of shift mechanism, trigger plate used when the safety gear hits the safety contact, boots lining of the sliding boots, one of the two parts of the impacting door locks, door hanging wheel, door block wheel, door guide shoes, etc. [10-13]

4 Conclusions

Explosion-proof elevator is a kind of special lifting equipment used in special environment. This article briefly describes the necessity of explosion-proof elevators in the design and supporting of explosion protection. Some of the design and supporting requirements described in this article are optional and can provide a useful reference for the design and use of explosion-proof elevators, which will therefore help to enhance China's development, production and use of explosion-proof elevator and promote the healthy and rapid development of explosion-proof elevator industry.

References

1. Heyan Wang, Liyong Ma, Biao Li and Sun Tao. *A Study on Running-in Quality Evaluation Method of Power-shift Steering Transmission Based on Performance Parameter Testing [J]* Lubrication Engineering 3 (2008): 025.
2. Jianguo Kong, Weirong Zhu, Wubiao Jiang. *Design of the Dual-PWM Control Energy Feed-back Elevator Drive System Based on DSP and IPM [J]* Electric Drive 8 (2007): 004.
3. General Administration of Quality Supervision, Inspection and Quarantine of the People's Republic of China (AQSIQ), China National Standardization Management Committee, GB3836.15-2000 *Electrical equipment in explosive gas atmospheres (Part 15 - electrical installation in dangerous places (except coal mines)*
4. General Administration of Quality Supervision, Inspection and Quarantine of the People's Republic of China (AQSIQ), China National Standardization Management Committee, GB3836.4-2010 *explosive environment Part 4: equipment protected by the intrinsically safe 'I'*
5. Hua F. U. and L. I. U. Yin-xia. *The Ex i Explosion-Proof Technology in the Fieldbus System [J]*. Electric Explosion Protection 3 (2003): 005.
6. Hongwu Wang, Shaohua Liu. *Application of PLC and Inverfer on Elevator [J]* Electrical Engineering 1 (2007): 013.
7. Q. P. Zhang and J. R. Wan, *Simulation of Generated Heat by Friction of Explosion-Proof Elevator Ropes*, Advanced Materials Research, Vol. 853, pp. 413-420, 2014

8. Hua Zhang, Zhao Wenzhu, and Wang Guodong. *Research on dynamic temperature measurement method based on transient forward and inverse heat transfer model.* Chinese Journal of Scientific Instrument 11 (2011): 016.
9. General Administration of Quality Supervision, Inspection and Quarantine of the People's Republic of China (AQSIQ), TSG T7011-05, *Test rules of elevator buffer types*
10. General Administration of Quality Supervision, Inspection and Quarantine of the People's Republic of China (AQSIQ), TSG T7012-05, *Test rules of elevator door lock types*
11. General Administration of Quality Supervision, Inspection and Quarantine of the People's Republic of China (AQSIQ), TSG T7010-2004, *Test rules of elevator safety gear types (draft)*
12. General Administration of Quality Supervision, Inspection and Quarantine of the People's Republic of China (AQSIQ), TSG T7009-05, *Test rules of elevator speed governor types*
13. General Administration of Quality Supervision, Inspection and Quarantine of the People's Republic of China (AQSIQ), TSG T7003-2011, *Rules of supervised inspection and periodic inspection of elevators—explosion-proof elevator*

Design and Performance Analysis of Contra-rotating Pumpjet Propulsor under Condition of Torque Unbalance

Xiao-er Wang[1,a], Zhen-shan Zhang[1] and Meng Zhang[1]

[1]*Naval University of Engineering Weaponry Engineering Department, 430033Wuhan, China*

Abstract. In order to balance the torque of front rotor and rear rotor of underwater vehicle, the analysis of the speed triangles at the inlet and outlet of the front and rear rotor has been done. Then, the thought of using contra-rotating pumpjet to achieve the objective was raised. The stator is installed behind the rear rotor so as to ensure the overall torque of the propulsor balance, at the same time, the stator can also support the shroud of the propulsor. the parameter design of the rotor and the stator has been carried out by using the three dimensional inverse design method. At last, the performance of the designed pumpjet propulsor is obtained when it is installed on the underwater vehicle By using computational fluid dynamics. The results show that the total torque of the propulsor is reduced to 1.8 N * m on the design point although the power difference ratio of the front rotor and the rear rotor is 20%. The torque ratio is also reduced from 4.6% to 0.4%, which is good to meet the propulsor balance requirement and verifies the 3-D design method of pumpjet is effective.

1 introduction

As the speed of underwater vehicle which propelled by counter-rotating propeller increasing, the cavitation incipient of the front and the rear propeller has begun to become a serious problem. The generation and collapse of the cavitation bubble at high speed will increase the noise sharply which of course will reduce the concealment of the vehicle. once the cavitation is serious, the propeller may even facing the danger of stall. On the other hand, the increase of the speed will increase the resistance of the vehicle so that the power of the engine will be bigger than before. In order to ensure the operation of the engine, it is common to use part of the power of the front shaft for the auxiliary engine of the vehicle, which makes the power received by the front and rear rotor no longer equal. When the engine power is increased, the difference between the power of the front and rear rotor will be further increased when compared with the low speed, so the design of contra-rotating propeller becomes more difficult under the condition of unbalanced shaft power.

In order to reduce the propeller noise and improve the speed of the vehicle, a counter-rotating pump jet propulsor has been designed which used on the "stingray-1" torpedo[1]. Some research on the Contra-Rotating axial flow pump has been done in Japan[2]. But it is needed on the design method and performance analysis of the contra-rotating pumpjet under the condition of unbalanced shaft power.

[a] Xiao-er Wang : hnwzwxe@163.com

Based on the three dimensional flow theory, the flow field calculation is completed after the iterative calculation between S1 flow surface and S2 flow surface is used to solve the continuous equation and the equation of motion in the pumpjet, then The blade shape is obtained by substituting the flow field calculation results into the blade equation. So that a new flow field will be calculated after the blades are obtained. The final blades can be obtained if the errors of the previous flow field and the last flow field meet the requirement. assumed that the are equal between the front and rear rotor, then the blades of the pumpjet propulsor will be obtained. Then, By means of computational fluid dynamics (CFD), numerical experiments were carried out on the designed pumpjet propulsor, the performance and the condition of torque balance for the propulsor under design speed were also can be analysed.

2 Speed triangles analysis

Assuming that the direction of the incoming flow to the pumpjet is axial, the velocity components inside the propulsor are analyzed as shown in Figure 1 (where subscript 1 indicates the inlet and subscript 2 indicates the outlet; The subscript f indicates the front impeller, the subscript r indicates the rear impeller). The rear rotor speed is 1.1 times the front rotor, But rotation direction in the opposite direction. Assuming that the flow field keeping same between the front rotor outlet and inlet of the rear rotor. The same applies to the rear rotor and the stator. So,

$$V_{m2f} = V_{m1r} \qquad V_{m2r} = V_{1s} \tag{1}$$

V_{w1f} and V_{w2f} respectively represent the front rotor inlet and outlet relative speed. V_{w1r} and V_{w2r} respectively represent the rear rotor inlet and outlet relative speed. V_{1su} and V_{1sm} Represents the velocity component of V_{1s} in the circumferential direction and the velocity component of V_{1s} in the axial direction. V_{2s} represents the final outflow velocity. As shown in Figure 1, because the tail of vehicle is conical, the circumferential velocity at the inlet of the front and the rear rotor is lower than the circumferential velocity at the outlet.

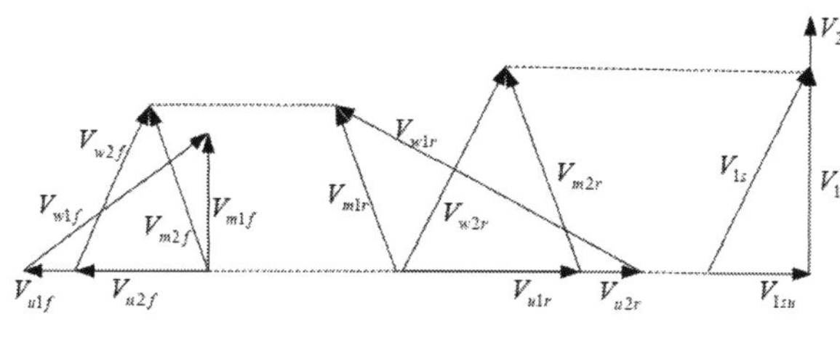

Figure 1. Inlet and outlet velocity triangles of contra-rotating pumpjet

3 Basic equations

Assuming that the flow of water in the pumpjet is steady, viscous and incompressible, the rotor comes in a non-swirling flow. Regardless of the thickness of the blade, the center plane of the blade is considered as a vortex sheet, and the vortex on the central plane is used to replace the effect of the blade on water. The strength of the vortex is controlled by the average circumferential circulation $2\pi r\bar{V}_\theta$ [3-4].

$$r\bar{V}_\theta = \frac{B}{2\pi}\int_0^{\frac{2\pi}{B}} rV_\theta d\theta \tag{2}$$

Where, B represents the number of blades, r represents radius, V_θ represents circumferential velocity.

Before the blade shape obtained, It needs to solve the following three equations based on S2 flow surface theory [3]:

(1) Continuity equation:

$$\frac{\partial}{\partial z}\left(H_2H_3\bar{V}_z\right)+\frac{\partial}{\partial r}\left(H_1H_3\bar{V}_r\right)=0 \tag{3}$$

Where, H_1, H_2, H_3 are the Lame coefficient in the curvilinear coordinate system, which are taken as 1, 1, r, respectively in this paper. $\bar{V}_r$ and $\bar{V}_z$ represents the radial average velocity and the axial average velocity, respectively.

(2) Motion equation

$$\frac{\partial \bar{V}_z}{\partial r}-\frac{\partial \bar{V}_r}{\partial z}=F(r,z) \tag{4}$$

$$F(r,z)=-\frac{\partial \bar{V}_\theta r}{\partial r}\frac{\partial \varphi_m}{\partial z}+\frac{\partial \bar{V}_\theta r}{\partial z}\frac{\partial \varphi_m}{\partial r} +\frac{1}{w^2}\left[\frac{\partial \bar{E}_r}{\partial r}\left(\bar{w}_z+\bar{w}_\theta\frac{r\partial\varphi_m}{\partial z}\right)-\frac{\partial \bar{E}_r}{\partial z}\left(\bar{w}_z+\bar{w}_\theta\frac{r\partial\varphi_m}{\partial r}\right)\right] \tag{5}$$

where, φ_m is the average circumferential S2 flow surface equation. represents the energy of unit mass of water at the inlet of the pumpjet propulsor[4].

(3) Blade equation

$$\bar{V}_\theta-\omega r=\frac{H_3\partial\varphi_m}{H_1\partial z}\bar{V}_z+\frac{H_3\partial\varphi_m}{H_2\partial r}\bar{V}_r \tag{6}$$

Where, ω represents the rotor angular velocity.

Formulas(3), (4) and (6) form the system equations used to obtain the flow field and blade. It can be seen that $\bar{V}_z, \bar{V}_r, \bar{V}_\theta$ and $r\bar{V}_\theta$ are unknowns in the system of equations. In order to close the system, the distribution of the $r\bar{V}_\theta$ along the meridian streamlines is required during the design process.

According to some test data, $\bar{E}_r$ distribution at the inlet can be obtained. As the viscosity of water in the pumpjet is not considered, the $\bar{E}_r$ along the streamline is constant. Streamlines will be determined by iterative calculations. The steps are as follows: (1) Given the meridian surface of the pumpjet, the position of leading edge and trailing edge of the rotor and stator. Angular velocity and other boundary conditions are also needed; (2) the initial streamline and velocity field can be obtained by solved equation (4) when its right side assuming to equal to zero; (3) Method of Characteristics will be used to solve the equation (6), in order to obtain the φ_m . (4) The distribution of $F(r,z)$ obtained from equation (5) and then equation (4) will be calculated when the result of $F(r,z)$ is Substituted in (4); (5) Repeat step (4) and step (5) until the error meets the requirement ; (6) Blades of pumpjet will be obtained after thickened according to the NACA data.

4 Pumpjet design

Before the design of pumpjet , the following parameters should be known. Rotating speed of the front rotor and the rear rotor, flow rate Q_d , the distribution of $\partial r\bar{V}_\theta/\partial m$, the number of front rotor, rear rotor and stator. In order to meet the shaft power of the front rotor and the rear rotor at the same time, keep the shroud of pumpjet taper, the rotating speed of rear rotor is slightly higher than the front rotor.

4.1 Meridional geometry of the pumpjet

Meridional shape of the pumpjet is closely related to the flow rate, efficiency, cavitation performance, so it needs to be determined according to the past experience, thrust requirement, and the geometry constraint of the underwater vehicle body. In this paper, Meridional shape of the pumpjet is plotted according to the actual size constraints of the tail of the navigating body and the required axial thrust force at the designed speed. The meridional geometry of the pumpjet is shown in Fig. 2.

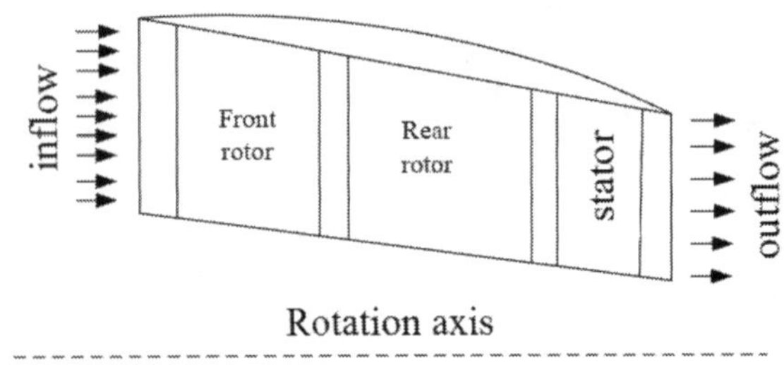

Figure 2. Meridional geometry of the pumpjet

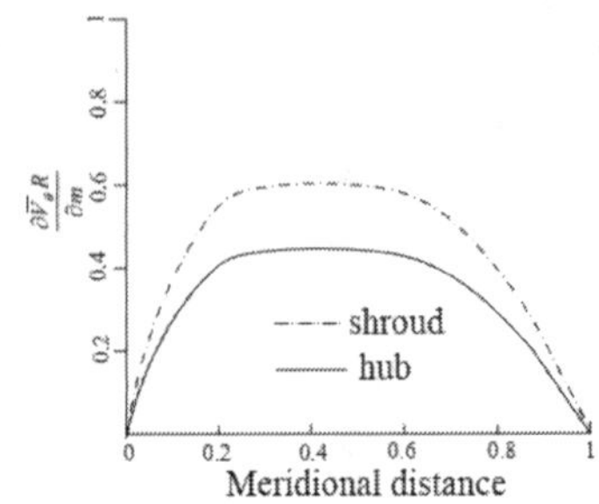

Figure 3. Blade loading distribution of the front rotor

4.2 Loading distribution of Pumpjet

Distribution of $\bar{V}_\theta r$, also known as loading distribution (the pressure difference between the blade pressure surface and suction surface), which has a significant impact on the efficiency and cavitation, so choose a reasonable loading distribution is key to design the blade of pumpjet.

According to Eq.(6), it can be seen that the loading on the blade is related to the partial derivative of $\bar{V}_\theta r$. Therefore, in order to facilitate the iterative calculation, the distribution of $\bar{V}_\theta r$ along the streamline line is given here. This method is effective to precise control of the loading, which makes low pressure zones can be prevented [5-6]. At the same time, the blade loading distribution needs to satisfy the following two conditions [7]:

(1) At the Exit,

$$\frac{\partial r\bar{V}_\theta}{\partial m} = 0 \tag{7}$$

This is because the pressure at the pressure surface should be equal to the pressure at the suction surface according to the kutta condition.

(2) Integral value from the leading edge to the trailing edge is equal to the circulation at the trailing edge, namely:

$$\int_0^1 \frac{\partial r\bar{V}_\theta}{\partial m} dm = V_{\theta_2} r_2 \tag{8}$$

where, $m = 0$ represents blade leading edge. $m = 1$ represents blade trailing edge, the subscript 2 represents the value at the outlet, and $V_{\theta 2} r_2$ is the speed moment at the outlet.

According to reference [5-6], we decided to take into account both the efficiency and cavitation of the pumpjet propulsor, using the intermediate uniform loading distribution form, that is, the maximum load appeared at 0.2 chord length, then keep it unchanged and gradually decreased at 0.6 chord length. The loading distribution at the hub and shroud of the front rotor and the rear rotor are showed in figure 3 and figure 4. The loads at the other radius cross sections are obtained by linear interpolation.

One of the roles of the stator is to convert the rotating flow discharged from the rotor into an axial flow to improve the efficiency. On the other hand it can plays the role of balance the propulsor's torque and support the shroud. In this paper, the main research about the stator is focus on whether the outflow from the pumpjet is straight, while avoiding the secondary flow. The stator loading distribution is based on the reference [7], as shown in Fig. 5, the loading distribution at the hub and the shroud is given, and the other radius sections obtained by linear interpolation.

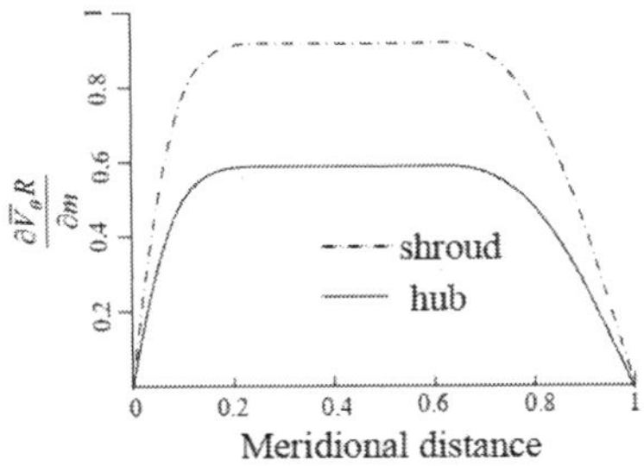

Figure 4. Blade loading distribution of the rear stator stator

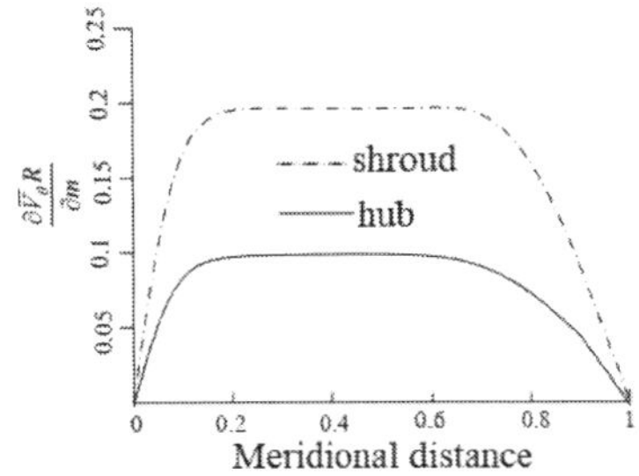

Figure 5. Blade loading distribution of the

4.3 blade thickening

After obtaining the blade without thickness, The final pumpjet geometry will be obtained through thickening the blade with NACA series airfoil thickness distribution rule.

5 Pumpjet propulsor performance caculation

After obtaining the geometry of the pumpjet, the pumpjet propulsor is placed on the tail of the underwater vehicle. Then, the numerical self-propelled calculation result can be analyzed.

5.1 Numerical model

In this paper, the central node control and finite volume method are used to solve the RANS equation. SST turbulence model to complete the system closed [8].

5.2 Mesh and boundary conditions settings

After obtaining the front rotor, the rear rotor and the stator of the pumpjet propulsor, the rotor mesh is completed by Turbo-Grid. J-type topology is selected in the process of generating mesh of the front and the rear rotor, and the tip clearance is taken as 0.5 mm. The grids are selected as 376,000 and 371,000. The stator selects H type topological structure, the quantity of the grid is chosen as 218,000. The computational domain size is selected according to reference [8], grids number of computational domain is 7.28 million. Boundary conditions are set to speed inlet and pressure outlet. The front and rear rotors and their hubs are set at relatively stationary wall boundary and their shrouds are set to absolute stationary wall boundary [8]. The interface between the calculation domain and the front rotor, the interface between the front rotor and the rear rotor and between the rear rotor and the stator are set as the frozen rotor mode.

5.3 Calculation results

The result of self-propelled of the fully attached underwater vehicle is shown in Fig.6. It can be seen that at design speed, the thrust of pumpjet is slightly larger than the resistance of vehicle. After the speed increases slightly, the thrust and resistance are basically balanced. The post-treatment shows that the thrust efficiency of the pumpjet propulsor at this time is 67.9%, which basically meets requirements. Under the design speed, the pumpjet has reduced the overall imbalance moment of the propulsor to 1.8 N * m by appropriately increasing the speed of the rear rotor and the balance of the stator under the condition that the front and the rear rotor have a power difference of 20%, The proportion of the total torque was reduced from 4.6% to 0.4%, basically completing the design goal of overall moment balance, which verified the validity of the design ideas and methods. The reference [2] shows that the efficiency of the pumpjet is lower under the condition of the rear rotor has a higher rotary speed than that the front and the rear rotors have a same rotary speed.

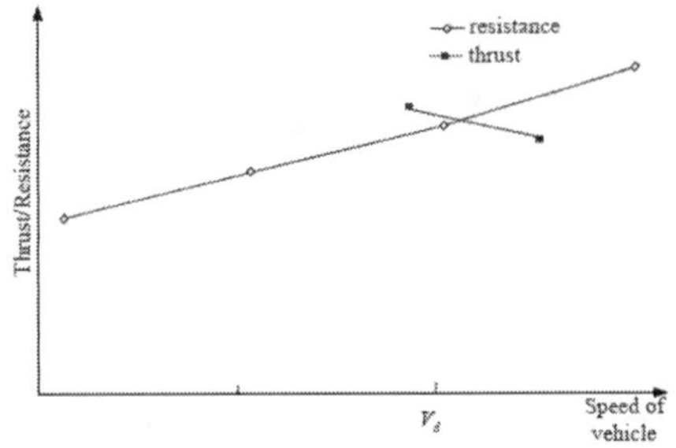

Figure 6. Speed evaluation of underwater vehicle

The pressure distribution of the pumpjet blade under the condition of the design speed is shown in Fig.7 and Fig.8. It can be seen from the figure that except for the leading edge, in which area the blades have a violent action with water, the pressure distribution of the pressure surface and the suction surface is smooth. There is no outstanding pressure fluctuation, which shows that the blade loading distribution is uniform. there is no pressure gradient larger area, to avoid the local low pressure area, reducing the risk of cavitation. This is also best demonstrated by the absence of a blue depression in each of the blades of the pressure contour in Figure 9.

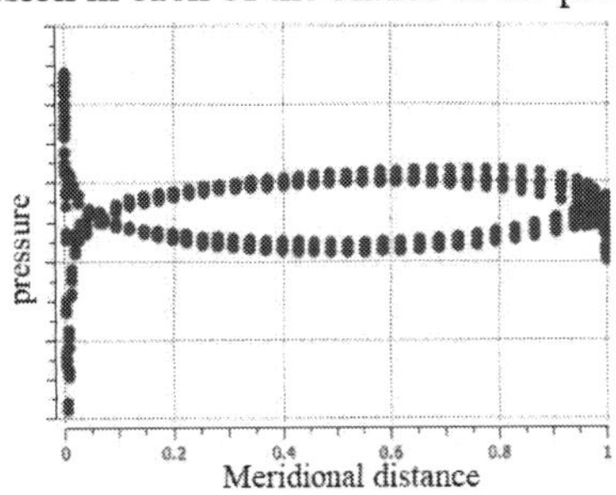

Figure 7. Pressure loading distribution of the front rotor

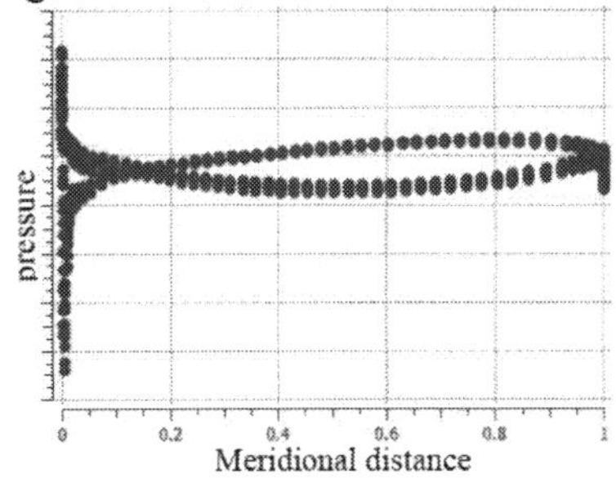

Figure 8. Pressure loading distribution of the rear rotor

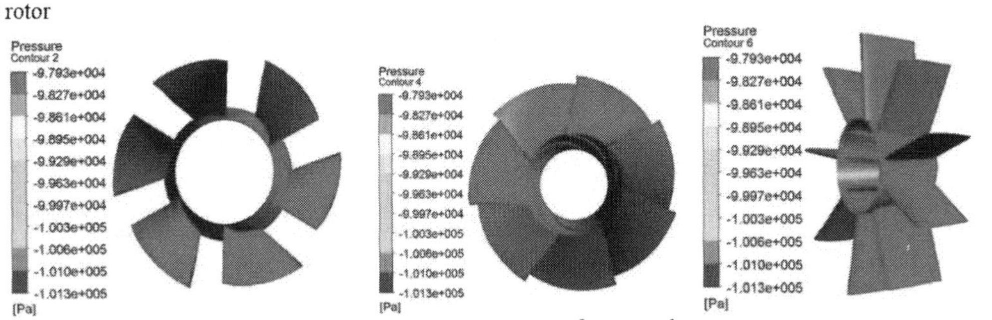

Figure 9. Pressure contour of rotor and stator

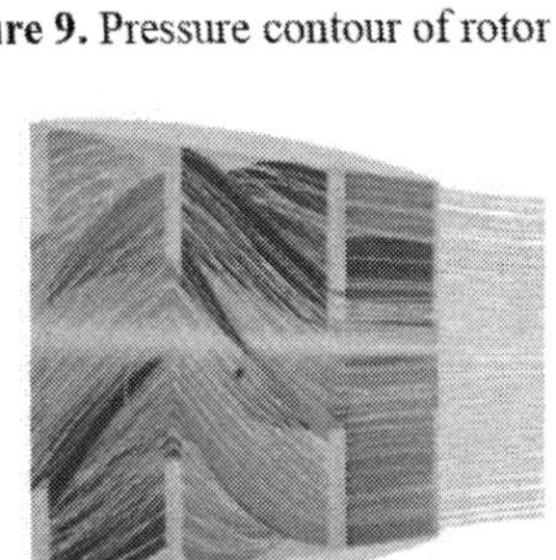

Figure 10. Three-dimensional streamline of the pumpjet

As shown in Fig.10, from the pumpjet internal flow chart, it can be seen that the entire flow is smooth without disorder. The flow around the pumpjet rotor and the stator blades at three different radius is observed. The vortex flow between the blades is not seen in Fig.11, and the final flow is smooth and straight, which indicates that the design of the stator is good.

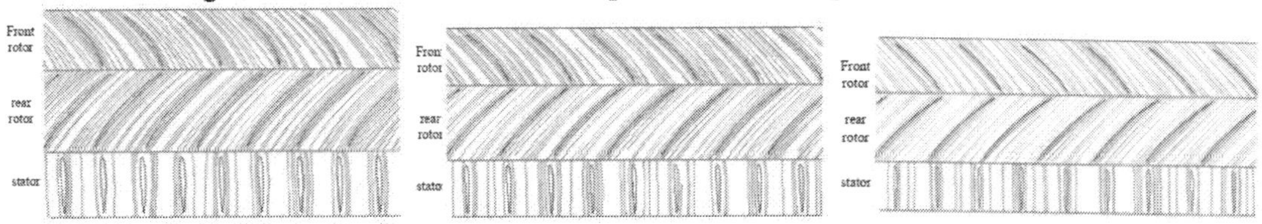

(a) 0.3R sectional drawing (b) 0.5R sectional drawing (c) 0.9R sectional drawing

Figure 11. Two-dimensional streamline at the different radius of the pumpjet

6 Conclusion

(1) A pumpjet of underwater vehicle was designed using the three dimensional inverse method after analyzing the speed triangle of the pumpjet. The design process has less assumptions about the internal flow of the pumpjet.

(2) The $r\bar{V}_\theta$ is used to control the loading of the blade, which well balances the entire propulsive torque. Meanwhile, the distribution of the velocity moment effectively controls the pumpjet efficiency and the cavitation of the blade. The method is simple and effective. The loading distribution, which selected according to the reference [8], effectively takes into account the efficiency and anti-cavitation performance of the pumpjet, and there is no obvious low pressure area on the rotor surface.

(3) Under the design speed, the thrust the pumpjet and the resistance of the vehicle is nearly balanced, and the thrust efficiency reached 68%, which verifies the thought of the pumpjet's structure is useful.

References

1. Wang Lan, The history of torpedo development in the world and its landmark study[J]. Underwater weapons, **2**:1(2015)
2. A. Furukawa, Y. Cao, K. Okuma, S. Watanabe, Experimental Study of Pump Characteristics of Contra-Rotating Axial Flow Pump, Proc. 2nd International Symposium on Fluid Machinery and Fluid Engineering, Beijing, China,245(2000)
3. Luo Xingqi. Full three-dimensional inverse and optimum design of Francis runner [D].Beijing: Tsinghua University(1995)
4. Peng Guoyi. Three-dimensional rotational flow inverse computation and optimum design of Kaplan runner[D]. Beijing: Tsinghua University(1996)
5. Duccio Bonaiuti, Mehrdad Zangeneh. Parametric Design of a Waterjet Pump by Means of Inverse Design, CFD Calculations and Experimental Analyses[J].ASME Journal of Fluids Engineering, **132**,1(2010)
6. Duccio Bonaiuti, Mehrdad Zangeneh. On the coupling of inverse design and optimization techniques for the multi-objective, multipoint design of turbomachinery blades[J].ASME Journal of Turbomachinery, **131**,1014(2009)
7. Zangeneh.M, Goto.A, Takemura T. Suppression of secondary flows in a mixed –flow pump impeller by application of three-dimensional inverse design method: part1: design and numerical validation [J]. ASME Journal of Turbo-machinery, **118**, 536(1996)
8. Jin Shuanbao, Wang Yongsheng, Ding Jiangming. Three-dimensional design and numerical experiment of mixed-flow waterjet with CFD[J]. Journal of Harbin Engineering University, **33**,1223(2012)

Surface Morphology and Bending Deformation of 2024-T3 Thin Sheets with Laser Peen Forming

Junfeng Wu[1,a], Shikun Zou[1], Yongkang Zhang[2] and Shuili Gong[1]

[1] *Science and Technology on Power Beam Processes Laboratory, AVIC Manufacturing Technology Institute, Beijing 100024, China*
[2]*School of Electro-mechanical Engineering, Guangdong University of Technology, Guangzhou, Guangdong, 510000, PR China*

Abstract. Laser peen forming (LPF) is a pure mechanical forming method through accumulated plastic strain, which has been successfully applied in wing components. Experimental investigation has been performed to understand the effect of process parameters such as constraint conditions, sheet thickness and laser energy on surface morphology and bending deformation of 2024-T3 thin sheets of dimensions of 76 mm × 19 mm (length × width). The research results indicated that bulges on the aluminum foil were generated at the bottom surface and not generated at the topmost surface. It was different for transition value of two-way bending deformations of thin sheets after LPF with different constraint conditions. Remain flat thicknesses of thin sheets after LPF were about 1 mm ~ 2 mm for 20 J, 25 J and 30 J. Arc heights and curvatures of 3 mm thickness sheets increased with laser energy and those of 2 mm thickness sheets only made little change. It was found that convex deformation, flat, concave deformation and laser deep drawing for thin sheets with different thicknesses after LPF.

1 Introduction

Laser peen forming (LPF), derivative laser shock processing (LSP), has been developed as a metal forming method for flexible shaping and adjusting. Because of some advantages such as non-contact, tool-free and high efficiency, LPF has been successfully applied for forming of B747-8 wing panel [1]. In addition, compressive residual stresses after LPF are generated both on two sides of plates, which would contribute to resist the stress corrosion crack and fatigue [2, 3].

Bend deformation of thin sheets with LPF has attracted much concern in the past several years because of demands of micro-electromechanical system (MEMS) fabrication and bending deformation mechanism. Hu et al. [4] addressed effect of process parameters on the bending deformation of aluminum alloy thin sheets with different thicknesses at cantilever state and proposed the forming mechanism about two-way bending deformations. Ocana et al. [5] presented the influence of process parameters on the net bending angle of thin sheet metals. Sagisaka et al. [6, 7] showed the changes of bending property with atmosphere, process parameters and elastic pre-bending on pure aluminum thin sheets. Hu et al. [8, 9] discussed the bending capability of thin sheet metals, residual stress distributions and the microstructure close to their opposite sides. Shen et al. [10] analyzed the surface modification of aluminum alloy thin sheets. However, surface morphology and bending deformation

[a] Junfeng Wu: wjf88813@163.com

of different thin sheets thicknesses are still little studied under different constraint conditions and different laser energies. Researched results would contribute to fundamental process experiment for large-scale thin sheets with LPF, which can be beneficial to the application of LPF in the contour design of aircraft wing skin panels.

This paper aims to study the surface morphology and bending deformation of 2024-T3 thin sheets with different thicknesses after LPF under different process conditions. Then surface morphology of aluminum alloy both at the topmost surface and at the bottom surface was analyzed. Arc heights and bending profiles of thin sheets were discussed. Furthermore, different bending deformation types were analyzed.

2 LPF experiment and measured methods

LPF, a derivative of LSP, is a new mechanical process to make the target bend deformation without heat effect, as shown in Fig. 1. Under high compressive shock loading, plastically deformated layer with high strain rates (up to $10^6 s^{-1}$) is generated in the surface layer. In order to balance plastic strain and plastic stress induced by LPF, the target bends and stretches. The convex or concave deformations depend on the plastically deformed layer depth at the cross-section [4, 11].

2024-T3 aluminum alloy thin sheets with different thicknesses were used to study the surface morphology and bending deformation after LPF. The dimensions of thin sheets were 76 mm ×19 mm (Length × width). The thicknesses are 3 mm, 2 mm and 1 mm, respectively. Thin sheets were wirely cut from one piece of plate and the length direction of thin sheets is parallel to the rolling direction of plate. Its mechanical properties of longitudinal direction are Tensile strength of 456.11 MPa, Yield stress of 370 MPa, Elongation of 21.7 % and Young's modulus of 73.29 GPa.

Thin sheets with different thicknesses were LPF by a YAG laser system with 1064 nm wavelength and overlapping square spots of 4 mm × 4 mm, as shown in Fig. 2. Free stated thin sheets without and with heel block were LPF with 40 mm shocked area in length direction. The LPF area is symmetrical to the center line in the length direction. Both scanning velocity V_y at the length direction and step interval Vx at the width direction are 3.4 mm. Laser process parameters are laser energy of 15 J/pulse, 20 J/pulse, 25 J/pulse and 30 J/pulse. Pulse width is 15 ns in Full Width at Half Maximum (FWHM). Repetition frequency is 1 Hz. The square spots were focused at the surface of thin sheets by a beam shaping device with a lens and a shaping lens. A water curtain of about 1 mm ~ 2 mm thickness was used as a confined layer. An aluminum foil of about 0.12 mm thickness was served as an absorbed layer.

Arc heights of thin sheets with LPF were measured by arc height tester with TSL-3A and a 31.75 mm×15.87 mm measured area. Bending deformation profiles of thin sheets along the length direction were measured by dial indicator and its support frame based on an optical platform and achieved around the Y-axis into the Z-direction. The optical platform was adjusted by electronic water balance meter adjustment (shaft level 300 mm, 1DIV=0.02 mm/m). The measurement path of bending deformation profiles was at the center line of the width direction, as shown in Fig. 2.

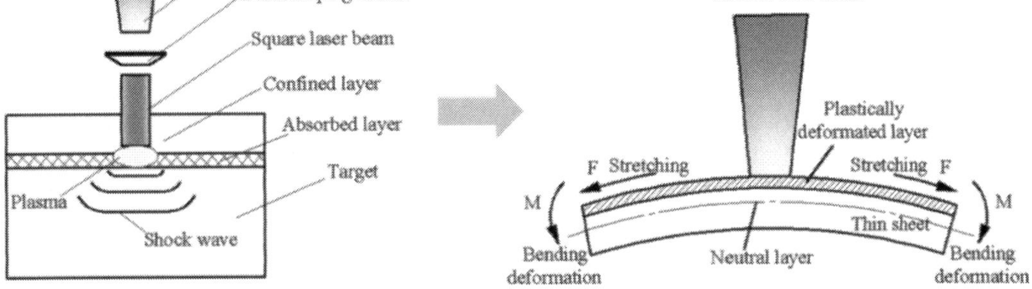

Figure 1. Schematic diagram of LPF process, (a) LSP, (b) LPF.

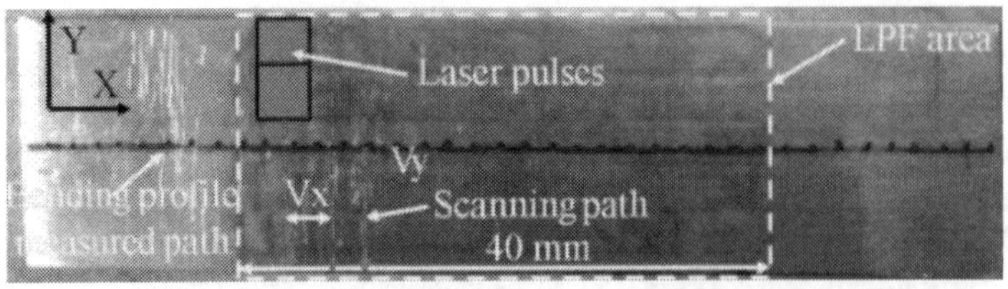

Figure 2. LPF area and bending profile measured path.

3 Results and discussions

3.1 Surface morphology of aluminum foil

Fig. 3 shows surface morphology of aluminum foil of thin sheets with different process parameters. It presents that bulge on the aluminum foil is generated at the bottom surface of thin sheets and bulge height increases with the decreasement of thin sheet thickness, as shown in Fig. 3(a)~Fig. 3(c). However, no bulge on the aluminum foil is generated at the bottom surface of thin sheets with heel block, as shown in Fig. 3(d). The reason may be ascribed to shock wave propagation into the heel block or bulge on the aluminum foil restricted by heel block [12]. In addition, it can be seen clearly that no bulge on the aluminum foil is generated at the topmost surface of thin sheets with different thicknesses, which indicates that no or little reflection wave and no negative effect are formed inside the thin sheet. The reason may be attributed to aluminum foil used for absorbed wave layer, which can absorb the laser shock wave and avoid the reflected wave to propagate into the thin sheet. Therefore, aluminum foil at the bottom surface should be used for thin sheets with LPF, which can effective avoid the negative effect and improve the mechanical property of thin sheets.

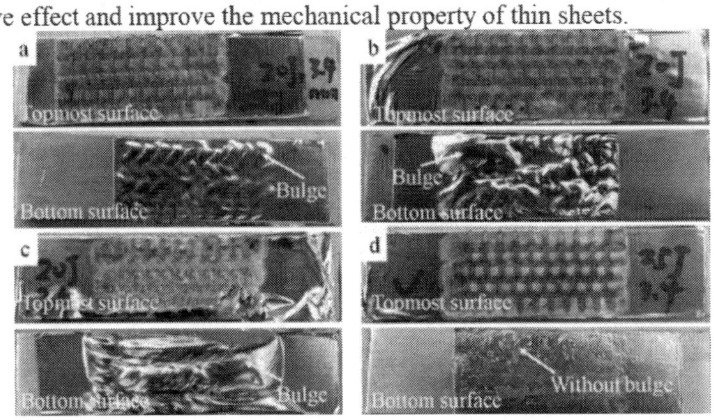

Figure 3. Surface morphology of aluminum foil of thin sheets with different process parameters, (a) 3 mm thickness sheet and 20 J laser energy and without heel block, (b) 2 mm thickness sheet and 20 J laser energy and without heel block, (c) 1 mm thickness sheet and 20 J laser energy and without heel block, (d) 1 mm thickness sheet and 25 J laser energy and with heel block.

3.2 Different constraint conditions

Fig. 4 shows arc heights of 1mm thickness sheets with LPF under different constraint conditions. It can be seen from Fig. 4(a) that the arc heights at the topmost surface of thin sheets are 0.075 mm for 15 J with heel block, -0.816 mm for 20 J without heel block, -0.064 mm for 25 J with heel block, respectively. It indicates that more large plastic deformation is generated on thin sheet without heel block compared to that of thin sheet with heel block. The reason may be attributed to plastic deformation along the depth of thin sheets is not restricted under the condition without heel block. In addition, 1mm thickness thin sheets produce two-way bending deformations with convex deformation of positive value and concave deformation of negative value, as shown in Fig. 4(b). About 25 J is the transition value of two-way bending deformations of 1 mm thickness sheets under the condition of heel block. And transition value of two-way bending deformations of 1 mm thickness sheets is less than 20 J under the condition without heel block.

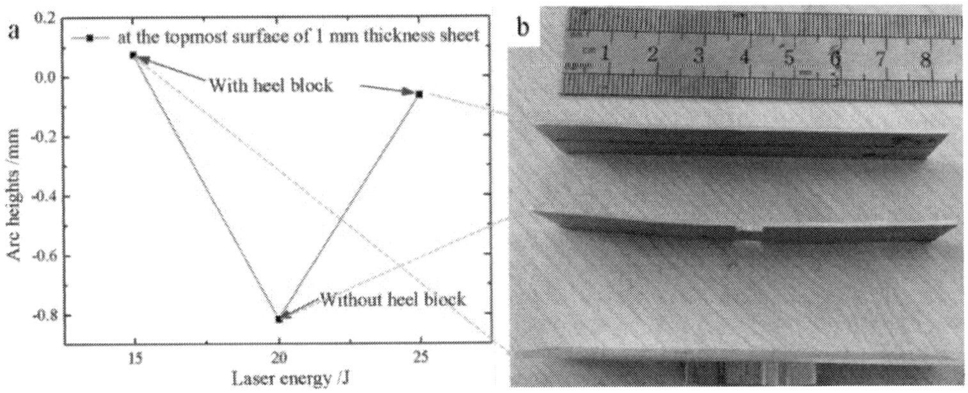

Figure 4. The arc heights of 1 mm thickness sheets with LPF under different constraint conditions, (a) arc heights value, (b) photos of deformed out of sheets.

3.3 Different sheet thicknesses

Fig. 5 shows arc heights of thin sheets with different thicknesses after LPF without heel block. It can be seen clearly in Fig. 5(a) that arc heights at the topmost surface of thin sheets after LPF with laser energy of 20 J are -0.816 mm for 1 mm thickness, 0.138 mm for 2 mm thickness and 0.206 mm for 3 mm thickness, respectively. It indicates that remain flat thickness of thin sheets after LPF with laser energy of 20 J is about 1 mm ~ 2 mm. The reason may be ascribed to the increasement of ratio of compressive residual stress layer/ sheet thickness. Fig. 5 presents that arc heights of 2 mm thickness sheets are less than those of 3 mm thickness sheets after LPF with 20 J, 25 J and 30 J. The reason may be ascribed to compressive residual stress layer reaching the bottom surface of thin sheets. Arc heights at the topmost surface of thin sheets are 0.13 mm for 2 mm thickness sheet and 0.221 mm for 3 mm thickness after LPF with laser energy of 25 J. They are 0.136 mm for 2 mm thickness sheet and 0.239 mm for 3 mm thickness sheet after LPF with laser energy of 30 J. Based on the above experimental data, remain flat thicknesses of thin sheets are about 1 mm ~ 2 mm for 20 J, 25 J and 30 J.

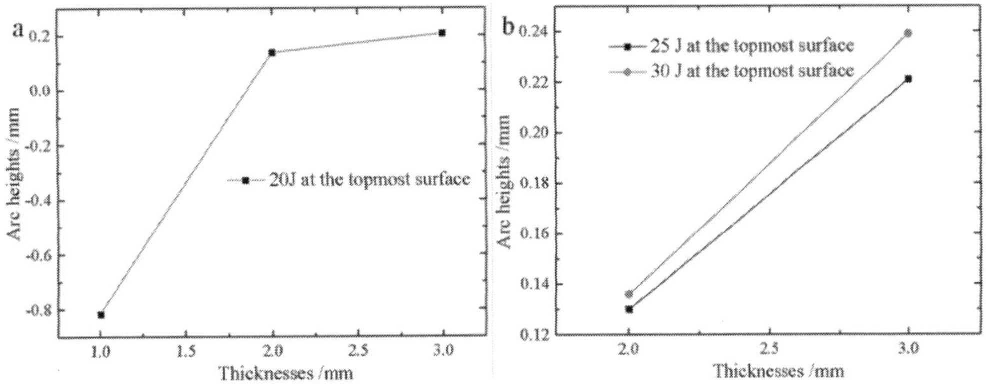

Figure 5. Arc heights of thin sheets with different thicknesses after LPF without heel block, (a) 20J, (b) 25J and 30J.

3.4 Different laser energies

Fig. 6 shows bending deformations of thin sheets after LPF with different laser energies. Bending deformations of 3 mm thickness sheets gradually increase with the increasement of laser energy, which may be attributed to high amplitude and deep compressive residual stress layer induced by LPF. Arc heights (*ha*) of 3 mm thickness sheets are 0.25 mm for 20 J, 0.29 mm for 25 J and 0.31 mm for 30 J in shocked area with chord length (*Lc*) of constant 40 mm, as shown in Fig. 6(a). According to the calculated formula of curvature radius [13], curvature radii (*R*) of 3 mm thickness sheets are about 800 mm for 20 J, 690 mm for 25 J and 645 mm for 30 J, respectively. Fig. 6(b) shows arc heights of 2 mm thickness sheets after LPF with different laser energies in arc height tester with chord length (*Lc*) of 31.75 mm. It indicates that arc heights of 2 mm thickness sheets make little change with the increasement of laser energy, which may be ascribed to compressive residual stress reaching to the bottom surface.Arc heights for 2 mm thickness sheets at the topmost surface are 0.138 mm for 20 J, 0.13 mm for 25 J and 0.136 mm for 30 J, respectively.

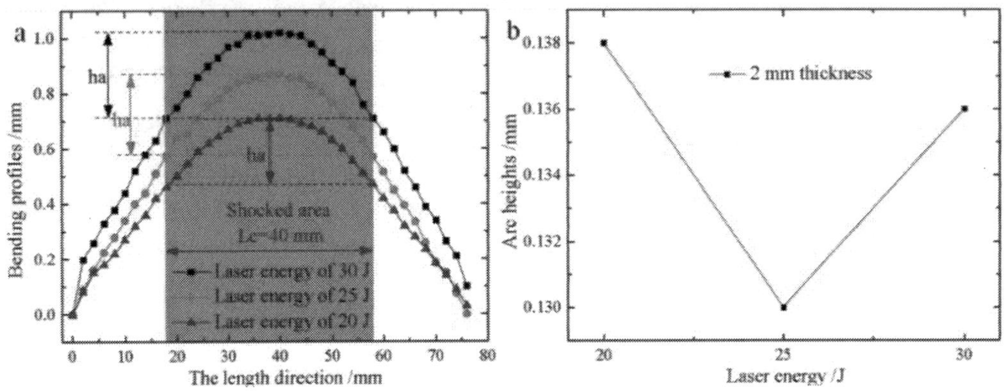

Figure 6. Bending deformations of thin sheets after LPF with different laser energies, (a) 3 mm thickness sheets, (b) 2 mm thickness sheets.

3.5 Bending deformation types

Bending deformation types of thin sheets with LPF are shown in Fig. 7. It can be seen that different bending deformation types induced by LPF are convex deformation, flat, concave deformation and laser deep drawing. The similar results have been reported [4,11]. When the depth of plastic deformation layer is small to sheet thickness, only elongation is generated in the surface layer. The curvature will emerge convex and increase to a maximum value. The convex curvature would decrease with the increasement of sheet thickness. Convex deformation of thin sheets will transform into concave deformation with the decreasement of sheet thickness. When compressive residual stress layer depth reach to the bottom surface, thin sheets keep flat. However, when sheet thickness is too thin or plasic deformation induced by LPF is not restricted by sheet, laser deep drawing is generated on thin sheets, as shown in Fig. 7.

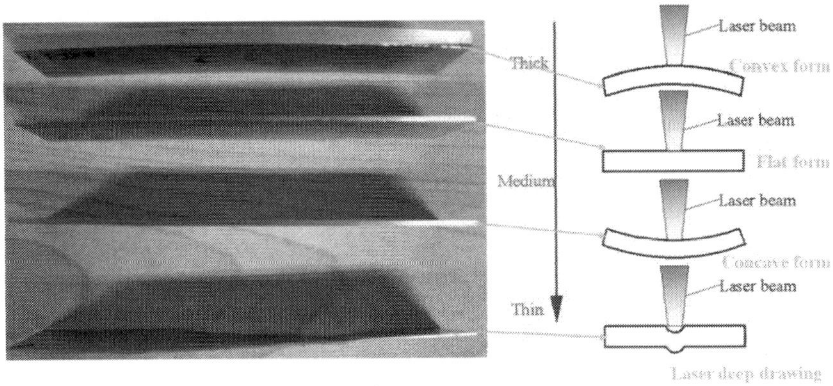

Figure 7. Bending deformation types of thin sheets with LPF.

4 Summary

LPF experiments have been performed to understand the effect of process parameters on the surface morphology and bending deformation of thin sheets with different thicknesses. The following conclusions have been reached:

(1) Bulges on the aluminum foil are generated at the bottom surface of 2024-T3 thin sheets with LPF and bulges heights increase with the decreasement of sheet thickness. No bulges on the aluminum foil at the topmost surface are generated. More large plastic deformation is generated on thin sheet without heel block compared to that of thin sheet with heel block.

(2) The arc heights at the topmost surface of thin sheets are 0.075 mm for 15 J with heel block, -0.816 mm for 20 J without heel block, -0.064 mm for 25 J with heel block, respectively. It is obtained by transition values of two-way bending deformations on thin sheets with LPF. They are about 25 J and less than 20 J for 1 mm thickness sheets with heel block and without heel block, respectively. Remain flat thicknesses of thin sheets are about 1 mm ~ 2 mm for 20 J, 25 J and 30 J.

(3) Arc heights (*ha*) of 3 mm thickness sheets with LPF gradually increase with laser energy of 20J ~30J. Curvature radii (*R*) of 3 mm thickness sheets with LPF become small with laser energy of 20J ~30J. However, arc heights of 2 mm thickness sheets with LPF make little change with laser energy of 20J ~30J. Different bending deformation types after LPF are convex deformation, flat, concave deformation and laser deep drawing with the decreasement of sheet thicknesses.

References

1. C. B. Dane, L. Hackel, F. Harris, D. Francis, Recent progress in laser technology for industrial laser peening, 4th International Conference on Laser Peening, Madrid, Spain 2013.
2. A. Telang, A. S. Gill, S. Teysseyre, S.R. Mannava, D. Qian, V. K. Vasudevan, Effects of laser shock peening on SCC behavior of Alloy 600 in tetrathionate solution, Corros. Sci. **90** (2015) 434-44.
3. C. Correa, L. R. D. Lara, M. Díaz, J. A. Porro, A. García-Beltrán, J. L. Ocaña, Influence of pulse sequence and edge material effect on fatigue life of Al2024-T351 specimens treated by laser shock processing, Int. J. Fatigue **70** (2015) 196-204.
4. Y. X. Hu, X. X. Xu, Z. G. Yao, J. Hu, Laser peen forming induced two way bending of thin sheet metals and its mechanisms, J. Appl. Phys. **108** (2010) 235-359.
5. J. L. Ocana, M. Morales, J. J. Garcia-Ballesteros, J. A. Porro, O. Garcia, C. Molpeceres, Laser shock microforming of thin metal sheets, Appl. Surf. Sci. **255** (2009) 5633-5636.
6. Y. Sagisaka, K. Yamashita, W. Yanagihara, H. Ueta, Microparts processing using laser cutting and ultra-short-pulse laser peen forming, J. Mater. Process. Technol. **219** (2015) 230-236.

7. Y. Sagisaka, M. Kamiya, M. Matsuda, Y. Ohta, Thin-sheet-metal bending by laser peen forming with femtosecond laser, J. Mater. Process. Technol. **210** (2010) 2304-2309.
8. Y. X. Hu, M. S. Luo, Z. Q. Yao, Increasing the capability of laser peen forming to bend titanium alloy sheets with laser-assisted local heating, Mater. Des. **90** (2016) 364-372.
9. Y. X. Hu, Z. Li, X. C. Yu, Z. Q. Yao, Effect of elastic prestress on the laser peen forming of aluminum alloy 2024-T351: Experiments and eigenstrain-based modeling, J. Mater. Process. Technol. **221** (2015) 214-224.
10. N. G. Shen, H. T. Ding, Q. H. Wang, H. Ding, Effect of confinement on surface modification for laser peen forming without protective coating, Surf. Coat. Technol. **289** (2016) 194-205.
11. Y. Sagisaka, Application of Femtosecond Laser Peen Forming to Sheet Metal Bending, J. Laser Micro/nanoeng. **7** (2012) 158-163.
12. Z. G. Che, S. L. Gong, Z. W. Cao, S. K. Zou, Y. Gao, H. Ma, Theory Analysis and Experiment Investigation of Laser Shock Processing on Titanium Alloy Blade, Rare Metal Mat. Eng. **40** (2011) 235-239.
13. A. Gariépy, S. Larose, C. Perron, P. Bocher, M. Lévesque, On the effect of the orientation of sheet rolling direction in shot peen forming, J. Mater. Process. Technol. **213** (2013) 926-938.

Study of Storage Shelves Deformation Using Finite Element Analysis during Seismic

Dyi-Cheng Chen[1,a] and Tsung-Ying Kuo[1]
[1] *Department of Industrial Education and Technology, National Changhua University of Education, Changhua 500, Taiwan*

Abstract. In recent years the government strongly promoted industry 4.0. Domestic enterprises highly developed of wisdom manufactured. Therefore, in the warehouse, the automatic storage system has gradually been taken seriously. The focus of this study is on the deformation of the shelves during the earthquake. First, this study used Inventor graphics software to model the shelf, the model into the Workbench for meshing and conditional constraints, then the completion of the model into ANSYS waiting for the solution. In earthquake conditions, in this study a scale 6-level seismic data is taken from the network PEER seismic database system. Using the ANSeismic program developed in Matlab, the seismic data downloaded from the network is transformed into ANSYS readable seismic code. The simulation results can be obtained by importing the seismic code into ANSYS and model simulation. In this study, we observed and analyzed the simulation results, through the shelf stress, strain and deformation to analyze the situation of the shelves, and improved the shelf structure.

1 Introduction

In recent years the government strongly promoted the industry 4.0. It is domestic enterprises a high degree of development of wisdom, with the application of new technologies such as internet of things, automatic warehousing, storage robots, mechanical vision and robotic arms. Industrial automation logistics system also will be produced. Which automatic storage system as part of the advanced manufacturing system [1]. Because of the easy operation, rapid processing, saving manpower and footprint and other advantages have gradually become a variety of industry management warehouse alternative [2].

Taiwan is located in the Pacific Rim seismic belt. The island has the frequency of frequent earthquakes. From the previous experience of the earthquake found domestic enterprises to be caused production disruption during the earthquake, mainly due to the destruction of production equipment [3]. The main body of the shelves is made of cold plates. Use bolts to assemble columns, beams and other parts, when the shelf by the role of goods, its structure of the external force and internal force to maintain a balance. The balance of the shelf is divided into the lasting balance state and limit balance state, when in the limit equilibrium state. As long as the application of an external force to balance the state, its structure will be destroyed and collapsed [4].

[a] Corresponding author : dcchen@cc.ncue.edu.tw

Based on the above, this study hopes to use finite element analysis to simulate the deformation of the shelves in the 6 earthquake. Observe the simulation results improved the structural strength of the shelves, to reduce the loss of natural disasters for the industry.

2 Methodology

Finite element method (FEM) for an unknown continuum to be analyzed, its complexity is difficult to get the mathematical solution of the analysis. You can use finite element analysis. The analytical solution is obtained from the mathematical formula of the "partial", and the analysis results include an infinite number of "partial" components. The principle assumes that this continuum is divided into finite "partial", called the element, the element of the boundary point called the node. Each node carries a mathematical equation, called interpolation equation. The analytic behavior of the continuum is expressed by a finite interpolation function equation. As long as the parameters are assumed to be correct and within tolerance errors. The exact solution of the solution can be compared with the exact solution of this continuum [5].

3 Establishment of Storage Shelves Finite Element Model and Condition Setting

This study is based on the market commonly used iron plate cold bay assembly rack to model. Inventor3D graphics software built the model, as shown in Fig. 1. The shelves can put a total of 60 boxes, each box the highest load 50kg. In this study, the full storage rate of 80% of the load conditions. In the condition set, the shelf material is set to iron, Density of 7.9g / cm^3, Elasticity coefficient is E = 211GPa, Poisson's ratio of 0.3, and the Y-axis vertical contact with the ground bottom of the shelf area as a fixed end. In the condition set with fixed constraints, as shown in Fig. 2 [6].

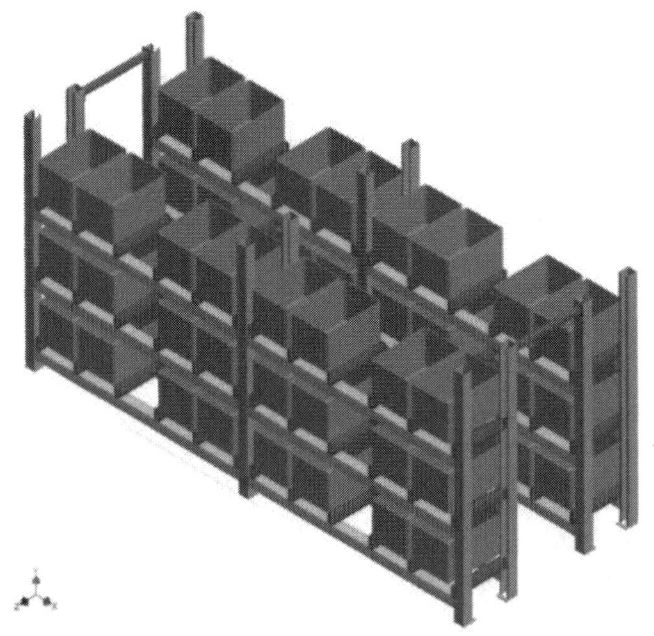

Figure 1. Storage shelves model

Figure 2. Model condition setting

4 Seismic Numerical Application

ANSeismic program is developed in Matlab. It can be in the PEER seismic database system to obtain real seismic data, and converted to ANSYS readable seismic code, to facilitate post-processing in ANSYS [7].

The quasi-simulated earthquake is a magnitude 6 earthquake, as shown in Fig. 3, the seismic data in the PEER seismic database system. This study needs the maximum value of 0.27516g, in the national 6 earthquake 0.25g ~ 0.4g within the definition.

Seismic waves are divided into transverse and longitudinal waves. Since the longitudinal wave propagation velocity is faster than the lateral wave, often the role of vertical seismic force on the shelves, the level of seismic power to reach. It can be considered that the longitudinal and lateral seismic forces are acting on the shelves, respectively [8]. The study is mainly to observe the shelves by the 6 level of seismic force when the deformation state.

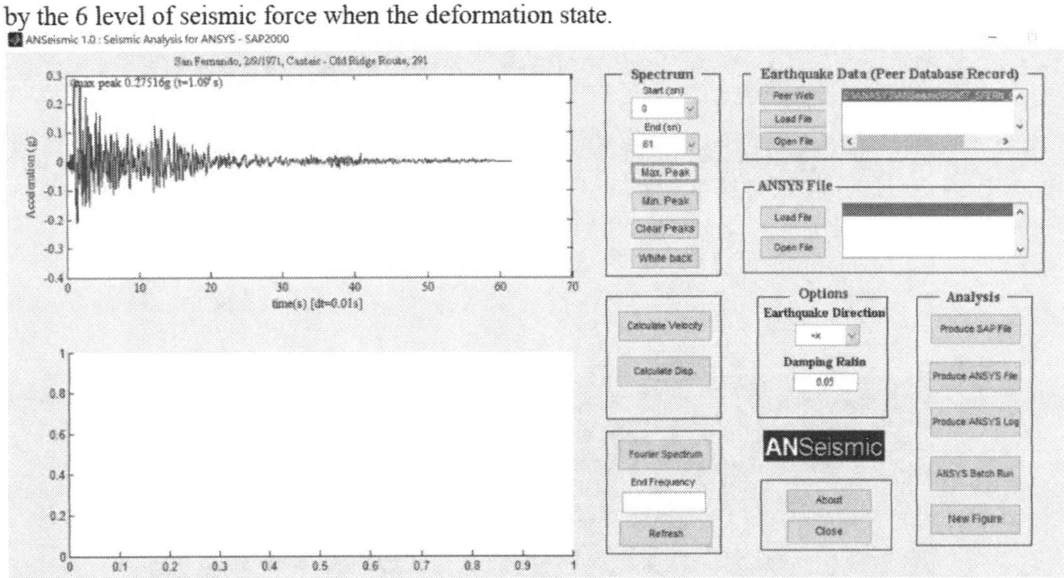

Figure 3. ANSeismic program main window

5 Analysis of Simulation Results

As shown in Fig. 4, the shelves in the X axis 1.09 seconds when the earthquake simulation deformation diagram, observation of this Fig. 4 can be found, the deformation of the shelf increases with height, deformation also gradually increased. In the top of the head and tail position of the maximum deformation is 0.302E-04 mm. The minimum amount of deformation at the bottom is zero.

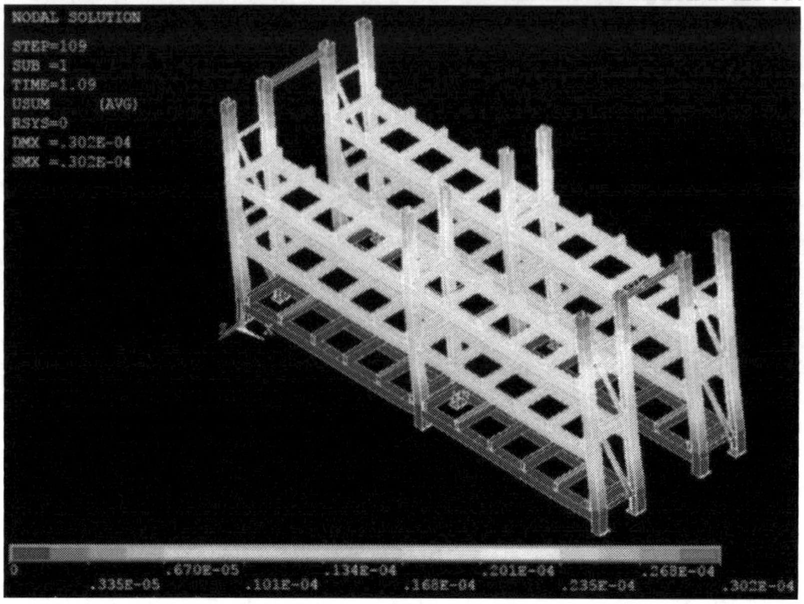

Figure 4. X-axis storage shelves deformation

As shown in Fig. 5, the shelves in the Z-axis 1.09 seconds when the earthquake simulation deformation diagram, observation of this Fig. 5 can be found, the deformation of the shelf increases with height, deformation also gradually increased. And in the same height of the middle of the deformation of the shelves are greater than the head and tail of the deformation, the maximum value of 0.545E-04 mm. The minimum amount of deformation at the bottom is zero.

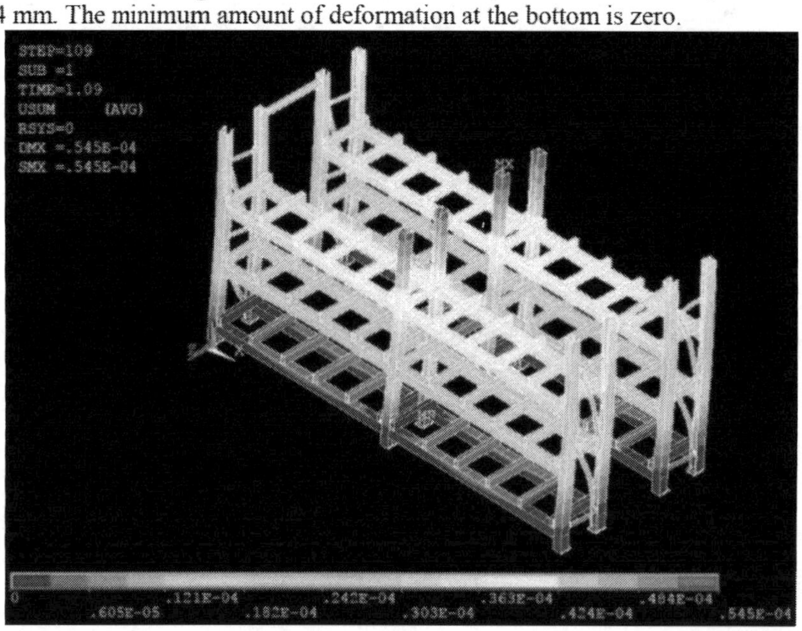

Figure 5. Z-axis storage shelves deformation

As shown in Fig. 6, the shelves in the X axis 1.09 seconds when the seismic simulation of the strain diagram, observation of this Fig. 6 can be found, the shelf strain are concentrated in the columns, and the greater the bottom of the larger strain, the maximum value of 0.431E-05, beam at the junction with the column at a higher strain value of 0.959E-06. The minimum value of the overall shelf strain is 0.172E-08.

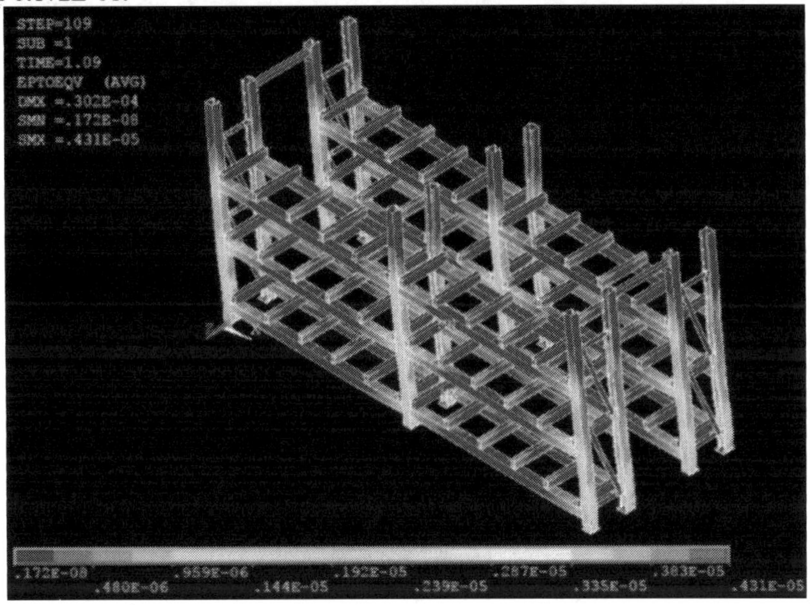

Figure 6. X-axis storage shelves strain

As shown in Fig. 7, the shelves in the Z-axis 1.09 seconds when the seismic simulation of the strain diagram, observation of this Fig. 7 can be found, the shelf strain are concentrated in the columns, and the greater the bottom of the larger strain, the maximum value of 0.264E-04. The strain value of the beam at the junction with the column is higher, 0.294E-05, Z-axis column between the rods also have strain occurred, the closer the bottom of the rod strain the greater the strain. The minimum value of the overall shelf strain is 0.562E-08.

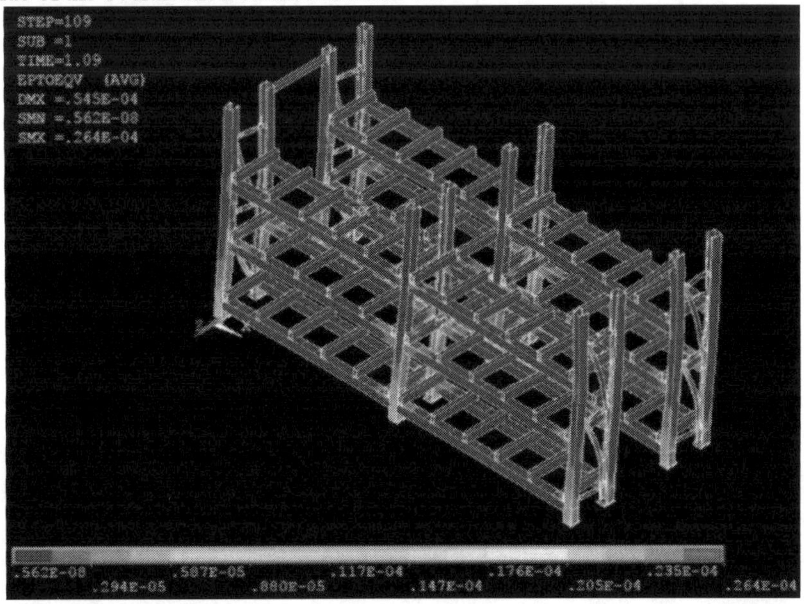

Figure 7. Z-axis storage shelves strain

As shown in Fig. 8, the shelves in the X axis 1.09 seconds when the earthquake simulation stress diagram, observation of this Fig. 8 can be found, the shelf stress are concentrated in the column, and the greater the bottom of the larger stress, the maximum value of 861595 Pa. The stress value of the beam at the junction with the column is 383103 Pa. The minimum value of the overall shelf stress is 309.569 Pa.

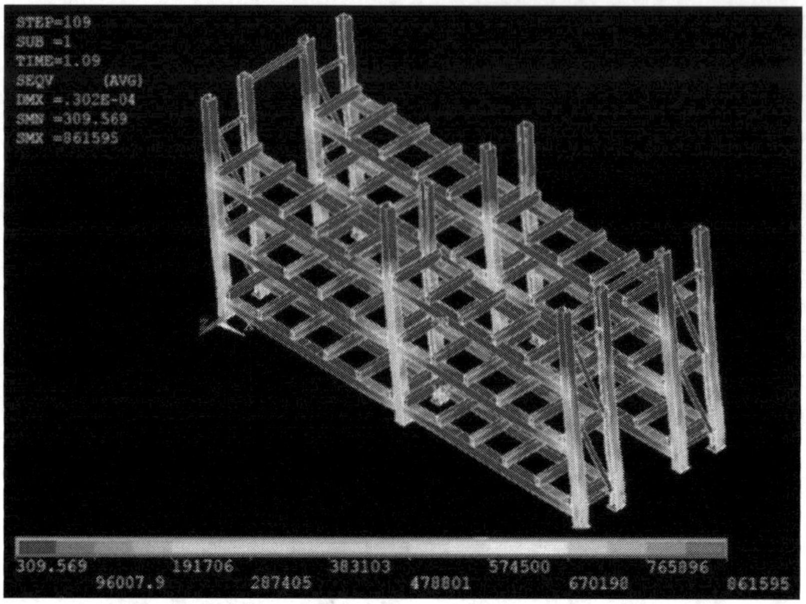

Figure 8. X-axis storage shelves stress

As shown in Fig. 9, the shelves in the Z-axis 1.09 seconds when the earthquake simulation stress diagram, observation of this Fig. 9 can be found, the shelf stress are concentrated in the column, and the more the bottom of the larger stress, the maximum value of 0.528E + 07 Pa. The stress value of the beam at the junction with the column is 586.930 Pa, Z-axis column between the rods also have stress occurred, the closer the bottom of the rod stress is greater, the overall shelf stress value of the minimum value of 625.499 Pa.

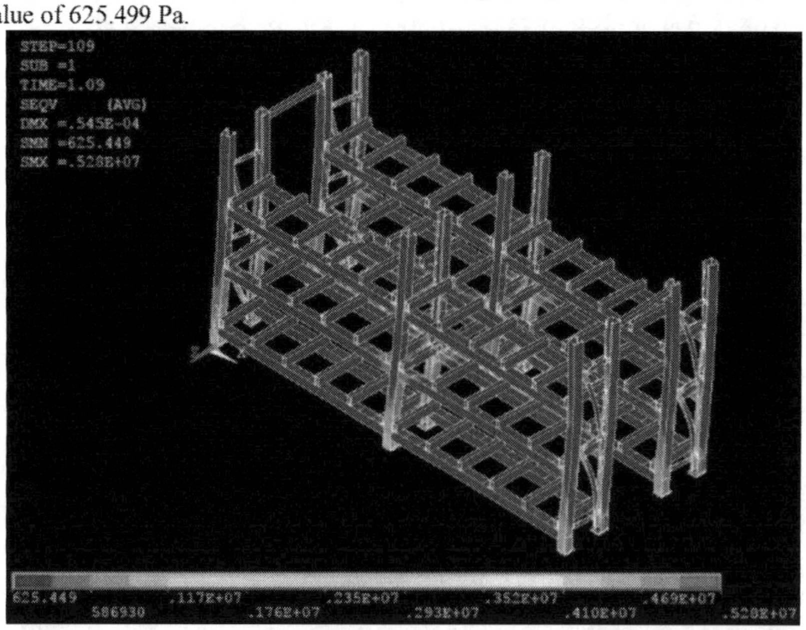

Figure 9. Z-axis storage shelves stress

6 Conclusion

This study used Inventor graphics software to model the shelf, the model into the Workbench for meshing and conditional constraints, then the completion of the model into ANSYS waiting for the solution. In earthquake conditions, in this study a scale 6-level seismic data is taken from the network PEER seismic database system. It can be seen that when the earthquake occurred. Shelf stress and strain are concentrated in the junction of the column and beam. If you want to increase the strength of the shelves to withstand the earthquake, it can strengthen the junction. In order to it improved the higher height, the greater shelf deformation problem. If necessary, consider increasing the fixed condition.

References

1. J.P. Van Den Berg, A.J.R.M. Gademann, International Journal of Production Research, *Simulation study of an automated storage/retrieval system*, **38**, 1339-1356, 6 (2000).
2. K.J. Vis Roodbergen, F.A. Iris, European journal of operational research, *A survey of literature on automated storage and retrieval systems*, **194**, 343-362, 2(2009).
3. J.C. Wang, J.F. Chai, C.H. Lin, Z.Y. Lin, C.M. Chang, S. Chang, Journal of Architecture, *The shockproof study for automation stocker system and high-raised floor*, **89**, 177-193, (2000).
4. H. Zhao, Y. Ding, X.Y. Wang, S.L. Sheng, Journal of Taizhou Polytechnic College, *FEA on stub columns of new steel storage racking*, 3 , 70-72 , (2012).
5. J.Q. Liu, C.H. Chue, *Engineering Applications of Finite Element Analysis with ANSYS* (Taichung, Taiwan：Tsang Hai Publishing, 2006).
6. X. Zhang, X. Zhao, Electric Technology and Civil Engineering, *Numerical analysis of seismic performance of frame structure with semi-rigid connections,* 5216-5218, (2011).
7. A. Şahin, International Congress on Advances in Civil Engineering, *ANSeismic – A simple assistant computer program for implementing earthquake analysis of structures with ANSYS and SAP2000*, **9** , 27-30 , (2010).
8. X. Weng, G.Q. Cheng, Q.F. Gao, Hoisting and Conveying machinery, *Finite element analysis of special rack structure*, **6**, 36-38, (2004).

Impact Loads on the Occupant under the Protection of an Inversion Tube Energy Absorber during a Helicopter Crash

Yunhua Luo[1, a*], Mengqi Wang[1,b], Tao Yan[2,c] and Lingjie Wang[2,d]

[1] *School of Materials Science and Engineering, Huazhong University of Science and Technology, Wuhan 430074, China*
[2]*Avic Aerospace Life-Support Industries .LTD*

Abstract. The objective of this paper is to investigate the impact loads on the occupants' head, neck, and spine under the protection of an inversion tube energy absorber during a helicopter crash landing. Due to the high vertical acceleration, the head, neck and spine are the most vulnerable parts of a body, so that an energy absorber is needed to dissipate the kinetic energy of the occupant and the seat to minimize the impact loads. In this paper, an inversion tube was adopted as an energy-absorbing device. The occupant injury conditions were evaluated by a numerical simulation. The result indicates that the impact loads on occupant's head, neck and spine are below the regulated thresholds under the protection of the energy absorber when the helicopter crash at a speed of 12.81m/s in vertical direction. As a consequence, the design of the occupant protection system has been proven reliable.

Introduction

Occupant protection is one of the most basic requirements for the helicopter seat structure design. During a crash landing, the impact loads may cause fatal injury on occupants' head, neck and spine [1]. To attenuate the impact loads on the occupant, the seat must process enough energy-absorbing capacity to dissipate the kinetic energy of the occupant and the seat before it bottoms out. Amount of work has been done by many researchers to develop various methods to meet the requirements of occupant safety. Michle Guida et al [2] proposed an innovative method of integrating anticrash energy absorbers made of a new composite material into landing gears to absorb the kinetic energy of the seat and the subfloor of a helicopter. S.A. Galehdari et al [1] developed a graded honeycomb shock absorber to attenuate the impact loads on the occupant to an acceptable level. Zhang [3] designed a sliding seat for preventing the neck from severe injury. However, there is no better way of attenuating the impact loads on the occupant than using energy-absorbing devices. The inversion tube is one of the most commonly used energy absorber for its excellent performance in absorbing energy, high manufacturability and low cost [4].

Considerable investigations have been conducted on the physical properties of the inversion tube energy absorber, however, little information exists on the impact loads on the occupant with an inversion tube installed on the seat. This study investigated the injury conditions of occupants' head, neck and

[a] Corresponding author : yhluo@hust.edu.cn, [b]1406472367@qq.com, [c]2453926883@qq.com, [d]353229291@qq.com

spine during a helicopter crash by a numerical simulation conducted by MSC.Dytran, an explicit code, associated with the articulated total body（ATB）program.

MSC.Dytran is good at solving nonlinear transient impact problems, while the ATB program is commonly used for computing the gross body response in various dynamic environments. The process of this simulation is not a procedure for running one program, then transferring information and running the other, these programs run concurrently and exchange data as the other proceeds. The ATB model used in this simulation consists of 17 rigid segments connected by 16 joints and takes on a tree-like structure [5]. The mass and inertia moment of each segment are equivalent to human counterparts, and the properties of the dummy model joints are similar to the corresponding human articulations. Besides, some shell elements with negligible density are covered on the dummy to provide a correct body contour for a contact calculation.

In this study, much endeavor was made to obtain the impact loads on the occupant and evaluate occupant injury conditions. The following proceeds as follows: the inversion tube used in this occupant protection system is introduced in section 2, section 3 shows how the whole model of the seat/occupant system was established, the results were showed and discussed to evaluate the occupant injury conditions in section 4, in the final section, conclusions were drawn and prospects were made for future research.

1. Inversion tube energy absorber.

The inversion tube is one of the most significant parts of the shock-mitigation system. The kinetic energy of an occupant/seat system can be dissipated rather than stored by the plastic deformation of an inversion tube. Figure 1 shows the process of the inversion tube deformation under the axial compressive force. It is obvious that the deformation part constantly keeps circular in the process of deformation from A to A'. Therefore, the deformation force keeps fixed, which provides the occupant with a steady deceleration process.

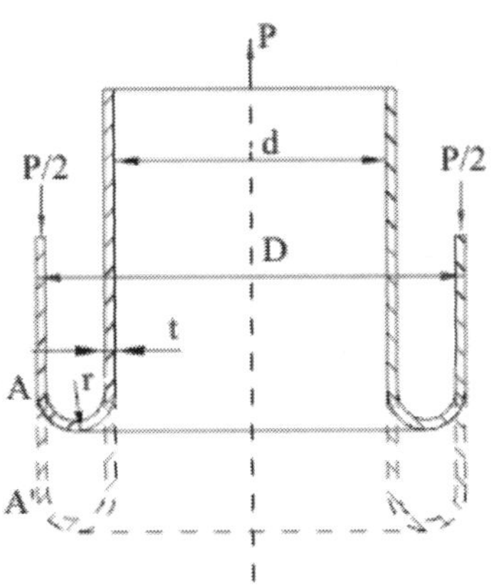

Figure 1. The deformation process of an inversion tube

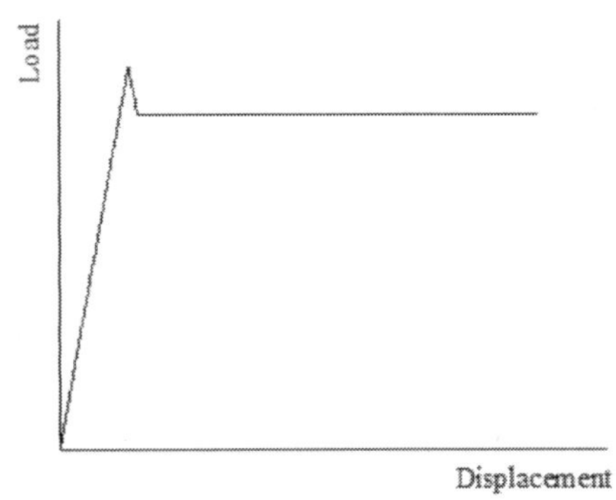

Figure2. Load-displacement curve of an inversion tube

Figure 2 shows the Load-displacement curve of an inversion tube. The load rises rapidly to its peak value, after that the deformation force descends a little and then keeps fixed until the inversion tube strokes out. The deformation force of the inversion tube can be depicted as the following equation:

$$P = \frac{\pi D \sigma_s t}{\sqrt{3}} \left[ln\frac{D}{d} + \ln\left(1 + \frac{t}{2r}\right)\right] \quad (1)$$

Where t, r, d and D respectively represent the thickness, knuckle radius, inside diameter and outside diameter of an inversion tube, showed as Figure 1. And σs is the yield stress. In this research, 1Cr18Ni9Ti was selected as the material of the inversion tube and its physical properties are showed in Table 1. 1Cr18Ni9Ti is known for its high plasticity, tenacity, and weldability.

Table 1. The properties of 1Cr18Ni9Ti

Name	Density (kg/m^3)	Elastic modules (GPa)	Position ratio	Yield stress (MPa)
1Cr18Ni9Ti	7850	202	0.3	330

2. Finite element model

The numerical model of the seat/occupant system showed in Figure 3 was established in MSC.Patran, a pre-processing software. The redundant seat structures that exert little influence on simulation results were taken out to improve the computational efficiency. The seat consists of 5 major parts: an energy absorber, two guides, a seat bucket, belts and a pedal. The inversion energy absorber is used to prevent the occupant from severe injury. The guides transform occupants' motion to vertical movement relative to the subfloor. The simplified seat bucket is used to define a contact between the seat and the dummy. The belt is designed as a 4-point belt to offer the occupant a symmetric constraint. The pedal constraints the foot and legs from a large movement. Excluding the dummy, the complete model includes 9788 nodes, 4200 HEXA elements, 1024 QUAD elements and 142 beam elements.

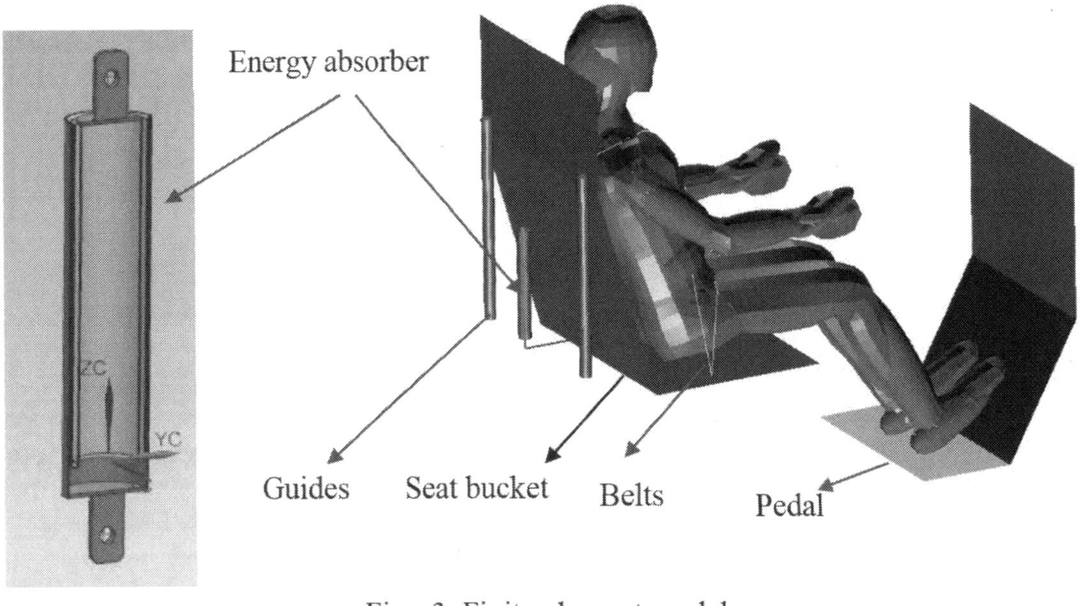

Figur3. Finite element model

A FAA 50th-percentile Hybrid-III dummy was selected in this model to maximize the effectiveness over the range of occupant weight. The hybrid-III dummy has a large improvement over the hybrid-II dummy on the characteristics of the neck, chest, and knee and became the only test dummy regulated by the FMVSS208 (Federal Motor Vehicle Safety Standards) . The dummy was seated in normal upright position and located in the middle of the seat to keep a symmetric model.

The initial velocity of the whole model was set at 12.81m/s for a statistical fact that the initial vertical velocity is smaller than 12.81m/s among 95% helicopter crash accidents. The seat input deceleration pulse is a triangle shape with a peak value of 30G and a span time of 0.087s. The boundary conditions are not complicated, however, two points must be noted in the process of preparing the simulation:

One is the contact between the bushes and the pillars. Both of the guides and the pillars are rigid bodies, so that the pillars are easy to penetrate the bushes, which may cause the inversion tube to deform unstably. Two methods are suggested to prevent this penetration:

(1) Set a small penetration check distance.

(2) Define a large contact force scale factor.

The other needed to be noted is the time step. Too small time step may result in too long computational time or even quitting from the simulation. The critical time step can be calculated by the following equation:

$$\nabla t_{cri} = \frac{SL}{C} \tag{2}$$

Where S is the time step factor, whose default value is 0.66, C is the sound speed of this material, and L is the smallest dimension in this model. Therefore, a small dimension in the model will generate a small time step and result in solving difficulty or even bringing the simulation to a halt. In the model of this simulation, the smallest dimension is the thickness of the inversion tube which must be controlled at an acceptable level.

3. Results and discussion

3.1. Occupant responses

The process of occupant response in the simulation is presented in figure 4. The result indicates that there is no large movement on the body except hands, which may be attributed to the belt constraint. The hands move largely so that they are easy to collide with the helicopter instruments and get fractured. Besides, it is found that the inversion tube deformed stably until it bottoms out.

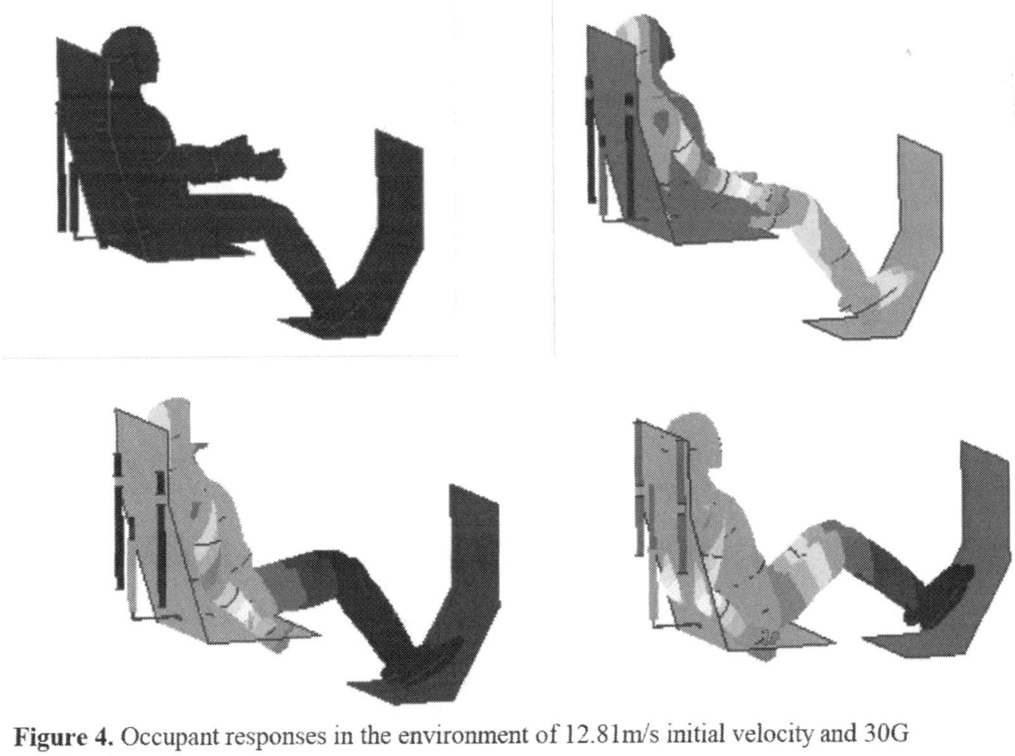

Figure 4. Occupant responses in the environment of 12.81m/s initial velocity and 30G deceleration pulse

3.2. Head injure criteria

The National Highway Traffic Safety Administration（NHTSA）has established several head injure evaluation criterions, foremost among them is the Head Injure Criteria（HIC）that has been accepted by FMVSS208 as a head impact tolerance specification. It regulates that the HIC value should be less than 1000, or the head will suffer a severe injury. The HIC value can be calculated by the following equation:

$$\mathbf{HIC} = \mathbf{max}[(t_2 - t_1)\{\frac{1}{t_2 - t_1}(\int_{t_1}^{t_2} a(t)dt)^{2.5}\}] \qquad (3)$$

Where t_1 and t_2 respectively represent the starting and ending time, during which time the HIC value reaches to its maximum value. Figure 5 shows the HIC value of the occupant during the crash. The maximum HIC value is 325, occurring at 35ms and under the HIC threshold. As a result, the occupant head is safe under the protection of the alleviation system.

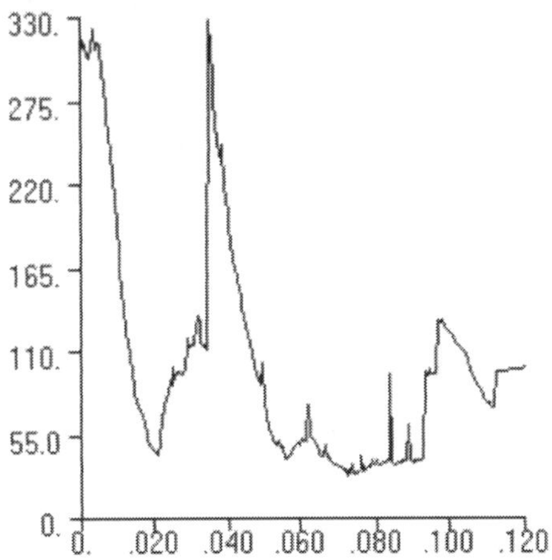

Figure 5. The HIC curve

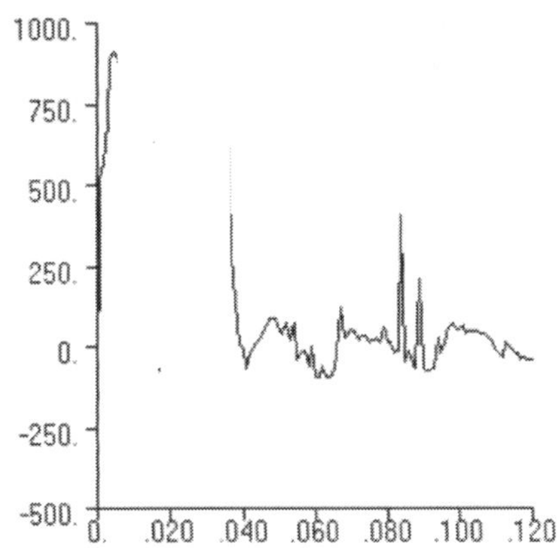

Figure 6. The neck load of the occupant

3.3. Neck load

The neck is a vulnerable part of a body and the impact load on the neck should be under 1100N or the neck may get injured. From figure 6, it can be found that the maximum neck load is 909N at 0.007s, below the neck load threshold. Therefore, the neck will not be objected to a severe injury.

3.4. Spine load

The spine is the main segment that supports the body weight. It consists of many movable sclerotomes and the motion of a spine is a result of the associated movement of all sclerotomes. The spine will bent when it is objected to a large compressive force.

Chandcer and Shanahan developed a lumbar load criterion that the maximum spine load should be no more than 6675N, which is now integrated into the Federal Aviation Regulation（FAR）[6].

A spine consists of many segments, among which the lumbar spine bears the maximum load, 50%~60% body weight. At the circumstance of 30G impact acceleration pulse in vertical direction and without any energy-absorbing device, the maximum lumbar spine load can be roughly calculated by the following equation:

$$F=0.6m50^{th}a \qquad (4)$$

Where m50th is the mass of a 50th-percentile occupant. The mass of a 50th-percentile hybrid III dummy is 76.23kg. From Eq.4, we can obtain that the lumbar force is 13447N under the impact acceleration of 30G, which is far more than the spine load-carrying capacity. As a consequence, an inversion tube energy absorber is needed to attenuate the lumbar force to an acceptable level.

Figure 7 shows the lumbar compressive force of the dummy. The result indicates that the lumbar compressive force rises rapidly to its peak value, 4855N, at 0.006s, and then decreases sharply to 181N at 0.017s, after that it keeps no more than 3000N until the end. It can be concluded that the spine will not get injured for the inversion tube alleviates the impact load on occupant spine.

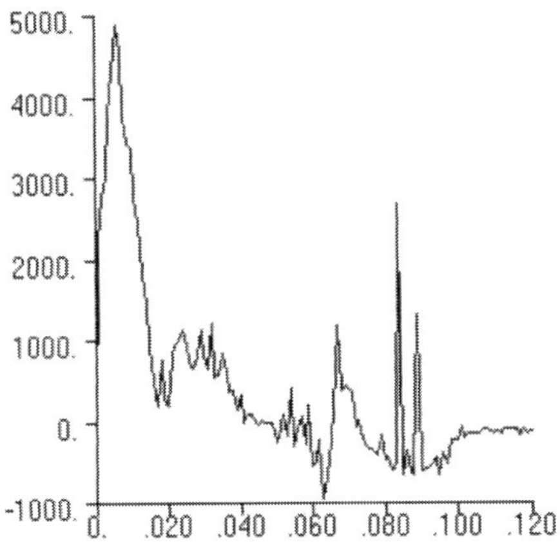

Figure 7. The lumbar compressive spine load

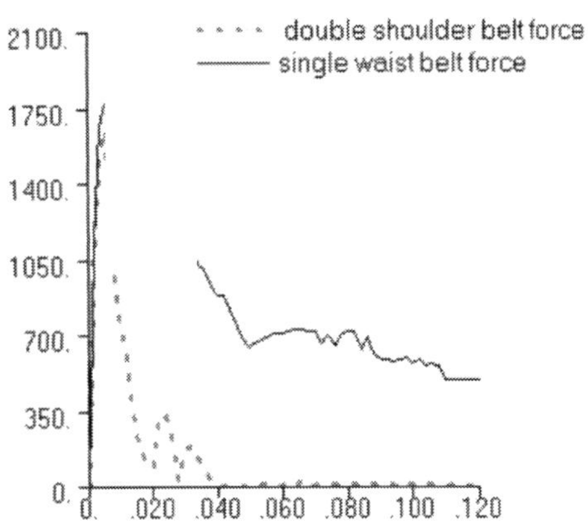

Figure 8. Seat belt load

3.5. Seat belt load

The seat belt consists of two shoulder belts and a waist belt. The CCAR23.562 regulates that the double shoulder belt load should be no more than 8900N while the single waist belt load should be less than 7790N. The shoulder belt load and waist belt load are showed in Figure8. The maximum shoulder belt load is 1645N and the maximum waist belt load is 2024N. The load on both of the two kinds of belts meets the requirements of CCAR23.562. Besides, it is found that both the two kinds of belts play a significant role on preventing the dummy from strenuous movement from the beginning to 40ms, after that only the waist belt works and takes effect.

4. Conclusions

To evaluate the reliability of the impact-alleviation system design, the occupant injure severity was investigated by a numerical simulation. It can be concluded that the inversion tube design is reliable, for that the HIC value, impact loads on the neck and the spine and belt force are all under the regulated thresholds under the protection of the shock-mitigation system.

This study is also committed to providing a reference for the inversion tube energy absorber design. The inversion tube parameters, such as the knuckle radius, outside diameter, inside diameter, thickness, length, material type and so on, can be changed and analyze the impact loads on occupant under the protection of different inversion tubes to optimize the energy-absorbing energy design. Besides, two things must be paid attention to:

(1) Owing to the fact that other seat structures, such as the cushion, take effect on attenuating the impact loads on the occupant but they were neglected in this simulation, the impact loads on occupant in this simulation results are a little larger than the loads in the real environment.

(2) This simulation only takes the velocity and the acceleration in the vertical direction into account. But the real helicopter crash environment is much more complicated. The model processes the velocity and the acceleration in the lateral and forward direction during a helicopter crash.

This research is just a preliminary investigation on the occupant protection system during a helicopter crash. To further this study, more factors, such as the cushion, initial velocity and acceleration in the forward and lateral direction should be considered.

Acknowledgements

The work described in this paper is financially supported by the Avic Aerospace Life Support Industries .LTD. under grant number 20142979004. The authors wish to gratefully thank this support.

References

1. S.A. Galehdari and H. Khodarahmi, Design and analysis of a graded honeycomb shock absorber for a helicopter seat during a crash condition, International Journal of Crashworthiness, 2016 21(3), 231-241.
2. Michele Guida, Francesco Marulo , Bruno Montesarchio and Massimiliano Bruno, Innovative Anti Crash Absorber for a Crashworthy Landing Gear, Appl Compos Mater , 21(2014) pp:483–494.
3. Xiaowei Zhang and Qing Zhou, An energy-absorbing sliding seat for reducing neck injury risks in rear impact—analysis for prototype built, Traffic Injury Prevention, 2016, 17(3), 313-319.
4. Kai Yang, ShanqingXu, JianhuShen, ShiweiZhou and YiMinXie, Energy absorption of thin-walled tubes with pre-folded origami patterns: Numerical simulation and experimental verification, Thin-Walled structures, 103 (2016), pp:33–44.
5. Huaining Cheng,Annette L. Rizer, Articulated Total Body Model Version V User's Manual, Biodynamics and Protection Division Human Effectiveness Directorate Air Force Research Laboratory,1998. Engineering, 2(4), 2012, 562-577.
6. Rasoul Moradi, Hamid K. Beheshti and Hamid M. Lankarani, Lumbar load attenuation for rotorcraft occupants using a design methodology for the seat impact energy-absorbing system , Central European Journal of Engineering, 2（4）, 2012, 562-577.

Research on the Dimensional Accuracy Measurement Method of Cylindrical Spun Parts Based on Machine Vision

Xiao Gangfeng[a], Zhong Xingwei[b], Shen Weixian[c], Xia Qinxiang[d*] and Chen Weiping[e]

School of Mechanical and Automotive Engineering, South China University of Technology, Guangzhou 510640, China

Abstract: A new method for measuring the dimensional accuracy of the cylindrical spun parts based on machine vision was proposed to overcome the artificial deviation and low efficiency of manual measurement. The image acquisition system of machine vision was built up. The methods of image processing and edge extraction of cylindrical spun parts were studied. The straightness and ovality of the cylindrical spun parts were obtained by the proposed new method. The results showed that the edge contour of the cylindrical spun parts extracted by Canny edge detector is better than Sobel and Prewitt edge detector. The dimensional accuracy of the cylindrical spun parts can be obtained accurately by the proposed measurement method based on machine vision. The relative errors of the straightness and ovality between the machine vision and the manual measurement are less than 10%.

1 Introduction

The cylindrical spun parts are one of the most important structural components which are widely used in aviation, aerospace, automotive, defense, energy and electronics industries [1]. The cylindrical parts manufactured by power spinning have inherent advantages, such as good dimensional accuracy and high material utilization [2]. The dimensional accuracy of the spun parts is usually obtained by manual measurement. However, the artificial deviation is existed and the efficiency is low if the manual measurement was used.

Machine vision technology is simulated the human visual function by machine and obtained the information from the image objects. It has advantages such as non-contact, high accuracy and high degree of automation. Therefore, machine vision technology has been employed in several manufacturing application including measurements, assembly and inspection [3]. Babu et al. [4] utilized a machine vision system to determine the surface roughness of the milled surfaces. The result indicates that the surface roughness could be estimated with a reasonable accuracy using machine vision. Ralf et al. [5] built an automatic visual inspection for small stampings with free-form surfaces. The results showed that the typical defects of cutoffs, incomplete stampings, cracks and scratches, etc. on both sides of the parts are inspected successfully. Li et al. [6] proposed a machine vision detection method to measure the geometrical size and shape-position of step shaft. The results showed that the length, radius, roundness error of the step shaft can be measured automatically. Liu et al. [7] proposed a non-contact detection method based on machine vision to measure the geometrical size and shape-position error of the ball nut. The results showed that the proposed method is feasible and

Corresponding author : [a]xiaogangfeng1111@163.com, [b]470741286@qq.com, [c]616674484@qq.com, [d]meqxxia@scut.edu.cn, [e]mewpchen@scut.edu.cn correct. However, no literature was reported about the dimensional accuracy measurement of spun parts based on machine vision.

The measurement method of dimensional accuracy based on machine vision can realize the online measurement, which lay the foundation for the intelligentialize of spinning technology. In the present work, a new measurement method of dimensional accuracy for cylindrical spun part based on machine vision was developed and experimentally tested. The image acquisition system of machine vision for cylindrical spun parts was built up and the image processing method was investigated in details. The straightness and ovality of the cylindrical spun part were automatically calculated.

2 Building up of the machine vision system

The machine vision technology mainly consists of image acquisition, image processing and data processing [3]. The image acquisition system built up to obtain the images of cylindrical spun parts is shown in Fig.1. It consists of four main components: LED light source, CCD camera, computer and software-MATLAB. Two LED light source is located above the spun part and irradiated it from two sides to obtain the clear image of the cylindrical spun parts and eliminate the shadow. The image is transmitted to the computer, and then processed by software-MATLAB. Fig.2 shows the image of cylindrical spun part obtained by the image acquisition system.

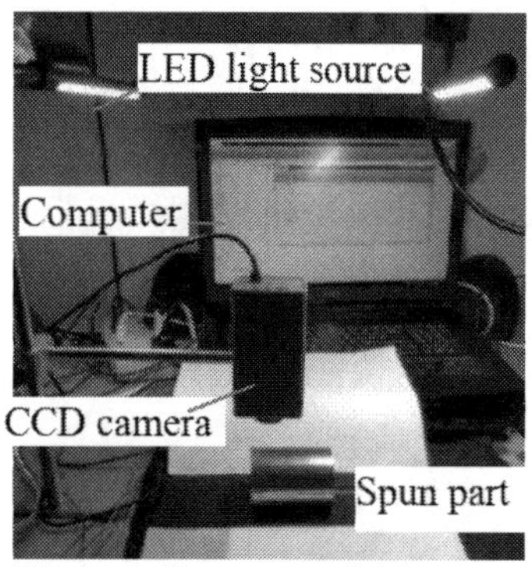

Figure1. Image acquisition system of machine vision for cylindrical spun parts

(a) Vertical view (b) Lateral view

Figure 2 Image of the cylindrical spun parts

A standard block with the length L of 25mm is selected as the calibration object to acquire the actual length that a pixel presents. The image of the standard block is obtained by the image acquisition system, and the number of the pixels M is calculated. Therefore, the length that a pixel presents $k= L/ M$. For the image acquisition system built up in this study, k is 0.1975mm/pixel.

3 Image preprocessing and edge extraction of the cylindrical spun parts

3.1 Image preprocessing

The digital image obtained by image acquisition system is RGB color image, which described the color of the pixel by intensity superposition of red, green and blue color. However, the edge extraction of color image is difficult. Therefore, the color image should be transformed to grayscale image. The greyscale value of the each pixel in the image is defines as the average of the color intensity of the three primary colors, as shown in Eq.1[8]. Fig.3 shows the vertical and lateral views of greyscale image of cylindrical spun part, respectively.

$$f(i,j) = (R(i,j)) + G(i,j) + B(i,j)) / 3 \quad (1)$$

where the $f(i, j)$ is the greyscale value of the arbitrary point in the image. $R(i, j)$, $G(i, j)$, $B(i, j)$ are the intensity of the red, green and blue color respectively.

(a) Vertical view (b) Lateral view

Figure 3. Greyscale image of cylindrical spun part

The noise is generated during the image acquisition and transmission, which makes the image not clear. Therefore, the de-noising processing is necessary. The photon noise and current noise should be considered during the image processing of the cylindrical spun part. These noises can be eliminated by spatial filtering. The method of noise elimination mainly includes mean filtering and median filtering. The edge of the image after media filtering is clearer than that of mean filtering, which is benefit for the edge extraction [3]. Therefore, the media filtering is selected. Fig.4 shows the image of cylindrical spun part after media filtering.

Figure 4. Cylindrical spun part image after spatial filtering

Image binarization is set the grayscale value of the pixel in the image as 0 or 255 [3]. The purpose of binarization is separated the objet and background of the image. The approach is converted the grayscale image to binary image, as shown in Eq. 2 [8].

$$g(x,y)=\begin{cases}1 & f(x,y)\geq T\\ 0 & f(x,y)<T\end{cases} \tag{2}$$

where $g(x, y)$ is value of the pixel in the image after binarization, $f(x, y)$ is the greyscale value of the arbitrary pixel point in original image, T is the threshold value.

Generally, the gray histogram is used to show the overall gray distribution of the grayscale image. Fig.5 shows the grey level histogram of cylindrical spun part image. It shows that the grayscale value of the background is less than 50 and that of the cylindrical spun part is more than 50. Therefore, the threshold value is selected as 50. Fig.6 shows the cylindrical spun part image after binarization.

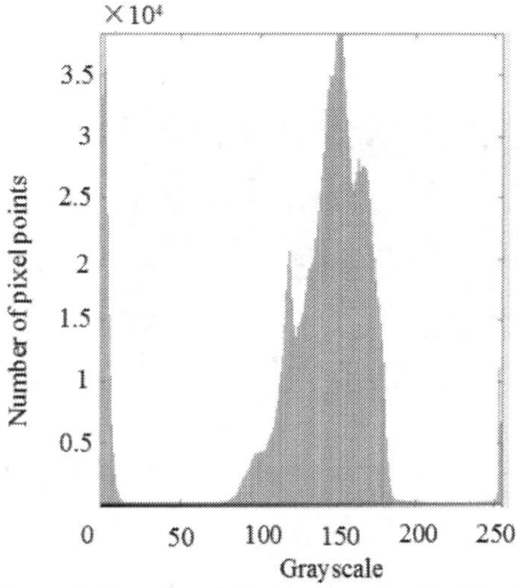

Figure5. Grey level histogram of cylindrical spun part **Figure 6.** Binary image of cylindrical spun part

3.2 Edge extraction

Accurate extracting the edge information of the cylindrical spun part is the key to measure its dimensional accuracy. The grayscale gradient of the binary image is small parallel to the edge of the spun part, and that is large along the direction perpendicular to the edge of the spun part. Therefore, the criterion of edge detection and extraction can select the grayscale gradient reaching maximum or its second derivative is zero.

For continuous image functions, the grayscale gradient in arbitrary point (x, y) can be described as follows [8]:

$$\mathrm{G}[f(\vec{x,y})]=\begin{bmatrix} G_x \\ G_y \end{bmatrix}=[\frac{\partial f}{\partial x},\frac{\partial f}{\partial y}]^T \tag{3}$$

The magnitude and direction of the grayscale gradient are as follows [8]:

$$G[f(x,y)]=\sqrt{\left(\frac{\partial f}{\partial x}\right)^2+\left(\frac{\partial f}{\partial x}\right)^2} \tag{4}$$

$$\alpha(x,y)=\arctan(G_x/G_y)$$

where $\alpha(x,y)$ is the angle between the gradient vector and the X-axis.

The edge detectors mainly include Sobel edge detector, Prewitt edge detector and Canny edge detector. Fig.7 shows vertical view of the edge contour of the cylindrical spun part extracted by Sobel edge detector, Prewitt edge detector and Canny edge detector respectively. It shows that the contours of the spun parts extracted by Sobel edge detector are simple contours with a single pixel width. The contours extracted by Prewitt edge exist partial contours with multi-pixel. The most complete contours can be extracted by Canny edge detector. Therefore, the Canny detector is selected for edge extraction in this study. Fig. 8 shows the lateral view of the edge of cylindrical spun parts extracted by Canny detector.

a) Sobel detector b) Prewitt detector c) Canny detector

Figure7. Vertical view of edge contour of cylindrical spun part extracted by different detectors

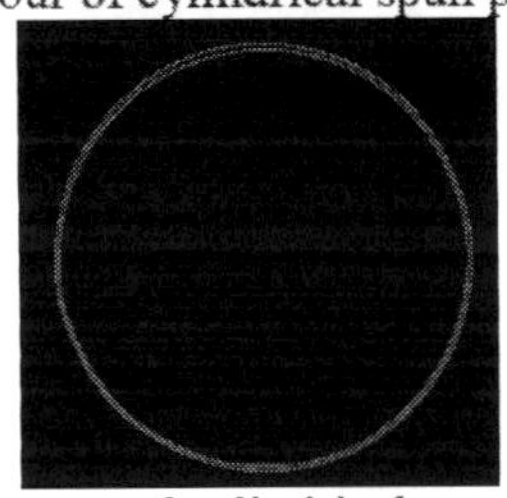

Figure 8. Lateral view of Edge contour of cylindrical spun part extracted by Canny detector

4 Method of dimensional accuracy measurement

4.1 Straightness calculation

The straightness e_s is defined as the distance of the two nearest parallel planes, and the arbitrary generatrix on the outer surface of the workpiece is located between these two planes.

The date processing method of obtaining the straightness consists of the following stages: firstly, the starting point of straight line is obtained by Hough transform line detection, and the end point is calculated by the straight slope and the dimension of the spun part. Then the contour line is retrieved out by loop algorithm and saved as single pixel contour point set. Finally, the contour line of the spun part is obtained by linear fitting of the contour points based on least square method, and the distance from each point to the fitting straight line is calculated as shown in Eq.5. e_s of the cylindrical spun part equals to the difference between the maximum and minimum of the distance.

$$d = \frac{kx + b - y}{\sqrt{1 + k^2}} \tag{5}$$

where k is the slope of the fitting line, b is the intercept of the fitting line.

The straightness of the cylindrical spun part are 0.19 mm and 0.21mm calculated by the proposed method and manual measurement, respectively. The relative error is only 9.5%.

4.2 Ovality calculation

Ovality e_o refers to the degree of the cross section of the workpiece approaching to the theoretical circle. e_o is defined as the radius difference of the two nearest concentric circles, and the arbitrary point on the outer surface of the workpiece is located between these two concentric circles. The center of the concentric circles is determined by the least squares circle method.

The date processing method of obtaining the ovality consists of the following stages: firstly, the outer circle contour with single pixel of the cylindrical spun part is retrieved by loop algorithm and saved as outer contour point set(x, y). Then, the center of the outer contour is obtained by least squares fitting of the point set (x, y), which acted as the polar point in the polar coordinates. Finally, the Cartesian coordinate point set (x, y) is converted to polar point set (θ, r) by the function of [theta, rho]=cart2pol(x, y) in software-MATLAB. e_s of the cylindrical spun part equals to the difference between the maximum and minimum of the polar distance .

The average values of the outer circle radius calculated by the proposed method and manual measurement are 29.00 mm and 29.02 mm, respectively. The ovality of the cylindrical spun part calculated by the proposed method and manual measurement are 0.77mm and 0.70mm calculated by the proposed method and manual measurement, respectively. The relative errors are both less than 10%. It indicates that the dimensional accuracy measurement method based on the machine vision can be considered as reliable.

5 Summary

A new method of measuring the dimensional accuracy of the cylindrical spun parts based on machine vision is proposed in this study. An experimental investigation is carried out to draw the following conclusions:

(1) The Dimensional accuracy of the cylindrical spun parts can be obtained accurately by the proposed measurement method based on machine vision.

(2) The relative errors of the straightness and ovality between the machine vision and the manual measurement are less than 10%.

Acknowledgments

This work was supported by the Natural Science Foundation of Guangdong province (Grant No.2017BQ019), and the special project for public welfare research and ability construction of Guangdong province (Grant No. 2016A010102011)

References

1. G.F. Xiao, Q.X. Xia, X.Q. Cheng, L. Hong, New forming method of manufacturing cylindrical parts with nano/ultrafine grained structures by power spinning based on small plastic strains, Sci. China Technol. Sci. 59(2016) 1656-1665.
2. Q.X. Xia, G.F. Xiao, H. Long, X.Q. Cheng, X.F. Sheng. A review of process advancement of novel metal spinning, Int J Mach Tools Manuf, 85(2014):100-121.
3. S. Carsten, U. Markus, W. Christian, Machine vision algorithms and applications, Beijing, 2008.
4. G.D. Babu, K.S. Babu, B.U.M. Gowd, Evaluation of surface roughness using machine vision, International Conference on Emerging Trends in Robotics and Communication Technologies. IEEE, (2011) 220-223.
5. L. Ralf, O. Alexander, S. Peter, S. Jörg, Automatic visual inspection system for small stampings with free-form surfaces, Proc Spie, 4301(2001) 9-19.
6. Z.T. Li, J. Liu, G.H. Sun, D.W. Shao, Multi-diameter of geometrical size and shape-position error detection by machine vision, Tool Engineering, 42(2008) 77-80 (in Chinese) .
7. Q.M. Liu, L.S. Wang, X.W. Chen, G.F. Li. Ball nut detection by machine vision, Journal of Jilin University (Engineering and Technology Edition), 36(2006) 534-538.(in Chinese)
8. R.C. Gonzalez, R.E. Woods, S.L. Eddins, Digital image processing using MATLAB, Beijing, 2013.

Molecular Dynamics Simulations of Crater Formation Induced by Laser Ablation on the Surface of α-Fe Substrate

Jenn-Kun Kuo[1], Pei-Hsing Huang[2,a], Shih-Kai Chien[3], Kuang-Yao Huang[3] and Kun-Tso Chen[3]

[1]*National University of Tainan, Department of Greenergy, 70005 Tainan, Taiwan*
[2]*National Pingtung University of Science and Technology, Department of Mechanical Engineering, 912 Pingtung, Taiwan*
[3]*ITRI SOUTH Laser and Manufacturing Technology Center (LAMC), Laser Manufacturing System Department, , 73445 Tainan, Taiwan*

Abstract. In this study, we employed molecular dynamics (MD) simulations to investigate ablation induced by the application of pulsed laser irradiation to a thin α -Fe substrate. We observed several mechanisms underlying the removal of material, including ultrafast melting, cluster ejection, and thermal vaporization. We also examined the effects of laser fluence on the resulting surface morphology as well as the amount of material ablated and deposited around the craters

1 Introduction

Laser cutting has been shown to outperform conventional oxyacetylene cutting with regard to speed and quality while enabling non-contact processing. This technology has been adopted in numerous industrial applications, as a good compromise between efficiency and accuracy. The removal and cutting of materials using laser ablation involves two primary mechanisms: vaporization and melt ejection [1-4]. Achieving a high-quality laser cutting surface requires a basic understanding of the underlying processes; however, observations of laser cutting are limited to extremely short time periods. Furthermore, this process involves complex interactions among a variety of phases. In this study, we used numerical simulations to directly observe the transient phenomena associated with laser cutting and ablation [5-8].

Laser-induced structural changes occur on extremely short timescales, falling somewhere between the natural oscillation cycles of an atom or molecule and several picoseconds. These changes are induced by a complex combination of processes, including the transmission of light energy, carrier dynamics, heating, and mechanical processes. As a result, the fundamental mechanisms underlying laser ablation have yet to be fully elucidated, which has hindered the further development of this technology for manufacturing [9-12]. In this study, we adopted molecular dynamics (MD) at the microscopic scale to explore structural metamorphoses induced by light, heat, and mechanical processes. We used simulations to conduct a qualitative study on the interactions between lasers and metal targets followed by experimental verification of our results. Our overall objective was to enhance the precision laser cutting and thereby improve the quality of laser cut surfaces.

---

[a] Corresponding author : phh@mail.npust.edu.tw

2 Numerical Methods

MD simulation was used to investigate the ablation of a thin α-Fe substrate. Laser energy applied to the substrate decayed exponentially in the direction of incident light in accordance with the Beer-Lambert law. The energy of the photons was transformed into kinetic energy (in the irradiated atoms) within a characteristic time (τ_{eq}) [8]. Time τ_{eq} corresponds to the relaxation time between electrons and the lattice, which typically takes between hundreds of femtoseconds to several tens of picoseconds for metals, depending on the properties of the target [1,10,11]. Here, we adopted a typical relaxation time of $\tau_{eq} \sim 10$ ps. The laser energy deposited to depth z within the substrate is expressed using the equation of S(z) as follows:

$$S(z) = I_0(1-R)D_p^{-1}e^{-z/D_p} \tag{1}$$

where I_0 is the intensity of the laser radiation, R denotes the reflectivity of the substrate, and D_p is the optical penetration depth. We accounted for effects of electron thermal diffusion as an increase in the effective depth of penetration by the laser, wherein depth was estimated using the electron thermal diffusion length [2]. The physical parameters used in the simulation are listed in Table 1.

Table 1. Physical parameters of α-Fe used in the simulation.

a_0 (nm)	R	D_p (nm)	Δt (fs)	r_e (nm)
0.286595	65%	18.5	5	2.481987

The EAM potential energy model developed by Wadley et al. [13] was used in the MD simulations to describe interactions between atoms. The EAM model is in good agreement with the basic material properties (e.g., lattice constant, elastic constant, bulk modulus, sublimation energy, and heat of solution) of the substrate [14]. The total energy Etot in this system of atoms is defined as follows [13]:

$$E_{total} = \sum_i F_i(\rho_{h,i}) + \frac{1}{2}\sum_{i,j(i\neq j)} \phi_{ij}(r_{ij}) \tag{2}$$

$$\rho_{h,i} = \sum_{j\neq i} f(r_{ij}) \tag{3}$$

where the summation includes the total number of atoms in the system. Embedded energy, F_i, is a function of the host electron density $\rho_{h,i}$ induced at the site of atom i by all other atoms in the system. Pair potential, ϕ_{ij}, is a function of the distance r_{ij} between atom i and atom j. re is the equilibrium distance between the nearest neighbor atoms [14]. The evolution of atomic dynamic trajectories is governed by Hamilton's classical equation based on Newton's second law of motion. The force acting on atom i can be calculated according to the gradient of potential energy as follows:

$$\begin{aligned}\vec{F}_i &= -\nabla_{\vec{r}_i} E_{tot} \\ &= -\sum_{j,j\neq i}\left[\frac{\partial F_i(\rho_i)}{\partial \rho_i}\frac{\partial f_j(r_{ij})}{\partial r_{ij}} + \frac{\partial F_j(\rho_j)}{\partial \rho_j}\frac{\partial f_i(r_{ij})}{\partial r_{ij}} + \frac{\partial \phi_{ij}(r_{ij})}{\partial r_{ij}}\right]\frac{\vec{r}_i - \vec{r}_j}{r_{ij}}\end{aligned} \tag{4}$$

The potential energy parameters used in the EAM model are detailed in Ref. [13]. Before initiating the simulation, the atoms in the system were randomly assigned atomic velocities appropriate to room temperature (300 K) for a given simulation, in accordance with Maxwell–Boltzmann distribution [15]. Periodic boundary conditions were applied to the atoms in the x direction of the computational domain. Free boundary conditions were assumed in the y and z directions, under which Fe atoms were permitted to travel straight along the two directions without any restrictions. Ablation was simulated at a pulse duration in which the electron-phonon relaxation time was 10.0 ps where the photonic wavelength of λ = 800 nm. The optical parameters of the material used in the calculations at this wavelength were as follows: reflectivity R = 65% and optical penetration depth = 18.5 nm [2]. The laser pulse was focused at the center of the simulation system with an even spatial distribution of intensity along the x-y plane. The ablation process was performed under vacuum. Gear's fifth-order predictor-corrector algorithm [15] was used to estimate the positions, velocities, and accelerations of the atoms, as well as the interaction forces between atoms. This algorithm features excellent numerical stability and provides a satisfactory compromise between time consumption and precision [16].

3 Results and Discussion

Before initiating the MD simulations, we adopted the embedded atom method (EAM) to describe the interactions among metallic atoms. To ensure accuracy, we performed a heating and melting point experiment on 10×10×10 a_0 Fe-BCC crystal material (a_0 represents the lattice constant of α-Fe). The λ-function [15] was used to identify the degree of order in the periodic arrangement of the lattice. A λ-function value approaching 1 indicates that the substrate is in solid state, a lower λ-function value refers to a less ordered state, and a λ-function approaching 0 means that the substrate is in a completely molten state. The simulation results in Fig. 1 clearly illustrate the changes in the λ-function and crystal structure as the heating time was increased. λ-function rapidly decreased with an increase in temperature. When the temperature rose to between 1,200 K and 1,400 K, the λ-function tended to remain constant. This period falls within the stage of allotropic transformation in the α-Fe structure. At a temperature of 1,850 K, the crystal began melting, such that the λ-function value fell to 0. At room temperature, the atomic polymerization energy of the crystal was −4.28 eV/atom, which is consistent with calculation results obtained via quantum mechanics [15]. These results prove the accuracy of the EAM and MD simulations used in this study.

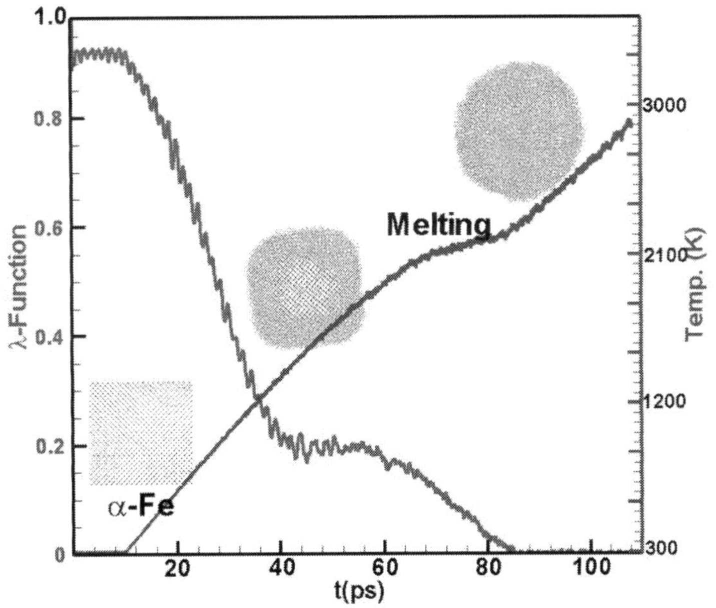

Figure 1. Time evolution of temperature and relationship between λ-function and crystal structure in molecular dynamics simulations of α-Fe undergoing heating.

Figure 2. illustrates the ablation state resulting from the application of a laser spot with diameter of 4.0 nm and laser energy of 600 mJ/cm^2 to a 50×50×50 a_0 (14.3×14.3×4.29 nm) Fe-BCC crystal. The colors of the atoms in the figure indicate their potential energy values. Under laser irradiation, the rapid increase in temperature led to ultrafast melting in the substrate at the location of the laser spot. At time t = 17 ps, phase eruption and cluster ejection were observed in many atoms within the surface layer. Once a certain number of the melted atoms had been ablated, they did not undergo complete desorption, and neither did they lift-off from the substrate surface. Thus, this process led the formation of a crater deposit.

Figure 2. Ablation state resulting from the radiation of laser fluence of 600 mJ/cm^2 to Fe-BCC crystal.

Figure 3. presents changes in average temperature in the substrate when heated using a laser of various energies. The lattice heating process relies on the intensity of electron-phonon scattering; therefore, we assumed that the electron-phonon relaxation time was $\tau_{eq} \sim 10$ ps in this study. As shown in Fig. 3, the average temperature of the substrate increased rapidly from the onset of heating. After being heated using laser energies of 300 mJ/cm^2, 600 mJ/cm^2, and 900 mJ/cm^2, the average temperature in the substrate respectively peaked at 2,385 K, 3,064 K, and 3,962 K before gradually declining.

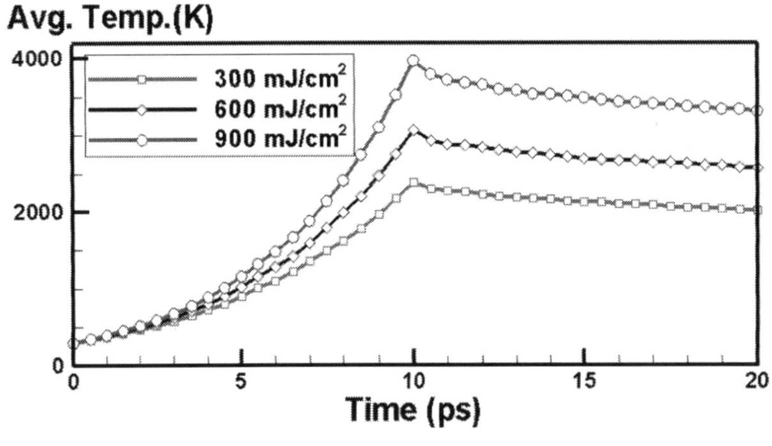

Figure 3. Time-evolution of average temperature in α-Fe substrate.

During the laser heating process, the surface and the interior of the target underwent rapid heating and swelling, which caused the eruption of melted as well as semi-melted material from the top and bottom surfaces. Laser energy of 900 mJ/cm^2 was sufficient to penetrate the substrate and form micropores, whereas laser energy of 300 mJ/cm^2 was insufficient to penetrate the substrate and merely formed a small pit. Melted material that did not undergo complete desorption or lift-off from the substrate surface (i.e.. residual slag or residue) created a crater-like structure, which severely reduced the quality of the surface (Fig. 4).

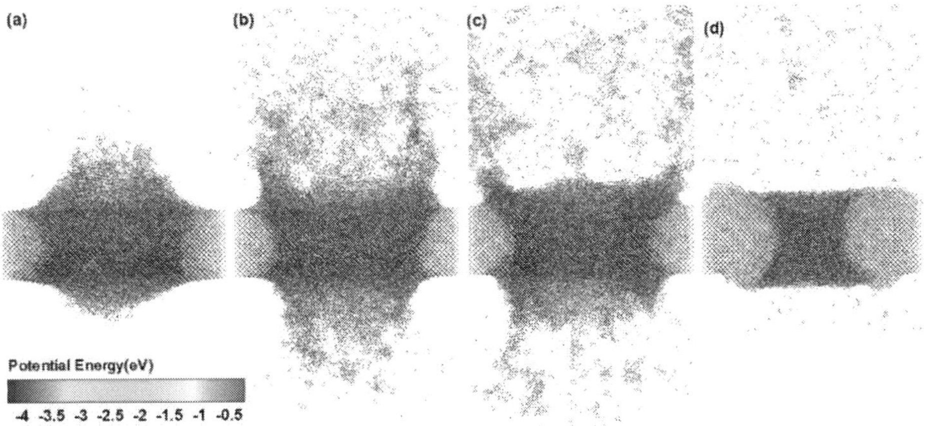

Figure 4. (a)-(d) Transient states of α-Fe crystal irradiated at a laser energy of 900 mJ/cm^2 for t =10 ps, 15 ps, 20 ps, and 50 ps.

As shown in Fig. 5(a), the application of a laser with energy of F≤ 600 mJ/cm^2 resulted in a larger number of atoms being ablated from the top surface (N_a). At this energy level, deformations also formed on the bottom surface, due to stress induced by the thermal shock wave. Nonetheless, the fact that the laser did not penetrate the target meant that almost no atoms were ablated from the bottom surface, as shown in Fig. 5(b). In contrast, higher energy levels enabled the laser to penetrate the substrate, which resulted in the formation of micropores (Fig. 4). In these cases, the number of atoms passing through the bottom surface exceeded the number of atoms ablated from the top surface. Nonetheless, a portion of the metal slag lacked the ejection or lift-off power required to overcome the attraction of the metal bonds. These atoms did not undergo complete desorption or lift-off from the target surface in the final stage, such that N_a ultimately decreased. These atoms eventually accumulated on the top and bottom surfaces of the target to form uneven craters and bumps.

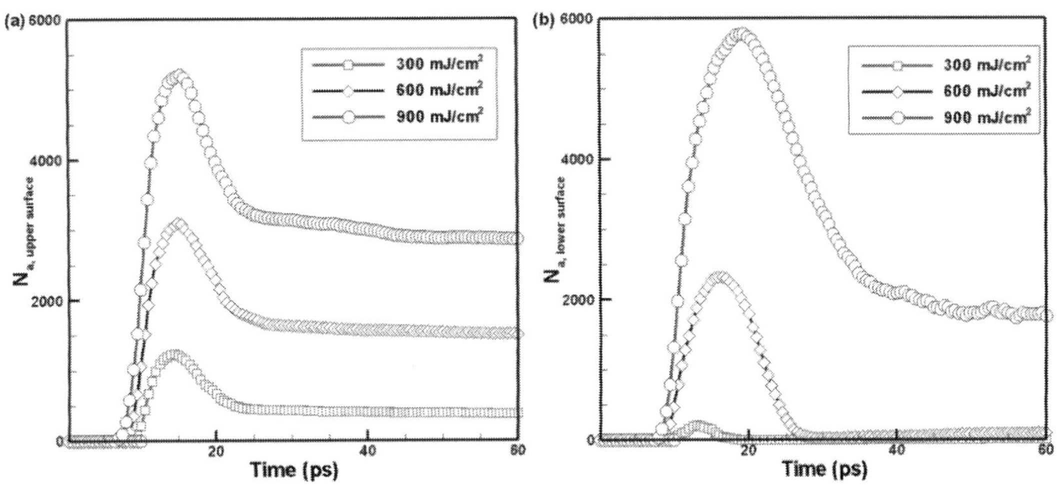

Figure 5. (a)-(d) Number of melted atoms that underwent desorption from (a) top surface and (b) bottom surface of substrate during laser heating.

4 SUMMARY

In this study, EAM molecular dynamics was used to simulate the laser ablation of α-Fe surfaces in order to elucidate the mechanisms underlying the formation of craters. Our results could be used to guide the design of process parameters with the objective of improving the quality of laser-cutting surfaces.

Acknowledgements

The authors gratefully acknowledge the support provided for this research by the ITRI SOUTH Laser and Manufacturing Technology Center under grant C106-078 and by the Ministry of Science and Technology, R.O.C. under grants MOST 106-2221-E-020-014 and MOST 106-2622-E-020-004-CC3. We would also like to thank the National Center for High-performance Computing (NCHC) of National Applied Research Laboratories (NARLabs) of Taiwan for providing computational and storage resources.

References

1. P.-H. Huang, H.Y. Lai, Phys. Rev. B **77,** 125408, 12 pages (2008)
2. N.N. Nedialkov, S.E. Imamova, P.A. Atanasov, G. Heusel, D. Breitling, A. Ruf, H. Hugel, F. Dausinger, P. Berger, Thin Solid Films **453–454**, 496–500 (2004)
3. P.-H. Huang, H.Y. Lai, J. Appl. Phys. **108**, 123504, 10 pages (2010)
4. H.Y. Lai, P.-H. Huang, Comput. Mater. Sci. **41**, 498-507 (2008)
5. S.E. Imamova, P.A. Atanasov, N.N. Nedialkov, F. Dausinger, P. Berger, Nucl. Instrum. Methods Phys. Res. Sect. B-Beam Interact. Mater. Atoms **227**, 490-498 (2005)
6. P.-H. Huang, H.Y. Lai, J. Chin. Soc. Mech. Eng. **31**, 477-483 (2010)
7. H.Y. Lai, P.-H. Huang, Appl. Surf. Sci. **254**, 3067-3073 (2008)
8. N.N. Nedialkov, P.A. Atanasov, Appl. Surf. Sci. **252**, 4411–4415 (2006)
9. H.Y. Lai, P.-H. Huang, J. Chin. Soc. Mech. Eng. **28**, 577-583 (2007)
10. N. Nedialkov, P. Atanasov, S. Imamova, J. Phys. D **36**, 638(2004).
11. P.-H. Huang, H.Y. Lai, Nanotechnology, **19**, 255701, 11 pages (2008)
12. H.Y. Lai, P.-H. Huang, T.H. Fang, Appl. Phys. A-Mater. Sci. Process. **86**, 497-503 (2007)
13. H.N.G. Wadley, X. Zhou, R.A. Johnson, M. Neurock, Prog. Mater. Sci. **46**, 329 (2001)
14. P.-H. Huang, J.K. Kuo, Appl. Phys. A-Mater. Sci. Process. **103**, 1083-1092 (2011)
15. J.M. Haile, *Molecular dynamics simulation: elementary methods*, Wiley-Interscience, New York. (1997)
16. P.-H. Huang, Compos. Sci. Technol. **72**, 599-607 (2012)

AUTHOR INDEX

AUTHOR INDEX